Law for Professionals

The *Law for Professionals* book series contains publications in all fields of the law aimed at professionals and practitioners. The book series is dedicated to publish high quality books that provide in-depth insights into the workings and the practical functioning of the law, international organizations or regional legal frameworks.

The authors contributing to the book series are experienced legal professionals and practitioners whose expertise combines theory and practice.

John C. Balzano

China Food and Drug Law

Policy, Practice, and Future Directions

Springer

John C. Balzano
Westport, CT, USA

ISSN 2662-141X ISSN 2662-1428 (electronic)
Law for Professionals
ISBN 978-3-031-61900-7 ISBN 978-3-031-61901-4 (eBook)
https://doi.org/10.1007/978-3-031-61901-4

This Springer imprint is published by the registered company Springer Nature Switzerland AG
The registered company address is: Gewerbestrasse 11, 6330 Cham, Switzerland

To Anthony, Sebastian, and Grace.

Acknowledgements

This book has been over 20 years in the making in some respects, so there are many colleagues, both present and former, to thank. Too many to name here. Thanks to my mentors, such as Frances Foster, John Haley, Jerome Cohen, Jamie Horsley, Paul Gewirtz, Ira Belkin, and other colleagues at the China Law Center, who got me started in this area and encouraged me to keep writing. I have also worked over the last 10 years with wonderful colleagues at Covington and Burling and a variety of individuals in the food and drug regulatory field, generally in China, the USA, Europe, and Japan, who have taught me a great deal. Thanks also to Andrea Thompson for her excellent editorial assistance and to the team at Springer Publishing. Despite the help and influence of all of these individuals, the opinions expressed in this book are mine alone, as are any errors.

Any thank you would be incomplete without thanking my immediate and extended family for all of their support and patience during the many late nights and early mornings, while I tried to work, write, and be a son, brother, husband and dad.

Contents

1 Introduction and Methodology 1
1.1 Scope and Methodology 1
1.2 Timeline of China FDD Regulation 2
1.3 Safety, Public Health, and China's Global Impact 2
1.4 Health Care Reforms and Access to Care and Products 4
1.5 Expansion Efforts to Develop Innovative Drug and Device Industries 4
1.6 Government and Legal Reform 6
1.7 Chapters to Come 7
References 8

2 Agency Jurisdiction Over Food and Drug Law 9
2.1 Overview of the Current Agency Structure 9
2.2 History of Food and Drug Regulatory Jurisdiction in China 10
2.3 Making of the NMPA 11
2.4 The Structure of NMPA 12
2.5 NMPA-Affiliated Centers and Provincial Counterparts 13
2.6 Provincial-Level Medical Product Associations 14
2.7 Food Agencies 14
2.8 Other Relevant Functions of SAMR 15
2.9 The Ministry of Agriculture and Rural Affairs 16
2.10 China Customs Administration 17
2.11 National Health Commission 18
2.12 National Healthcare Security Administration 19
2.13 The Ministry of Science and Technology 20
2.14 China National Intellectual Property Administration 20
2.15 Ministry of Public Security 21
2.16 Other Ministries and Agencies 22
2.17 China's Court System and Food and Drug Enforcement 23
References 23

3 Administrative Law Concepts and Enforcement Powers 25
3.1 Overview of Other General Legal Concepts 25
3.2 Constitutional Law and Legislation 26
3.3 Administrative Licensing and Related Policies: Improvement of Efficiency 27
3.4 Public Supervision as a Regulator 29
3.5 Open Government Information 30
3.6 Administrative Penalties and Compulsory Measures 31
3.7 Administrative Penalty Procedure 31
3.8 Administrative Compulsory Measures 32
3.9 NMPA and Provincial Administrative Rules 33
3.10 Administrative Reconsideration 36
3.11 Administrative Litigation 37
3.11.1 Vellpharm v. State Food and Drug Administration [now NMPA] (2015) 38
3.12 Administrative and Civil Enforcement in Practice 45
3.12.1 How Do Enforcement Issues Begin? 45
3.12.2 Investigations and Establishment of Cases? 46
3.12.3 What Evidence Is Needed? 47
3.12.4 When to Seek Reconsideration or Appeal? 48
3.12.5 When to Consider Settlement? 49
References 50

4 Common General Principles of China Food and Drug Regulation 53
4.1 Separation of Imported and Domestic Pathways 53
4.1.1 Imported Product Pathway 53
4.1.2 Domestic Pathway 54
4.1.3 Differences Between the Imported and Domestic Pathway 54
4.2 Evolution of the System of Pre-Approvals 57
4.2.1 Reduction of Pre-Approvals/Licenses 57
4.3 Creation of True Product Licenses 58
4.3.1 History of Non-Product Specific Licenses 58
4.3.2 Evolution of the Product License System 59
4.3.3 Restriction of Supply Chain Through the Product Life Cycle 60
4.3.4 Specialized Facility Licenses 61
4.4 Requirement for Local Testing and Clinical Data 62
4.5 Reliance on Foreign Approvals 62
4.6 Use of Technology 63
4.6.1 Websites and Online Databases 63
4.6.2 Online Submission Systems 64
4.7 Reliance on Conformity Standards Versus Safety and Effectiveness Data 65

4.8 Use of Experts . . . 66
4.9 Internationalization . . . 67
4.10 Intellectual Property . . . 70
References . . . 72

5 Pharmaceutical Regulation in China . . . 75
5.1 Introduction . . . 75
5.2 Content of China's Drug Market . . . 76
5.3 China's Drug Paradigm and Policy Shift . . . 78
5.4 Drug Definitions and Categorization in China . . . 80
5.4.1 Definitions . . . 81
5.4.2 Registration Categories . . . 83
5.5 Generic Small Molecule Drugs . . . 86
5.6 Biosimilars . . . 91
5.7 Standards of Approval . . . 92
5.8 Overview of the Drug Development and Marketing-Authorization Process . . . 95
5.9 Non-Clinical Development . . . 98
5.10 Registration Applications and Fees . . . 99
5.11 Clinical Development . . . 101
5.11.1 The Clinical Trial Application . . . 103
5.11.2 NMPA/CDE Meetings . . . 104
5.11.3 Clinical Trial Sites . . . 106
5.11.4 Good Clinical Practice Reform in China . . . 108
5.11.5 Human Subject Protections and Informed Consent . . . 112
5.11.6 Ethics Committees . . . 113
5.11.7 Management of Investigational Product and Quality Considerations . . . 114
5.11.8 Clinical Trial Monitoring and Pharmacovigilance . . . 115
5.11.9 Verifications . . . 116
5.12 Additional Aspects of Clinical Development . . . 117
5.12.1 International Multi-Center Trials . . . 117
5.12.2 Investigator-Initiated Clinical Trials . . . 118
5.12.3 Real-World Evidence . . . 120
5.12.4 Special Approvals in Medical Tourism and Other Pilot Zones Based on Clinical . . . 121
5.13 Expedited Programs . . . 122
5.13.1 Urgently Needed New Drugs Marketed Abroad . . . 124
5.13.2 Breakthrough Applications . . . 125
5.13.3 Conditional Approval . . . 125
5.13.4 Priority Approval . . . 126
5.13.5 Special Approval . . . 127
5.13.6 Orphan Drugs . . . 128

5.14 Marketing Application . . . 128
5.14.1 Drug Masterfiles . . . 129
5.14.2 Registration Testing . . . 129
5.14.3 Pre-Approval Verifications . . . 130
5.14.4 Marketing-Authorization Holder ("MAH") and Its Agent ("MAH Agent") . . . 131
5.14.5 One Drug One Registration One MAH . . . 132
5.14.6 MAH Transfers and Technology Transfers . . . 132
5.14.7 MAH Amendment and Renewal Processes . . . 134
5.15 Drug Labeling and Naming . . . 135
5.15.1 Drug Names . . . 135
5.15.2 Drug Labels . . . 136
5.16 Over-the-Counter Drugs . . . 137
5.17 Vaccines and Preventive Biologics . . . 138
5.17.1 Vaccine Post-Market Surveillance and Injury Compensation . . . 141
5.18 Other Specially Regulated Drugs . . . 144
5.19 Role of Drug Patents and Non-Patent Exclusivities . . . 146
5.19.1 New Drug Monitoring Period . . . 148
5.19.2 Early Forms of Patent Linkage . . . 149
5.19.3 New Patent Reforms . . . 150
5.19.4 Patent Linkage . . . 151
5.19.5 Patents and Procurement . . . 153
5.19.6 "Bolar" Exemption . . . 154
5.20 Drug Manufacturing . . . 154
5.20.1 Manufacturing Local and Pathway Determination . . . 155
5.20.2 Drug Manufacturing License and GMP Compliance . . . 157
5.20.3 Marketing-authorization Holders and Manufacturing Licenses . . . 158
5.20.4 Manufacturing of APIs/Bulk Chemicals and Drugs and Ingredients for Export . . . 160
5.20.5 Counterfeit Drugs . . . 161
5.20.6 Supply Chain Aspects . . . 163
5.21 Drug Distribution . . . 164
5.21.1 Drug Importation and Export . . . 165
5.21.2 Public Health Aspects of Drug Distribution . . . 167
5.21.3 Distribution to User Entities and Pharmacies . . . 168
5.21.4 Anti-corruption in Drug Distribution . . . 169
5.21.5 The Two-Invoice System . . . 171
5.21.6 China's Health Care System, Pricing and Reimbursement, and Procurement . . . 172
5.21.7 Prescription of Drugs By Healthcare Providers . . . 174

5.22 Advertising and Promotion 175
5.22.1 Drug Advertising 175
5.22.2 Drug Information 176
5.22.3 Medical Representatives 176
5.22.4 Patient Support Programs 177
5.23 Post-Market Surveillance 178
5.23.1 Pharmacovigilance Systems and Individual and Periodic Event Reporting 178
5.23.2 Adverse Drug Reaction Reporting 179
5.23.3 Quality Issue Reporting 180
5.23.4 New Risk Mitigation Systems and Post-Market Studies Under the Drug Administration Law 181
5.23.5 Annual Reporting 181
5.23.6 Drug Dossiers 182
5.23.7 Recalls 183
5.24 Perspectives on Drug Inspections and Regulatory Enforcement 184
5.24.1 Inspections 184
5.24.2 Penalties and Compensation 186
References 188

6 Medical Device Regulation 203
6.1 Introduction 203
6.2 Early Device Regulation 204
6.2.1 The 2014 MDSAR 205
6.2.2 The 2017 Revision and Implementation of the Innovation Opinion 206
6.2.3 2021 MDSAR 207
6.3 Medical Device Definition and Classification 208
6.3.1 Medical Device Definition 208
6.3.2 Device Classification 209
6.3.3 Medical Device Standards 213
6.4 Medical Device Registration and Record-Filing Process 216
6.4.1 Medical Device Pre-Clinical and Clinical Evaluations 216
6.4.2 Use of Foreign Data and Real-World Evidence 221
6.4.3 Other Studies on Medical Devices 222
6.4.4 In Vitro Diagnostics 222
6.4.5 Laboratory Developed Tests 223
6.4.6 Research Use Only IVDs 224
6.5 Medical Device Registration Process 225
6.5.1 Division of Labor Within NMPA 225
6.5.2 Registration Requirements 225
6.6 Technical Evaluation Process 231

6.7 Naming and Labeling . . . 233
6.7.1 Medical Device Naming . . . 233
6.7.2 Labels/Instructions for Use . . . 233
6.8 Device License Renewals . . . 235
6.9 Amendments to Pending Applications . . . 236
6.9.1 Change of License Holder Name Amendment . . . 236
6.9.2 New Amendment Regime Under the 2021 MDSAR . . . 237
6.9.3 Special Pathways . . . 238
6.10 Post-Market Obligations . . . 239
6.10.1 Adverse Event Regime . . . 240
6.10.2 Who Is Required to Report Adverse Events? . . . 240
6.10.3 What Is a Reportable Event? . . . 241
6.10.4 Targeted Monitoring of Medical Devices . . . 242
6.10.5 Medical Device Re-Evaluation . . . 242
6.10.6 Recalls . . . 243
6.11 Medical Device Distribution . . . 245
6.11.1 Definition of Distribution . . . 245
6.11.2 Distribution License System . . . 246
6.11.3 Basic Requirements for Becoming a Licensed Device Distribution Enterprise . . . 247
6.11.4 Medical Device Good Supply Practice . . . 247
6.11.5 Traceability and Emerging Unique Device Identifier (UDI) System . . . 248
6.11.6 Two-Invoice System . . . 249
6.11.7 Distributor Self-Inspections . . . 250
6.11.8 Distributor Risk-Tiered Inspections . . . 250
6.11.9 In-Market Sampling Inspections . . . 251
6.12 Medical Device Manufacturing . . . 251
6.12.1 Manufacturing License and Filing System . . . 252
6.12.2 Medical Device GMP Implementation . . . 253
6.12.3 PTR and Release . . . 255
6.12.4 Contract Manufacturing . . . 255
6.12.5 Refurbished Devices . . . 256
6.12.6 Manufacturing Inspections . . . 257
6.13 Medical Device Use Responsibilities . . . 259
6.14 Export of Medical Devices . . . 260
6.15 Digital Health and Medical Devices . . . 261
6.16 Penalties and Enforcement . . . 264
References . . . 265

7 Biosecurity and FDD Research and Development . . . 275
7.1 Introduction to Biosecurity . . . 275
7.2 China's 1998 "Interim" Human Genetic Resource Regime . . . 276
7.2.1 Emergence of the Current Regime . . . 277

7.3 Collaborators and Collaboration 280
7.3.1 Other Entities 281
7.4 Data Transfer 282
7.5 HGR Collection Permits 285
7.6 Intellectual Property Sharing 285
7.7 Application Delay and Other Difficulties Impeding Development 286
7.8 Sample Transport 288
7.9 Site Inspections, Self-Inspections and Surveys 288
7.10 Enforcement and Penalties 289
7.11 The Biosecurity Law and Surveillance of National Human and Biological Resources 292
References 293

8 Food Regulation in China 295
8.1 Introduction 295
8.2 History of Food Law and Regulation 296
8.3 General Principles of Food Regulation 297
8.4 Food and Food Ingredients 301
8.4.1 Mitigating Risk of Using Unapproved Ingredients . . . 303
8.4.2 Food Related Materials and Food Contact Substances 305
8.5 General Foods Versus Special Foods 306
8.5.1 Health Foods 307
8.5.2 Health Food Applications 310
8.5.3 Health Food Ingredient and Claim Reform 311
8.5.4 Health Food Manufacturing 312
8.5.5 Health Food Distribution 312
8.5.6 Health Food Labeling and Advertisement Restrictions 313
8.6 Infant Formula 313
8.7 Foods for Special Medical Purposes 316
8.8 Food Labeling 317
8.9 Food Manufacturing 318
8.10 Food Transportation and Distribution 320
8.11 Food Importation and Export 321
8.12 Food Regulation Enforcement 324
8.12.1 How Does Enforcement Start? 324
8.12.2 Food Litigation 325
8.13 Administrative and Criminal Sanctions 330
8.13.1 Criminal Law and Food Safety 331
8.14 Food Safety Accidents 332
8.15 Food Recalls 332
References 334

9 Cosmetics Regulation in China 341
9.1 Introduction to the Cosmetics Space 341
9.2 Definition of Cosmetics and Product Classification 344
9.3 Non-Special Use (“General Cosmetics”) Versus “Special Use” 346
9.3.1 Non-Special Use (“General”) Cosmetics 347
9.4 Children’s Cosmetics 348
9.5 Marketing-Authorization Holder System 349
9.6 Application Requirements 350
9.7 Animal Testing 356
9.8 Certificates of Free Sale 357
9.9 Cosmetic Ingredients 357
9.10 Cosmetic Manufacturing 359
9.11 Manufacturer Facility Licenses 360
9.12 Cosmetics GMP 361
9.13 Cosmetics Distribution 362
9.14 Cosmetics Claims and Advertising 363
9.15 Cosmetovigilance (Post-Market Surveillance) 367
9.15.1 Adverse Reactions 367
9.15.2 Recalls 368
9.15.3 Cosmetic Safety Re-Evaluation 369
9.15.4 Testing and Inspections 369
9.15.5 Enforcement 371
References 371

10 Advertising and Promotion for Regulated Products 375
10.1 Introduction 375
10.2 General Principles of Product Promotion 376
10.3 Entities Subject to this Law 377
10.4 What Constitutes an Advertisement 377
10.5 Intersection with Labeling 380
10.6 General and Specific Content Requirements 380
10.7 Overlapping Claims and Categories 384
10.8 Other Product Restrictions 386
10.9 Advertising and Promotion Enforcement 386
References 387

Introduction and Methodology 1

1.1 Scope and Methodology

This book explores the development of China's system for regulating foods, human pharmaceuticals, medical devices, and cosmetic products ("FDD regulation").[1] It analyzes the evolution of China FDD Regulation and its trajectory, including its balance between a science-based regulatory approach for approving innovative and consumer-safe products and China's political and economic policies.

It provides my perspective as a foreign practitioner who has been observing and interacting with this system in China for over twenty years and has watched China grow into an increasingly influential regulator in these product spaces, in part by adapting various concepts from the United States, the European Union, Japan, India, Argentina, and other countries to its own circumstances. My observations in this book consider China both in comparison to other countries with substantial regulatory systems and on its own as a regulator with a unique approach to these areas.

This book is intended as a resource for both foreign and Chinese practitioners, policymakers, and students in this area to gain an overall understanding of the basic structure of China's FDD regulatory system and the interpretation of those regulations in practice, and to provide significant historical context, with particular emphasis on developments over the past ten years, which is the most intense period of reform.

In some cases, the information in this book relies not on a specific law or regulation but on experiences within the Chinese system, including knowledge of customary practice and conversations with regulators, key opinion leaders, practitioners, academics, and other stakeholders in this area about their experience doing business in China. There are many unwritten rules or practices in China FDD regulation that are constantly evolving. My understanding continues to develop.

[1] It does not cover traditional Chinese medicine, or animal food and health products.

J. C. Balzano, *China Food and Drug Law*, Law for Professionals,
https://doi.org/10.1007/978-3-031-61901-4_1

I have been fortunate to work with colleagues in China, the United States, Japan, and Europe, many of whom have worked with me on projects that contributed to the knowledge I present in this book. They are thanked in the acknowledgement section. Nevertheless, the views in this book are mine, formed from my own reading of the laws and regulations in the original language, interactions with regulators and practitioners in China, and participation on various policy debates in China and the U.S. Translations of quotes from provisions of Chinese law or excerpts of cases, unless otherwise expressly indicated, are by me. Any errors are mine alone.

This book will not be a replacement for an original Chinese-language volume on these issues, but I hope that those in China will also find it useful, much as I have found listening to and reading Chinese commentators' observations of the United States helpful.

1.2 Timeline of China FDD Regulation

As with all stories, this one must begin somewhere. Although an examination of China FDD Regulation could stretch back deep into China's vast 5000-year past by examining traditional eating and Chinese medicine, or even to the 1950s and 60s when the issues were more about food and drug security (and shortages), this story starts much more recently. In the twenty-first century, China—after nearly twenty years of legal reform, beginning in the early 1980s—began to develop regulation necessary to be on par with other major markets in health foods, innovative medical products, and trendy cosmetics and lifestyle products.

Laws or regulations did cover these products during the early history of the People's Republic of China, dating back into the 1950s and, more substantially, to the early 1980s, when the first drug administration and food hygiene laws were enacted. However, it was the period between 2000 and 2004, when China entered the World Trade Organization (WTO), that the country began to revise existing foundational laws and regulations and the increasingly comprehensive China FDD regulatory regime began to take shape. China revised both its Drug Administration Law and Food Safety Law (formerly the Food Hygiene Law) and developed the first comprehensive framework regulation on medical devices, all within this period. In the second decade of the twenty-first century, China FDD regulation hit a fever pitch, with second and third comprehensive revisions to some of these laws and regulations.

1.3 Safety, Public Health, and China's Global Impact

Beginning in the early twenty-first century, food safety became a serious public health and political issue in China, with significant infant formula adulteration issues making international news in 2004 and 2008. In both cases, the effects were terrible, resulting in malnutrition, permanent health damage, and even death, for young infants. These and other similar issues marked the start of several years in which food safety in China was a domestic and international concern.

In 2007–2008, China made international news for its distribution of melamine-contaminated milk and pet food and antifreeze-tainted toothpaste. In 2008, the scandals continued with adulterated heparin exported to the United States, sharpening the focus on China's role in the global supply chain of active pharmaceutical ingredients. Combine these incidents with a SARS outbreak in China and the global struggle against coronavirus (COVID-19), along with China's sheer economic and geographic size, and it is fairly easy to see why China's ability to effectively handle public health issues and regulated products has become an important issue both domestically and internationally.

Different articles, books, and government reports (some of which continue to emerge) recorded China's problems during the first decade of the twenty-first century, identifying it as a safety concern of other countries as a source of these imports.[2] Those materials have informed the views of policymakers and commentators, and shaped policies and writings in countries like the United States.[3] Indeed, it is possible to find books that argue that China's FDD regulatory issues are a concern for consumer products and the food[4] and drug supply. Certain incidents have become symbolic of an ineffective response by China of making examples of a few cases but not pursuing systemic reform, such as the execution, in 2007, of Zheng Xiaoyu, the head the State Food and Drug Administration, for corruption that ostensibly led to some of the food safety incidents described above. I've been asked: If our company commits a food safety violation, could our people go to jail? Is this how all incidents are handled? As this book will show, the day-to-day reality is different.

The COVID-19 pandemic brought new headlines, including a mixture of anger at the spread of the disease and an early admiration for China's implementation of a "COVID-Zero" standard. That admiration turned to concern as the policy stayed in place longer than expected. The pandemic also marked a test of China's transparency and ability to instill confidence in its own innovative medicines, such as vaccines, and devices. Trade between China and the rest of the world has continued and—sometimes in the face of gloomier trends in other areas—life sciences has remained a growing sector in China.[5]

[2]Gibson and Singh (2018), pp. 137–160.

[3]China Economics and Security Review Commission (2021), p. 294.

[4]Liu et al. (2019), pp. 4127–4139.

[5]See, e.g., Wu (2022).

1.4 Health Care Reforms and Access to Care and Products

Health care access for its citizens is a concern for China. It first introduced policies in 2009 that established goals of providing universal health care in China to replace the health care system supported by China's state-run command economy since 1980. China set out to offer insurance through state-run urban and rural plans and to increase coverage. What began with access to locally made generic medicines has grown to include more affordable access to cutting edge, innovative medicines and cell and gene therapies in disease areas such as oncology, infectious disease, and cardiovascular disease. The infrastructure of basic plans is expanding to include drugs for rare diseases through Critical Disease Insurance and collaborations with an emerging, but still fledgling, private insurance industry. Hospitals themselves have become sites for increasing collaboration with innovative drug and digital health companies, despite sometimes limited room under law and regulation for foreign investment in this area. Nor has this evolution been entirely about bricks-and-mortar hospital resources. The chapters that follow will discuss how China has an increasingly advanced regulatory regime for internet hospitals, e-diagnosis, and telemedicine.

Other policies aimed to improve nutrition and exercise and promote a healthier lifestyle, such as Healthy China 2030, in which China has set goals for attention to healthy eating and exercise habits and the adoption of digital health technology to monitor physical health.

1.5 Expansion Efforts to Develop Innovative Drug and Device Industries

China has embarked on an era of innovation. The government has put in place policies aimed at decreasing the time to market and, increasing marketing access by establishing a procedure for drug companies to apply for reimbursement of their products via the state insurance plans. China has proposed and, in some cases, begun to implement policies that *could* protect intellectual property ("IP") over drugs in similar ways to other markets, such as a means for pre-market resolution of patent disputes with generics and protection of clinical data (i.e., data exclusivity), although significant IP questions remain, including in the area of trade-secret protection.

In 2015, China established the Made in China 2025 policy, setting goals for moving China's production away from lower-tech manufacturing arrangements and developing capacity in high-tech fields, such as human pharmaceuticals and medical technology. At around the same time, China issued its Thirteenth Five-Year Plan (a blueprint for industrial production), with a focus on innovation in various areas, including drugs and medical devices. As the following chapters will show, these high-level policies significantly strengthened China's resolve to reform regulation in these spaces, permitting the dismantling of several legal and regulatory restraints that had made the development and marketing approval processes slow and unproductive.

These reforms have not been entirely government-led. China has an increasingly sophisticated drug and device sector, thanks both to government subsidies and to the substantial investments in research and development and manufacturing from innovative companies and venture capital firms. Companies make investments on their own or through partnerships with either emerging Chinese biotech companies or the innovative arms of existing Chinese private and state-owned companies in the areas of drugs, food, and medical devices.

Companies from around the world conduct clinical trials on new medicines in China's state-led hospitals, work with healthcare key opinion leaders in China on collaborations, and these ventures provide valuable insight into international practices and can serve as the training grounds for important business, legal, regulatory, and quality talent in China. China's large drug manufacturers—such as Qilu, Hengrui, and Hisun—have begun to develop their own innovative drug divisions, in addition to their generic and active-pharmaceutical-ingredient businesses.

Reforms in 2015–2017 have driven other private-sector innovative drug companies, some of which have substantial operations in the United States and are listed on the Hong Kong or foreign stock exchanges, to raise capital as China-focused innovators. An increasing number of companies are pursuing data in China, and the role that those data play in marketing applications overseas is increasing. In 2018, the drug Roxadustat for the treatment of anemia obtained its first worldwide approval in China—taking advantage of new policies that rewarded China-first or simultaneous drug marketing submissions in China and other jurisdictions to reduce what had become known as a substantial "drug lag" of China approvals, sometimes over ten years behind other countries.

On January 15, 2020, the United States and China signed a new Phase One trade agreement. Although broad in scope, it also provides for several specific measures to protect the intellectual property of medicines, including methods to resolve patent disputes prior to generic marketing approval (patent linkage), patent term restoration, and stronger commitments to avoid conditioning approval on technology transfer. These efforts are meant to help resolve the dearth of IP protection for innovative pharmaceutical products in a country with, some estimates, a 75% invalidation rate for innovative drug patents.[6] China has made good on *some* drug-related reforms to date, including new drug-patent term extension and implementation of the patent linkage systems, which are explained in the chapters that follow. It remains to be seen how this trade deal, like the WTO accession that led to the initial push in the drug space, will drive additional reforms or lose force as China turns in different directions.

Thus, we have two policy eras close in time that have combined to inform China's regulations today: policies for safer, more affordable, and regularly available food, drugs and medical devices and policies that aim to position China as a leader in

---

[6]Lu (2018).

cutting-edge food and health care technologies such as genomics, digital health, and personalized medicine and nutrition.

1.6 Government and Legal Reform

Starting in the early years of the twenty-first century, the policies above were the basis for government (i.e., administrative agency) and legal reform in the FDD area. The progress has at times seemed slow and uneven, but, overall, there has been substantial reform in these areas.

The reorganization of central and local administrative agencies is a regular feature of government in China. Therefore, the agency (or agencies) in charge of FDD regulation in China have taken different forms, but the trend has been toward more specialized agencies with more technical experts and greater capacity to review and approve FDD products.

In 2003, China created a State Food and Drug Administration, from the agency that had previously been the State Drug Administration. This began a process of forming a unified regulator with comprehensive authority over China FDD regulation. In 2013, China established an agency directly under the State Council (the chief executive organ of government), the China Food and Drug Administration (CFDA), with authority over the primary aspects of development, registration, production, distribution, and certain aspects of use of foods, human drugs, medical devices, and cosmetics.

CFDA's creation began a more serious level of FDD regulation in China, one that reformed regulation at a rapid pace and promised to implement reforms that China had contemplated on some level since its entry into the WTO. Although China does not have one uniform law like the Food, Drug and Cosmetic Act in the United States, it has framework laws or regulations that govern each of the individual areas of regulation. In 2013, CFDA began or appeared to accelerate reform of each of these pieces of legislation. This was not only because of its increased standing as an agency, but also because CFDA had new leadership under its forward-thinking commissioner, Bi Jingxuan. As discussed, certain prior SFDA leaders were caught up corruption scandals, but Commissioner Bi ushered in a new era of serious regulatory and technical leadership that drove many reforms.

CFDA's authority was somewhat diminished in 2018. At that time, during a routine government reorganization, CFDA became (in part) the National Medical Products Administration (NMPA), which was a subordinate bureau of the State Administration for Market Regulation (SAMR). SAMR is a super-ministry created from various other ministries, including those regulating businesses, consumer protection, standardization, product quality, advertising, intellectual property, and certain aspects of antitrust. The government removed the food regulatory function from NMPA and transferred it directly to SAMR. Throughout this book, NMPA (rather than CFDA) may be used even when the developments pre-date the official establishment of NMPA.

Shortly after the reorganization, Commissioner Bi resigned because of a scandal involving vaccine manufacturers that were not meeting minimum good manufacturing practice standards. Once formed, NMPA issued a document stating that NMPA would continue to implement the same rules and documents that CFDA had put in place, but the pace of reform that existed from 2015–2017 in the drug and device spaces slowed.

As explained in Chap. 2, NMPA is not the only agency with authority over FDD regulation. China's Communist Party, its chief executive State Council, the National People's Congress legislature, the Supreme People's Court, various ministries, including NMPA's parent, the State Administration for Market Regulation, the health care regulator, the National Health Commission, and other ministries and commissions under the State Council, and certain provincial governments have significant influence in this area of regulation.

1.7 Chapters to Come

This book is organized into general and specific chapters. The first chapters discuss government organization and the overall administrative processes governing rulemaking, penalties, and appeals from agency decisions into which FDD regulation fits. The fourth chapter discusses the common features of each area of FDD, because the similar structure and organization (e.g., the general separation between imported and domestically made products) of these regulatory spaces give insight into China's overall legal and industrial policy. The next set of chapters is dedicated to a specific regulated product or products, drugs (including vaccines), medical devices (including digital health products), cosmetics, and foods (including dietary supplements and infant formula). The primary purpose of each chapter is to describe the regulatory structure for that product, including regulation of the research and development stage, pathways for pre-approvals (if any), distribution within China or abroad, advertising and promotion, and the monitoring and resolution of post-marketing safety and effectiveness issues.

These regulatory structures do not evolve in a vacuum; therefore, the chapters also set forth relevant context and case examples that have helped to shape the regulations. For example, drug regulation has developed alongside China's expanding government-run health care system, including national and provincial-level insurance plans. It has also been influenced by drug regulation in places like the United States, the European Union, and in many respects, Japan. Food regulation has developed in part based on the contamination scandals that occurred from 2004–2008 and, in some respects, by COVID. Medical device regulation and its focus on upgrading quality systems and inspections for domestically made products has been influenced by the heavy reliance on imported devices in hospitals. This book details these events and issues as part of the context that has shaped FDD regulation.

References

China Economics and Security Review Commission (2021) US and China Links in Healthcare and Biotechnology. In Annual Report, Chapter 2, Section 3. https://www.uscc.gov/sites/default/files/2020-12/Chapter_2_Section_3%2D%2DU.S.-China_Links_in_Healthcare_and_Biotechnology.pdf.

Gibson R, Singh JP (2018) China Rx: exposing the risks of America's dependence on China for medicine. Prometheus Books, Connecticut

Liu Z, Mutukumira AN, Chen H (2019) Food safety governance in China: from supervision to coregulation. Food Sci Nutr 7(12):4127–4139

Lu Z (2018) Over 75% patents are invalidated? New drug research and development is afraid to fall into the Paradox of Chine Style Innovation. China Intellectual Property Magazine, Issue 137. http://www.phirda.com/artilce_18053.html?cId=1.

Wu Y (2022) China's Biopharma Industry, Market Prospects and Investment Paths. China Briefing. https://www.china-briefing.com/news/china-booming-biopharmaceuticals-market-innovation-investment-opportunities/.

Agency Jurisdiction Over Food and Drug Law

2

2.1 Overview of the Current Agency Structure

The agency that currently holds most of the jurisdiction over food, drug device, and cosmetic regulatory issues in China, is the super-ministry of the State Administration for Market Regulation (SAMR), into which the former China Food and Drug Administration (CFDA) and several other agencies merged in 2018. Instead of CFDA, there is now a subordinate bureau under SAMR that regulates drugs, cosmetics, and medical devices, referred to as the National Medical Products Administration (NMPA). SAMR is a sprawling combination of this and other areas of consumer protection, including advertising and promotion, quality assurance, antitrust, standardization and accreditation, and intellectual property. Reforms in 2023 removed the National Intellectual Property Administration (the agency that makes intellectual property policy and resolves disputes) from SAMR's control and made it a separate agency under the State Council. SAMR supervises NMPA, and sets procedures for rule promulgation and administrative procedures, and SAMR issues administrative rules in the spaces NMPA regulates, although NMPA may draft the rules. NMPA itself issues lower-level guidelines and normative documents.

The food functions previously under CFDA are now directly under SAMR. SAMR has regulatory jurisdiction over prepackaged food, including dietary supplements, infant formula, and food formulated for special medical purposes that previously belonged to CFDA. These functions are housed in six separate food-related divisions directly under SAMR. Despite this reshuffling, SAMR and NMPA still enforce most of the same rules and documents that CFDA promulgated.

The following chapter describes the history and the current structure and the role of other agencies in the FDD area.

J. C. Balzano, *China Food and Drug Law*, Law for Professionals,
https://doi.org/10.1007/978-3-031-61901-4_2

2.2 History of Food and Drug Regulatory Jurisdiction in China

China has regularly reorganized its central agencies since the 1980s, doing so with the stated goal of making governance more efficient. The agencies went from previously running sectors of the economy—e.g., a ministry of coal and electricity—to regulating those sectors, e.g., the State Development and Planning Commission became the National Development and Reform Commission. The penetration of the state, i.e., the government, into the economy and social sphere is generally very strong, and it still manages the assets of its own set of state-owned-enterprises (SOEs) through the State-Owned Assets Supervision and Administration Commission of the State Council. Some drug and device companies have either whole or part state ownership. The local industry associations have agency sponsorship, as do the branches of foreign trade associations that operate in China. This sponsorship is a form of supervision of the activities of these associations.

The trend in agency regulation of the FDD area in China has been consolidation of regulatory functions for products under increasingly specialized agencies. China has adopted a life-cycle approach to food and drug regulation, meaning that regulations create a system to supervise the product from pre-development work through post-market monitoring. NMPA is now organized with this approach in mind, with different divisions and their sub-divisions devoted to development, registration, and post-market surveillance. NMPA relies on its local counterpart agencies for investigation and enforcement of many of its rules against local companies.

A combination of laws and policy documents establish NMPA's and other agencies' jurisdiction. These do not always match. In other words, the National People's Conference (NPC) and the State Council do not always amend the applicable law or regulations when an agency's scope of jurisdiction changes. Instead, they may change an agency's jurisdiction via a plan that briefly discusses what the new scope of agency jurisdiction. They also provide an administrative document that gives localities guidance on how to structure their local agencies in accordance with the arrangement at the central level. Different provinces do that in different ways, with some perhaps already having adopted a new arrangement under a pilot program and others following more slowly behind. This is what happened when different provinces established market supervision bureaus before SAMR was established at the national level of government.

Sometimes the law governing an area will not be amended to include the correct agency until years later. For example, the Cosmetic Hygiene Management Regulation (CHMR) did not include CFDA or NMPA until 2019. From 1989 until 2019, the CHMR stated that the health authority under the State Council was in control of cosmetics regulation. In contrast, the Drug Administration Law was amended in 2015 to include CFDA as the governing agency. Certain adjustments were made in the law to make it clear that CFDA had comprehensive jurisdiction.

Various agencies have controlled this area. Arguably, drug regulation can be traced back to the Pharmaceutical Administration under the Ministry of the Chemical Industry in the 1950s. In 1978, China established the State Drug Administration directly under the State Council. Thereafter it was bounced back and forth between an agency directly under the State Council and an agency subordinate to the State Economic Commission until 2003, when the government moved the food functions that resided within the Ministry of Health under a newly established State Food and Drug Administration. SFDA was also briefly subordinated to the Ministry of Health from 2008 to 2013.

Other agencies have played scattered roles. For example, the Ministry of Commerce was involved in the regulation of slaughterhouses.[1] Regulation of device importation resided under the General Administration for Quality Supervision, Inspection, and Quarantine (which housed the Standardization Administration of China—now under SAMR), through which products received the China Compulsory Certificate (CCC) mark of quality to enter China. The Ministry of Agriculture has always had supervisory authority over raw agricultural products and animal drugs. Cosmetics regulatory functions were originally given to the Ministry of Health under Cosmetics Hygiene Management Regulation.

Most of the responsibility for regulated products moved to what ultimately became the China Food and Drug Administration (CFDA) in 2013, which was a standalone agency until 2018. CFDA contained most responsibility for drugs, medical devices, cosmetics, and many functions of food. Food remained the most fragmented area with certain important functions, such as food safety risk assessment and standard making, residing in other ministries, such as the National Health and Family Planning Commission (formerly the Ministry of Health) and the Ministry of Agriculture.

2.3 Making of the NMPA

Although CFDA had a substantial number of responsibilities, it did not have substantial resources compared to other areas of focus in the China government. For example, it had only about 350 personnel (although much more in the provinces) and—although its budget is not public—charged modest fees of a few thousand dollars for applications, when agencies in other countries and regions have over 10,000 personnel and charge in the millions of dollars in review fees. Although important as a regulator of critical everyday products that all of China's vast population, CFDA was also not necessarily thought of as a significant agency in the Chinese government, like the Ministry of Commerce, Ministry of Finance, or the Ministry of Industry and Information Technology.

Some of these aspects of CFDA changed. For example, in June 2015 the government allowed it to charge much more substantial fees for reviewing license applications for drugs and medical devices, raising them from a few thousand to tens or hundreds of thousands of dollars. Later that year, the central government committed to expanding the number of reviewers of product applications in

[1] Ministry of Commerce (2008).

CFDA's Center for Drug Evaluation (CDE) and Center for Medical Device Evaluation (CMDE), from fewer than one hundred to several hundred. The numbers of the central agency also stand in stark contrast to the 10s of thousands of the personnel in the medical products administrations throughout the different provinces.

However, NMPA, like CFDA, has remained small compared to other agencies. It is not yet clear that NMPA's inclusion under the super-ministry umbrella of SAMR has lifted its ability to better deal with enforcement or application reviews, as its number of specialized personnel remains small.

2.4 The Structure of NMPA

NMPA's divisions correspond with different product spaces and functions related to the administration of the agency. Its general divisions include a Consolidated and Financial Planning Division, a Policy and Law Division, and a Science and Technology and International Cooperation Division. Like other agencies in the Chinese government, there is an internal party committee that ensures compliance with the Chinese Communist Party's (CCP) policies.

For both drugs and devices, NMPA has both registration divisions and supervision divisions. The registration divisions handle, among other things, marketing approval (i.e., issuance of licenses) for each of these products. For drugs, the registration division is broken down into specialized subdivisions for research, chemical drugs (both prescription and over-the-counter), Chinese and ethnic medicine, and biological products. Although the Center for Drug Evaluation (CDE), an affiliated center, but not a division approves clinical trial applications, the research division organizes and releases drug standards and supervises the implementation of good laboratory and clinical practice standards.

Registration and supervision for devices are subdivided by type of device, including powered devices, unpowered devices, and *in vitro* diagnostics, to name a few. Supervision divisions handle post-market surveillance matters for drugs, medical devices, and cosmetics.

There is no cosmetics registration division, but there is now a cosmetics supervision and administration division. This division has a general office and two subdivisions. The first subdivision handles the organization and implementation of product registration and filing activities and the second organizes post-market surveillance efforts. There is no center for cosmetics evaluation. Rather, a division within another affiliated center, the National Institute for Food and Drug Control, handles similar work to the CDE and the Center for Medical Device Evaluation (CMDE).

2.5 NMPA-Affiliated Centers and Provincial Counterparts

NMPA also has several public institutions (*shiyedanwei*), sometimes referred to as "centers" or "institutes," that are subordinate to NMPA (as they were to CFDA), but not divisions of it. The number of personnel in these centers has increased significantly, in addition to the 300–400 personnel that work for NMPA's divisions. For instance, there are estimated to be over 800 reviewers (over twice the number in NMPA itself) in the Center for Drug Evaluation (CDE).

These public institutions include the CDE and the Center for Medical Device Evaluation, (CMDE), which conduct the technical reviews of drug and device applications and issue recommendations to the registration divisions of NMPA as to whether it should approve a product. They must organize technical experts, including outside experts in some cases, to conduct these reviews and issue recommendations. These centers also handle the approval of clinical trial applications for drugs and certain devices and required amendment and renewal applicants for drug and medical device registrations.

Other important centers issue technical guidelines and technical documents to guide applicants, reviewers, and inspectors. The Center for Drug Revaluation and National Center for Adverse Drug Reaction Monitoring (one center with two names) creates policy for and, at national level, tracks and engages in risk analysis concerning adverse drug reactions and medical device adverse events. Individual drug or device adverse reaction monitoring centers also track and engage with stakeholders locally. The China Pharmacopeia Commission is responsible for China's Drug Code, which is a collection of standards covering drug products, ingredients, and testing methods. Compliance with some of these standards may also be required for devices or cosmetics. The Center for Food and Drug Inspections (CFDI) has played an increasingly pivotal role in organizing and guiding the technical policy behind manufacturing inspections, whether in China or, increasingly, through onsite or virtual inspections at overseas facilities that make imports destined for China, and clinical trial data inspections.

There is also the National Institute for Food and Drug Control (NIFDC), which, among other functions, oversees standard-drafting for devices and testing laboratories for drugs, devices, and cosmetics. "Registration" or sample testing remains an important part of development and marketing applications for all three types of products. This is testing according to general or product-specific standards to ensure safety and in some cases effectiveness. NIFDC has specialized testing divisions for chemical drugs, biologics, Chinese medicine, medical devices, *in vitro* diagnostics, cosmetics, and drug excipients and packaging materials.

NIFDC also plays a substantial role in cosmetic regulation, including developing standards and lists of banned and restricted ingredients (discussed in Chap. 9). As CDE and CMDE do for drugs and devices, NIFDC issues various guidelines and technical documents related to the registration and testing of cosmetics. In addition, as noted above, it houses the cosmetics safety and technical evaluation center that organizes expert reviews of special use cosmetics and new cosmetics ingredients, as well as reviews of filed dossiers of non-special use, imported cosmetics.

The chapters that follow discuss the functions of these various centers, as needed, in more detail.

2.6 Provincial-Level Medical Product Associations

The bureaucracy organized under provincial governments (or provincial-level municipalities such as Beijing, Shanghai, Chongqing, and Tianjin) includes provincial- and municipal-level food and drug agencies. The Provincial-Level Medical Product Associations (PMPAs) and MPAs at lower levels of local government have various responsibilities, including product-approval responsibilities, either fully approving applications, such as those for Class I and II medical devices, or serving as initial gate keepers for applications before passing them on to NMPA. The PMPAs and Municipal Medical Products Administrations (MMAs) also play a role in the review and approval of relevant facility licenses, including drug, device, and cosmetic manufacturing and drug and device distribution licenses or filings. They inspect these manufacturing and distribution facilities both prior to the license issuance and thereafter on basis of risk. They conduct spot checks on products by pulling them from circulation and testing them, and these local MPAs receive recall and adverse event and reaction reports for initial analysis. For medical devices and drugs, the local medical products administrations review and approve advertisements (see Chaps. 4 and 5).

The local MPAs are sometimes closely associated with the local Administration for Market Regulation departments, which play an important role in the investigation and enforcement of penalties for violations of various regulations, including those related to consumer protection. NMPA does not itself directly conduct various enforcement activities, but instead makes the policy on enforcement. It must rely more heavily on the local provincial bureaucracy in the local medical product administrations, which also have ties to local provincial governments. Fines and other penalties are issued at the local level.

2.7 Food Agencies

Under the earlier versions of the Food Safety Law referred to as the Food Hygiene Law, the Ministry of Health had the responsibility for supervising food safety. Another agency, the General Administration for Quality Supervision, Inspection, and Quarantine (AQSIQ) also had control over food import policy and even some domestic issues, such as food and food-additive manufacturing.

Many of these functions were consolidated under CFDA. Food was added to its name in 2003, and revisions of the Food Safety Law transferred many of the responsibilities, including that for supervising manufacturing (including health food, general food, and food additive manufacturing) and distribution of all foods to CFDA. Responsibility for making and proposing revisions to food safety standards and approving new food ingredients remained with the Ministry of Health, then referred to as the National Health and Family Planning Commission.

As of 2018, SAMR has direct jurisdiction over food and special foods (i.e., general processed foods, health foods, infant formula and foods formulated for special medical purposes). The responsibilities for these products are spread over six food-related divisions, including (1) the food safety coordination division, (2) the food safety inspection division, (3) the food manufacturing safety division, (4) the food distribution division, (5) the special food supervision division, and (6) the product quality and safety division.

A Center for Food Evaluation (CFE) is also under SAMR and plays a role in the evaluation of the various special food applications for approval. This includes health foods, infant formula, and foods formulated for special medical purposes.[2] CFE helps organize site inspections of facilities that make special and general foods.

Like CFDA, SAMR serves as the general office of the National Food Safety Commission, which has a mandate to analyze the overall food safety situation and make policy recommendations. A vice premier heads the Food Safety Commission. The membership consists of vice-minister-level officials from various ministries and commissions.

Based on its actions so far, SAMR appears to be focused more on enforcement, rather than rule-making. New food legislation was slow to emerge. SAMR did not finish various ongoing health-food reform that NMPA started after the 2015 amendment to the Food Safety Law until 2023 (Chap. 8). As discussed above and will be discussed below, the National Health Commission also plays a substantial role in food regulation, particularly in the areas of risk-assessment and standard making.

2.8 Other Relevant Functions of SAMR

SAMR has other relevant functions besides its supervision of NMPA and its direct food regulatory jurisdiction. Like its predecessor the State Administration for Industry and Commerce, SAMR still has jurisdiction over many areas that are relevant to these regulated products, including advertising, consumer protection, and antitrust. SAMR has jurisdiction to enforce the provisions in the Advertisement Law that pertain to drugs, devices, animal drugs, animal feed, and special foods. As will be explored below, the local MPAs have jurisdiction to approve drug, device, and health-food advertisements, and the local AMRs have jurisdiction to enforce the law by fining violators and seizing non-compliant advertising materials.

SAMR now also administers the Law for the Protection of the Rights and Interests of Consumers (LPRIC). The LPRIC governs multiple aspects of marketing products to Chinese consumers, including promotion, collection of consumer information and communications with consumers, complaints and reporting,

[2]Center for Food Evaluation, State Administration for Market Regulation (2023).

investigations of counterfeit products, and product recalls. SAMR has taken over the consumer product recall divisions that used to be part of AQSIQ.

The Law Against Unfair Competition (LAUC) is also partly administered by SAMR. This law governs other aspects of these regulated products, including misleading promotion. As we will discuss later, SAMR and its local counterparts frequently use the LAUC to restrict promotional activities and materials that may not be deemed traditional advertising. For example, the LAUC can be used to penalize off-label promotion (i.e., promotion beyond the approved indications) of drugs and medical devices. LAUC also governs certain key aspects of trade-secret protection.

SAMR's other functions are important to the products herein, such as product pricing, standard making, and third-party (e.g., laboratories or foreign facilities) accreditation. In the latter two respects, SAMR houses the Certification and Accreditation Administration of China (CNCA), which among various other things accredits laboratories that conduct product testing prior to approval, and the China National Accreditation Service for Conformity Assessment (CNAS), which evaluates various China and ex-China facilities for conformity with Chinese standards. This includes assisting with the registration of the meat and dairy plants that may supply China. The National Standards Administration is responsible for the promulgation of national mandatory and recommended standards, i.e., *guobiao* (GB) and *guobiao/tuijian* (GB/T), respectively.

2.9 The Ministry of Agriculture and Rural Affairs

The Ministry of Agriculture and Rural Affairs (MARA), formerly the Ministry of Agriculture, has authority over raw agricultural produce, up until it enters wholesale distribution and processing, and animals for slaughter.[3] This also includes the approval of varieties of genetically modified food and, perhaps in the future, cultured meat (i.e., meat produced by culturing animal cells). SAMR's food divisions and the food manufacturing regime take over once food is processed and packaged.

When the National People's Congress established CFDA in 2013, it issued a document describing the jurisdiction of CFDA and its division of labor. Below is an excerpt of that document, which still in many respects holds true for the split of jurisdiction on food-related matters:

> The Ministry of Agriculture is in charge of the quality supervision and management of edible agricultural products from the time of seeding/planning and raising meat to the time prior to wholesale, retail, or production by an enterprise. It is in charge of animal drugs, feed, feed additives, and the supervision and management of agricultural chemicals within its scope of management, fertilizers and other such things that influence product quality and use. When edible agricultural products enter the wholesale, retail, or production environments . . . the Food and Drug Administration shall supervise management. The [MOA] shall supervise and

[3] State Council of the People's Republic of China (2021).

administer quality and safety of the slaughter of livestock and fowl and the receipt and sale of fresh milk.[4]

Although animal food and health products are beyond the scope of this book, the MARA is also responsible for the safety of animal feed and feed additives,[5] including new feeds, which, of course, relates to the safety of the human food supply.[6] This includes pet food, which is regulated under the regime for animal feed. Pet food is becoming an increasingly specialized area of regulation as evidenced by, for example, the standards on dog and cat food and the proposal for a new pet food labeling regulation that were implemented starting in 2015 and the pet food rules issued in 2018.[7]

The MARA has primary jurisdiction over animal drugs, vaccines, and other biologics. These products are evaluated and approved for marketing by the Bureau for Animal Husbandry and Veterinary Medicine with assistance from the Institute for Veterinary Drug Control. The MARA's scheme for regulating animal drugs and biologics, including their manufacturing and distribution, is similar to the scheme employed for drugs by the NMPA, although animal drug regulation has not yet embraced the marketing authorization holder system that was adopted nationally in 2019 for drugs (Chap. 5).

2.10 China Customs Administration

China's Customs Administration, in its various forms and by other names, has always been the gateway for various FDD-regulated products and intermediates. One key function that it has regularly performed is to supervise the classification of goods coming into China to determine duties and required testing and inspection. It has been assisted in border testing work by local medical products administrations, which, for example, would know the appropriate testing for imported drugs for investigational use or commercial marketing.

China imposes a variety of standards, including quarantine standards, licensure requirements, and restrictions and prohibitions on animals and plants coming from countries or areas where there have been outbreaks of disease. China Inspection and Quarantine Bureaus (CIQs) in the various ports carry out testing and inspection of these products. CIQs used to be housed under the AQSIQ, but since 2018 CIQs are housed under the Customs Administration. Therefore, Customs now plays financial, trade, and public health roles previously held by AQSIQ.

---

[4] General Office of the State Council (2013).

[5] State Council of the People's Republic of China (2011).

[6] Ministry of Agriculture and Rural Affairs of the People's Republic of China (2022).

[7] General Administration of Quality Supervision, Inspection and Quarantine of the People's Republic of China (2014) Complete Pet Food—Dog Food; Complete Pet Food—Cat Food.

In 2013, AQSIQ's role in importation was described as follows, a description still helpful in understanding its role under Customs:

> The AQSIQ is responsible for the food safety of import and export products, and for their quality supervision, inspection, and supervision and management. Imported food products and food-related products must comply with China's national food-safety standards. The AQSIQ shall collect and store import and export food safety information and it shall inform the Chinese Food and Drug Administration in a timely manner. If food safety accidents produced abroad could have influence domestically within China or a serious food safety problem discovered in imported products, then the AQSIQ shall take risk warning and control measures in a timely manner, and it shall report to the Chinese Food and Drug Administration. The Chinese Food and Drug Administration shall take relevant measures in a timely manner.[8]

AQSIQ played a similar role for medical devices and cosmetics, whereas the State Drug Administration, then CFDA, and now NMPA, has more closely managed drug importation.

The CIQs have a great deal of discretion in carrying out their work to inspect, quarantine, and approve products as safe and of sufficient quality to enter and exit China. The procedures of local CIQs at different ports can differ on certain issues and, while some of these differences may seem insignificant, they can create very significant difficulties for companies that are trying to bring products in and out of China. Each CIQ likely posts an increasing amount of guidance online, which companies importing (and exporting) products should study closely. Different ports can handle issues in significantly different ways.

2.11 National Health Commission

The National Health Commission (NHC) has gone by two other names (the Ministry of Health and the National Health and Family Planning Commission), but its function as the chief health care regulator has remained constant. It regulates the hospitals and health care professionals (HCPs) that comprise China's health care system; it and local health commissions are responsible for issuing the licenses for these entities and individuals. This means that it also supervises investigator or HCP-led research inside of China's hospitals. As discussed below, the National Health Commission played a substantial role in the tide of research during the COVID-19 pandemic.

NHC has bureaus covering emergency responses to outbreaks, basic medicine, insurance, women's and children's health and population "monitoring" and "family planning." (Gone are the references to population control.) It used to play a larger role in drug reimbursement and procurement, but it still plays some role in drug circulation in health care, including by assisting with the scheduling of controlled substances. It houses the State Administration on Traditional Chinese Medicine

[8] General Office of the State Council (2013).

(TCMA).[9] NHC still manages the compilation of the Essential Drug List (with the help of the TCMA), which outlines which drugs medical institutions (i.e., hospitals and clinics) must have on hand and their reimbursement by insurance plans (see below).

The Centers for Disease Prevention and Control (CDC) are also affiliated with the NHC. The CDC acts not as an independent agency but as a public institution subordinate to the NHC. In addition to its role in infectious disease control, the CDC plays an important role in the supervision of distribution of vaccines throughout the country. It is responsible for vaccination work across China's provinces and plays a role in the administration of the compensation scheme for national inoculation program vaccines (compensated by the government) and non-inoculation program vaccines (compensated by the manufacturers).

A third important function of the NHC is its role in food safety. As discussed above, the predecessors of the NHC played a role in various other spaces, such as food and cosmetics. The NHC still retains its role in assisting with the drafting of food safety standards, including through the China National Food Safety Risk Assessment Center, which assists with the monitoring and assessment of safety threats that inform the need for the drafting and amendment of food safety standards. Food safety standards are the core regulatory tool for general foods, which are not pre-approved in China. A separate internal NHC bureau on food safety standards, monitoring, and evaluation supports this work. The NHC also approves two important categories of food ingredients, food additives (e.g., preservatives and flavors) and new food raw materials (e.g., non-additive substances that are "new" because they are not part of traditional eating habits, such as cultivated ginseng).

2.12 National Healthcare Security Administration

The National Healthcare Security Administration (NHSA) was established in 2018 and plays an important role in administering health insurance, including, perhaps most importantly, the government-run health plans that provide insurance coverage for most individuals in China. Specifically, NHSA is a specialized agency that articulates policies on reimbursement under the government health plans. Some of these functions used to reside under the Ministry of Human Resources and Social Security.

The NHSA runs the expert evaluation and negotiation processes that result in drugs being admitted to reimbursement lists for the state health insurance plans. This is also known as the process to admit drugs to China's National Reimbursement Drug List (NRDL), one of the primary positive lists of reimbursable drugs. Once a process shrouded in mystery, the NRDL negotiation process under NHSA is governed by publicly available regulations. The regulations governing these processes were established in 2019 and 2020. This has given more structure to the

[9] State Administration for Traditional Chinese Medicine (2023).

process and allowed for companies to plan for reimbursement as part of their overall launch plans for market.

The NHSA also plays an important role, working with other agencies, in structuring the procurement process through which medical institutions obtain drugs in different provinces. As discussed herein, it has worked to implement the volume-based approach that currently dominates health care procurement.

NHSA is related to China's insurance industry but there is another general insurance industry regulator. The agency governing the insurance industry, including the private commercial insurance industry, is the National Insurance Commission of China.

2.13 The Ministry of Science and Technology

Among other functions, the Ministry of Science and Technology (MOST) plays a substantial role in steering the direction of government-sponsored research. It administers laws designed to promote scientific development and to convert scientific discovery into technology, including the Science and Technology Promotion Law, Promotion of the Transfer of Scientific Achievements Law, and the Science and Technology Award Regulations. These laws and related policies include protection of the government's stake in this research, and it is responsible for certain policies encouraging intellectual property to remain in China and be held by Chinese parties long-term, i.e., "independent intellectual property," which is favored in grants and administrative approvals, such as for new drugs.

MOST is also responsible for administering funds for "key" drug and device research projects that can secure priority review with China's health authorities. It played an important role in directing state funds to targeted areas of research to address the COVID-19 pandemic.

Another important part and function of MOST was the Office of Human Genetic Resource Administration ("Office"). This Office—unknown to many stakeholders prior to 2015—became a key factor determining the direction of many types of research. The Office issues the approvals necessary for foreign companies and foreign-invested companies to collect and utilize the samples and data in research on human subjects. Every study conducted in Mainland China for a drug or a medical device is covered by this regime, and many companies have faced enforcement for non-compliance with the Office's often opaque and complicated dictates. As of May 2024, the National Health Commission administers the human genetic resource regime.

2.14 China National Intellectual Property Administration

Previously the State Intellectual Property Office, the China National Intellectual Property Administration (CNIPA) is relevant for this discussion because it has jurisdiction over patent prosecution and partial jurisdiction over the resolution of patent disputes. This includes invalidity and increasingly infringement disputes. CNIPA is where patent policy may be drafted, including relevant drug patent policy, though other parts of the government certainly have input. For example, CNIPA played a role in drafting the Patent Law, which contains an exemption from infringement for generic drug development, and CNIPA worked with NMPA to establish China's patent linkage system, which permits early resolution of patent disputes prior to the approval and/or marketing of potentially infringing generic drugs or biosimilars. CNIPA also drafts and promulgates other important rules and guidelines such as the Patent Examination Guidelines, which govern data supplementation during the drug development process and patent term adjustment and compensation to restore patent life lost during the patent and drug application processes. Once housed under SAMR, as of mid-2023, NIPA is now a standalone agency under the State Council.

2.15 Ministry of Public Security

The Ministry of Public Security (MPS) is China's police force, and it plays a role in prosecuting activities that rise to crimes under the FDD framework laws and regulations. For example, the Criminal Code permits prosecutions against those who cause harm by producing or distributing fake or substandard FDD products.

Criminal prosecutions for violations of food, drug, and medical device safety standards have increased, and the cooperation between NMPA and MPS has also increased, with a formal mechanism for NMPA and its local counterparts to refer cases for criminal prosecution. At the time of this writing, a concept called "biosecurity" is also increasingly involving the MPS in the biopharmaceutical area. An amendment to the Criminal Code, in 2011, increased penalties for those violations amidst the international controversy surrounding the safety of China's food supply, and amendments in 2020 revised those provisions and specifically criminalized human genetic resource violations (described in Chap. 7).

MPS's function in the food, drug, and device area is to receive information from food and drug regulatory authorities about potential criminal liability, investigate the potential violation, and refer cases to the procuratorate for formal prosecution. Under certain laws (e.g., the Food Safety Law and the Drug Administration Regulation), MPS has the power to detain responsible individuals for egregious violations.

All the FDD framework laws and regulations contain cross-references to portions of China's Criminal Code. When NMPA or the PMPAs (or SAMR or its local counterparts, in the case of food) suspect that a case may contain criminal behavior, they may make a referral of the matter to the MPS for investigation. The medical products authorities play an important role in compiling the basis and evidence for these cases using their expertise of the FDD regulatory system. MPS can also commence the cases on its own, and has the authority to investigate illegal narcotics.

The 2013 jurisdictional document establishing the functions of CFDA described cooperation between that agency and MPS in the following manner, which still holds true in many respects:

> The Chinese Food and Drug Administration and the Ministry of Public Security shall establish mechanisms for administrative enforcement and criminal judicial work. When the Chinese Food and Drug Administration discovers illegal food and drug behavior that relates to a crime, it shall, in accordance with relevant provisions, inform the Ministry of Public Security. The Ministry of Public Security shall quickly and expeditiously conduct an investigation, and it shall make a decision according to law to institute a case or not prosecute. If the Ministry of Public Security invites the Chinese Food and Drug Administration to conduct testing, expert analysis, and determination to assist it, the Chinese Food and Drug Administration shall render assistance.[10]

In 2014, MPS announced that it had established a special bureau to investigate food and drug violations, and, soon thereafter, multiple agencies issued regulations on cooperation on the discovery and investigation of food and drug criminal violations. These agencies included CFDA, the Supreme People's Court, the Supreme People's Procuratorate, MPS, and the Food Safety Office of the State Council.[11] These cooperative measures and their implications are discussed further in the chapter on enforcement.

MPS also plays a regulatory role. It works with the National Health Commission and NMPA to supervise the classification and distribution of controlled substances, including psychotropic and narcotic drugs. It assists with the assignment of risk-based classifications for these types of drugs, when applicable. Class I psychotropic drugs, for example, have a significantly narrower scope of distribution than do Class II.

2.16 Other Ministries and Agencies

There are various other ministries, local agencies, and quasi-governmental organizations that play a role in the FDD system. For example, the Ministry of Industry and Information Technology plays a role in the regulation of Internet and digital technology when it touches on digital health products, and in the regulation of internet websites that inform the public about health products and distribute them. As the value of health care data (including big data that informed tracking and control during the pandemic) has become more apparent, the China Cyberspace Administration plays a pivotal role in administering cyber and data security laws and addressing privacy aspects through China's new Personal Information Protection Law.

The Ministry of Ecology and the Environment administers the regime that registers new chemicals produced or imported into China, and its local counterparts issue the radiation safety licenses necessary for hospitals that use radiation technology, and manufacturing and distribution enterprises that produce or distribute

[10] General Office of the State Council (2013).

[11] China Food and Drug Administration (2015).

radiopharmaceuticals or radiation-emitting medical devices. It administers environmental standards that apply specifically to drug production facilities.

2.17 China's Court System and Food and Drug Enforcement

A discussion of the role of China's court system is beyond the scope of this book. However, while the courts should not be viewed as an area where agency action can be easily undone or policy made, nor do they have any law-making or, arguably, law-interpretation power, they are an important dispute resolution forum. As Chap. 3, on Administrative Law, will show, China's court system can also present a check on the vast administrative system.

First, the courts are a place where Chinese citizens can seek compensation for damage caused by defective products. Many laws contain provisions on compensation for damage caused by food, drugs, devices, and cosmetics. This includes the provisions on tort in the Civil Code and specialized laws, such as the Product Quality Law. As will be discussed in later chapters, China law can be plaintiff-friendly when the dispute is smaller (often to avoid larger social disorder) and allows consumers to sue any number of parties regardless of fault (distributors, manufacturers, hospital user entities) and does not permit cases to be dismissed via preliminary motion. The courts can also be a place to resolve patent infringement and validity cases.

The courts tend to avoid asserting jurisdiction over more technical matters, even if those matters are simple disputes about labeling rules or ingredient usage. For much more complex proceedings, such as those related to vaccine or medical malpractice issues, the court will order an expert identification (*jianding*) proceeding, or, if they do not, may try to find a way to avoid taking the case altogether. Expert functions are provided by medical associations or expert institutes, depending on the area.

The second function of the courts is to be a place to appeal the decisions of administrative agencies. As will be discussed in greater detail later, in the next chapter on administrative enforcement, China's Administrative Litigation Law and Administrative Reconsideration Law provide avenues for companies and individuals to challenge illegal decisions of administrative agencies. This has been a route to challenge various actions, from an improper decision on an application, to the exclusion of a party from an administrative proceeding, to an egregious and disproportionate administrative penalty.

The courts are also a place for government enforcement of orders for certain compulsory measures. Criminal cases proceed through the courts, although a substantial amount of work on the case is done by MPS and the procuratorate (i.e., the prosecutors).

References

Center for Food Evaluation, State Administration for Market Regulation (2023) Raw material use and compatibility. http://www.cfe-samr.org.cn/zyyw/bjsp/jsspyqzl/ylsyjpw/.

China Food and Drug Administration (2015) Measures on Linkage of Food and Drug Administrative and Criminal Enforcement. CFDA, et al. No. 271. https://www.nmpa.gov.cn/xxgk/fgwj/bmgzh/20151222120001856.html?type=pc&m=.

General Administration of Quality Supervision, Inspection and Quarantine of the People's Republic of China (2014) Complete Pet Food—Dog Food. GB/T 31216-2014. https://openstd.samr.gov.cn/bzgk/gb/newGbInfo?hcno=A8758F9C4633B2CCBEE9C82CC3F4A4D1.

General Administration of Quality Supervision, Inspection and Quarantine of the People's Republic of China (2014) Complete Pet Food—Cat Food. GB/T 31217-2014. https://openstd.samr.gov.cn/bzgk/gb/newGbInfo?hcno=9496ABE283E8EB1FF58DF961B2DBAC9B.

General Office of the State Council (2013) Printing and Distributing the Provisions on the Main Responsibilities, Internal Organizations and Staffing of the China Food and Drug Administration. http://www.gov.cn/zwgk/2013-05/15/content_2403661.htm.

Ministry of Agriculture and Rural Affairs of the People's Republic of China (2022) Measures on New Feed and Feed Additives. http://www.moa.gov.cn/govpublic/CYZCFGS/202201/t20220127_6387839.htm.

Ministry of Commerce (2008) Implementing Provisions on the Slaughter of Live Pigs. http://www.gov.cn/flfg/2008-08/04/content_1063909.htm.

State Administration for Traditional Chinese Medicine (2023) Main Page. http://www.natcm.gov.cn/.

State Council of the People's Republic of China (2011) Provisions on Animal Feed and Feed Additives. http://www.gov.cn/zwgk/2011-11/15/content_1993711.htm.

State Council of the People's Republic of China (2021) Regulations on the Management of Pig Slaughter. http://www.gov.cn/zhengce/content/2021-07/22/content_5626534.htm.

3 Administrative Law Concepts and Enforcement Powers

3.1 Overview of Other General Legal Concepts

FDD law and regulation in China fits within a system of constitutional law, general civil, administrative, and criminal laws. Constitutional law and legislation law govern the making of FDD laws and regulations. China has a Legislation Law that determines the ranking of different levels of legislation and how they are interpreted. The Legislation Law and other regulations governing the enactment of administrative regulations by the State Council (e.g., the device framework regulation, see Chap. 6, and the Drug Administration Law Implementing Regulation, Chap. 5) and rules made by the State Administration for Market Regulation (SAMR) or National Medical Products Administration (NMPA) (e.g., the Drug Registration Regulation, Chap. 5) set forth an important structure for "public participation" which allows entities and individuals to comment on drafts of laws and regulations. This system of public comment and stakeholder dialogue has become entrenched over the last 10 years and has developed into a platform for exchanges between agencies and various domestic and international industry associations and companies.

The civil law, i.e., the new Civil Code (a collection of many laws, tort, contract, and family), impacts the contractual relationships underlying clinical trials and manufacturing and distribution agreements, including their quality components, and the suits in tort by consumers or patients for injuries caused by regulated products. Such suits are also governed by specialized judicial interpretations and regulations governing liability for food safety, medical malpractice, and vaccine liability, for example.

Criminal laws also play a key role in enforcement, primarily as a powerful deterrent. There are specific crimes related to fake and substandard food, drugs, and medical devices, as well as false advertisements. In 2021, as will be discussed in Chap. 7, China added a specific crime related to mishandling human genetic resources. The criminal law can be used against entities and individuals and can make an example against certain behavior, including conduct that the government deems to have caused political, social, or economic harm to China's society.

J. C. Balzano, *China Food and Drug Law*, Law for Professionals,
https://doi.org/10.1007/978-3-031-61901-4_3

China FDD law and regulation is primarily a type of administrative law, and general administrative laws are important in its enforcement. China does not have one administrative procedure law. Instead, it has different laws and regulations governing administrative licensing, administrative penalties, administrative compulsory measures, administrative reconsideration, and administrative litigation (including challenging the decisions on agencies vis-à-vis individuals and entities on limited grounds), among others.

These general laws dictate the procedures for issuing and denying licenses, as well as important guidelines for administrative enforcement. For example, the Administrative Penalties Law states that a stakeholder may not be penalized twice for the same conduct. The framework rules in each space (Drug Administration Law, Food Safety Law, Medical Device Supervision and Administration Regulations, Cosmetics Supervision and Administration Regulation) set forth specific prohibited acts and related penalties, and SAMR and NMPA and the provinces have developed more specific regulations that implement the penalties in the framework laws and regulations and generally address procedures, such as for reconsideration and penalties.

3.2 Constitutional Law and Legislation

China's constitution and related constitutional implementing laws, primarily the Legislation Law, dictate the powers of different branches (legislative, executive, local government, court) to enact legislation, including the powers of different agencies to make rules. "Legislation" as it is meant here does not only mean the power to make laws. The National People's Congress and its Standing Committee (NPCSC) enacts most food and drug related laws, such as the Drug Administration Law (DAL), the Food Safety Law, and the Agricultural Product Quality and Safety Law.

The State Council is the chief executive and is organized as a council of ministers of individual agencies and commissions and individual state councilors. The State Council makes administrative regulations, which are only subordinate to laws. It has its own procedural rules on the making of those regulations, which include provisions on drafting, seeking comment, enacting and interpreting administrative regulations.[1] The key relevant framework administrative regulations in the FDD context are the Medical Device Supervision and Administration Regulation and the Cosmetic Supervision and Administration Regulation, but there are also implementing rules at the "regulation level," the Food Safety Law Implementing Regulations and the DAL Implementing Regulations.

The State Council's specific procedures for regulation and rulemaking suggest, but do not always require, that agencies issue rules for public comment. There is no clear requirement that agencies publicly account for their consideration and

[1] State Council of the People's Republic of China (2017).

acceptance or rejection of stakeholder comments. There is also no clear requirement that the common periods be longer than 30 days, and they often are not. NMPA itself may give as little as 10 days, even for early drafts of major legislation, such as proposed drafts of the Drug Administration Law. NMPA's immediate predecessor, the China Food and Drug Administration, had its own standards governing its rule-making processes, including the establishment of legislative projects, issuance of drafts, notice and comment procedures, and procedures for cancelling legislation;[2] now SAMR administers these.

3.3 Administrative Licensing and Related Policies: Improvement of Efficiency

An important element of China FDD regulation is the issuance of licenses for various aspects of product development, marketing, distribution, and manufacture. These licenses are often also referred to as "registrations." In some cases, a product or process considered lower risk might receive a record-filing certificate instead of a registration. A record-filing means that the company has submitted documentation to the government, and the government has issued a certificate to acknowledge that all required documentation has been submitted; however, a technical review is not conducted in advance of the approval (although in some cases the government may review after issuing the certificate).

In 2003, China issued an Administrative Licensing Law that set forth general principles on the issuance of administrative licenses, aiming to curb abuses of agencies requiring such licenses when not supported by laws and regulations and abusing discretion in issuing or denying such licenses without applying fair and neutral legal procedures. The Administrative Licensing Law became, at least in principle, an important part of China's administrative procedure.

The Administrative Licensing Law provides the circumstances under which a license can be established by an agency, and requires that they be based on national laws and regulations, or in some cases local regulations.[3] The administrative agencies that administer these licenses should provide an application form and instructions, and the required content should be directly relevant to the license issued.[4] Indeed, this idea of narrow tailoring and avoidance of broad requests for information that could capture unnecessary and commercially sensitive information has been a relevant issue both in trade agreements, such as the U.S.-China Phase One Agreement (2020), and later a draft guiding opinion by the Ministry of Justice to further refine the practices of requests for information during mandatory licensing processes. The draft guiding opinion stated, "An administrative licensing authority shall determine the scope of materials to be submitted by a license applicant

[2]China Food and Drug Administration (2013).

[3]Standing Committee of the National People's Congress (2019), Articles 14–15.

[4]Ibid, Article 29.

according to the principle of necessity, avoid requiring the submission of irrelevant materials, and effectively protect business secrets and confidential commercial information."[5] The draft guiding opinion has not been finalized, and tailoring of information requested to what's needed to review for safety, effectiveness, and quality remains an often-discussed issue between industry and NMPA.

In addition to the principles of the Administrative Licensing Law, the government also issues various "policies" related to administrative governance, licensure, and rulemaking. These policies are many, varied, and regularly changing.

One is provided here as an example. China has been pursuing reform of its administrative system for some time, striving to make it more effective and efficient at governance. In the food and drug space, China has pursued policies of reducing costly and time-consuming pre-approvals—which often offered little net benefit for safety of products on the market—and increasing post-market surveillance of products. We will see several examples of this phenomenon in the chapters to follow.

Some of the administrative policies that have been particularly relevant to the areas that NMPA (and now SAMR, with respect to food) oversees relate to administrative licensing. The State Council has continually conducted reviews of license requirements to either eliminate them or to decentralize approval to localities.[6] This policy has led to the abolishment or downgrading of several FDD-related pre-approvals. The State Council has also established a policy of strict control of new license requirements (limiting their creation), required agencies to issue lists of existing approvals and regularly evaluate those approvals for effectiveness, and abolished or revised license requirements that were not achieving their intended economic and social policy goals.[7]

Similarly, in 2018, the State Council introduced another policy referred to as "Separation from Licensing and Permitting." This policy sets out one of four outcomes:[8]

- Abolish the approval completely. This occurs when the agency realizes that the approval is unnecessary, and the market, industry organizations, and intermediary institutions that assist with self-regulation can regulate this area. Once the approval is abolished the entity can pursue its activity once it obtains a standard business license.
- Abolish the approval and change the process to a record-filing. This occurs when the government needs to keep track of materials associated with the product in order to regulate the product but a technical review to approve the materials is unnecessary. Therefore, an applicant files the materials and, once they are accepted, the applicant may market the products.

[5] Ministry of Justice (2020).

[6] See, e.g., State Council of the People's Republic of China (2013a, 2014, 2019a, 2020).

[7] State Council of the People's Republic of China (2013b).

[8] State Council of the People's Republic of China (2018).

- Maintain the approval but simplify the procedure to rely on a letter of intent. The agency produces a model text with the conditions for examination and approval. The applicant submits the materials and the letter, and the agency examines them on the spot and approves or rejects them. Once the product is on the market, the agency will continue to conduct inspections; if it discovers that the conditions in the letter have been violated, it will revoke approval and impose a punishment.
- Maintain critical approvals related to national security, public health, public safety and ecological safety, but make the process more efficient. These approvals are maintained but the process for obtaining them is streamlined, meaning there will be shortened timelines for approval, better standards and criteria for approving applications, decentralized approval processes (i.e., giving the responsibility to local levels of government, where there may be greater resources), and greater transparency in the approval process.

Record-filings, as noted above, should function more like a notification, although in some cases the record-filings are arguably overused where there should be no approval or just simple notification. Record filings may require the same or a similar level of material as a registration/licensure procedure (see, e.g., Chap. 9 discussion of general cosmetics); the difference is the time to market. Some have argued that as administrative licenses have been more strictly controlled and reduced in number, there have been "too many" record-filings introduced as substitutes.[9]

3.4 Public Supervision as a Regulator

The government considers consumer and other stakeholder complaints and whistleblower input a valuable form of enforcement. However, cases have shown that it draws the line at organized consumer group advocacy or concerted action, independent of government control. For example, in 2008 the government shut down the individual lawsuits about contaminated milk, in favor of top-down government-administered relief.[10] We will further discuss this case in the chapter on food regulation.

Still, reports to the government about perceived non-compliance are permitted and even encouraged. NMPA issued new regulations on complaints in 2016, setting forth uniform procedures for local counterparts to investigate, accept, and dispose of consumer claims. SAMR adopted these rules once the agencies were merged and revised them shortly thereafter.

Consumer complaints against food or drug companies for various alleged violations have been tolerated in China, sometimes based on little or no evidence. This is most evident in the phenomenon of the professional consumer. These are consumers that purchase products primarily for the purpose of suing the company. In

[9] Chen (2022).

[10] Balzano (2012), p. 46.

some cases, these are legitimate claims. The product does have a clear label or packaging defect, for example. And some professional consumers are quite sophisticated, even taking the form of former government officials.

Other professional consumer claims are not legitimate. Some use recycled boilerplate complaints downloaded from the Internet. Certain of these consumers make fraudulent claims by scrubbing off expiration dates or passing off expired product as the one they recently removed from the shelves. The consumers file these claims in order to produce revenue for themselves or their organizations either by obtaining small amounts per case or settling with the company under an agreement to stop the incoming stream of claims.

As we will discuss in the chapter on Food Regulation, and the central and local governments have pursued measures to restrict professional consumers. For example, the Supreme People's Court issued guidance that courts could dismiss cases where the primary motive of purchase was collection of damages or when the consumer knowingly purchased a defective product to collect compensation.

3.5 Open Government Information

Another important general area of administrative law is freedom of information or open government information. This area is governed by the Open Government Information (OGI) Regulations,[11] which require that agencies proactively release certain information about the administrative process and reactively release information in response to requests by stakeholders with appropriate standing. A company with an interest in a particular item of information may make a request to an agency like NMPA for release of information. There are three exceptions of information that may not be released: personal privacy information, commercial secrets (including trade secrets), and state secrets.

All three of these categories are relatively undefined, although NMPA has articulated in various regulations information that it will not release under these categories. This includes a rule in 2017 specifically on protection of confidential information in the drug device registration application review and approval processes,[12] and another rule that adds a number of safeguards for review of drug applications. NMPA has not always reconciled this promise to protect confidential information and the records of its technical review with the broad information it requires (sometimes more than in other countries or regions for similar processes) when creating a medical device product standard or registering a clinical trial, for example. When rules require that such information be submitted or disclosed (even publicly), NMPA may not always receptive to arguments the irrelevant confidential information should be redacted or otherwise protected. Requests to redact information and/or efforts to encrypt or password protect information may be rejected.

[11] State Council of the People's Republic of China (2019b).

[12] See, e.g., National Medical Products Administration (2017a).

If an OGI applicant files a request for release of information related to a drug or a device that is owned by another company, NMPA will have between 5 and 15 days to respond to the request. However, NMPA must seek consultation with the owner of the commercial information before it may release it to a third party. This process is not included in the 15 days. Even if the information owner objects to release, NMPA may still release it. Administrative appeals may not be a particularly effective vehicle for challenging these decisions by NMPA, except if the information holder manages to obtain a preliminary injunction from a court not to release the information.

3.6 Administrative Penalties and Compulsory Measures

Companies that make or distribute FDD regulated products may be subject to various types of administrative penalties for prohibited acts. The prohibited acts are set forth in the product specific framework laws and include acts such as making or distributing a product that does not have the requisite approval or is not compliant with good manufacturing practice standards or its registered specifications. Or, those penalties can be present in general laws, such as the Advertisement Law, which has penalties for the prohibited acts of false or misleading advertising content or failing to obtain approvals for drug, device or health food advertisements, or the Anti-Unfair Competition Law, which has penalties false or misleading publicity. There are other national laws that set general prohibited acts, which may be applied to regulated products in the absence of something more specific in the product law: Law for the Protection of the Rights and Interests of Consumers, Product Quality Law, and tort provisions of the Civil Code.

3.7 Administrative Penalty Procedure

The penalties for each prohibited act may include warnings and public criticisms, fines, disgorgement of illegal income, product seizures, license revocation, and debarment (or disqualification) of entities and individuals, and, in rarer instances, detention of individuals. Debarment means that the entity or the individual is disqualified from participating in product-related activities, such as the filing of product or facility applications. The penalties must be set by a law, regulation, or administrative rule.[13]

While the imposition of penalties is governed by the relevant law, there are also procedural and fairness requirements and measures under the general law on Administrative Penalties. The Law on Administrative Penalties includes the right to an administrative hearing prior to the imposition of administrative penalties[14] and

[13] Standing Committee of the National People's Congress (2021), Article 16.
[14] Ibid, Articles 64–65.

prohibitions on duplicative penalties for the same conduct.[15] It sets a statute of limitations of 2 years, starting from the time of commission or termination (in the event of continuous conduct), for most administrative violations, and 5 years for those actions that have consequences for human life and safety and financial security.[16]

It also contains principles of leniency and mitigation. For example, there should be no penalties for hypothetical violations or violations that produce no harm. And there are principles of mitigation for taking steps to reduce the harm of illegal conduct, confession of illegal actions, and good-faith cooperation with a government investigation.[17]

Once an administrative penalty is imposed, it can be enforced by the government, even if the defendant seeks reconsideration or an administrative appeal, because a court (which has the power to refuse to enforce orders during the pendency of appeals)[18] may not be a viable option to stop administrative authorities from using their local enforcement power to remove products from circulation. Some information about the case and the nature and extent of the penalty is published online for at least 6 months to several years.[19] The notice can be taken down if the penalty is revoked or withdrawn, e.g., if the company wins an appeal. This requirement of publication, even during the pendency of an appeal, has created difficulties for regulated companies because the penalties (even if lacking in justification or evidence) can generate consumer lawsuits that the company must then defend; it can be challenging to win those if the government has already concluded that the conduct is illegal.

3.8 Administrative Compulsory Measures

In 2011, in addition to the procedures for investigation and imposition of administrative penalties, China enacted a Law on Administrative Compulsory Measures that establishes procedures of fairness for various coercive measures. That Law defines compulsory measures as restriction of personal freedom and seizure and temporary control of property, including premises, money, or property. The law defines procedures for notifying and seizing persons or property, conditions for detention, and enforcement of obligations through courts.[20]

[15] Ibid, Article 29.

[16] Ibid, Article 36.

[17] Ibid, Article 32.

[18] Standing Committee of the National People's Congress (2011), Article 53.

[19] Ibid, Article 48.

[20] Standing Committee of the National People's Congress (2011), Article 2.

3.9 NMPA and Provincial Administrative Rules

The above-described laws set forth the national framework for constitutional law, civil law, and administrative law. But NMPA and local MPAs have also developed their own rules. For example, NMPA has developed rules on Food and Drug Administration Penalties (2014),[21] evidence in administrative cases, and appeals of product rejection and administrative reconsideration. Similarly, SAMR and the provinces may have more general rules that govern in certain cases. A province may have rules for determining when a case is sufficiently severe that local authorities should consider imposing elevated penalties called for in the various laws.

One important consideration in penalty adjudication is whether the conduct is "serious." As discussed above, and in the chapters that follow, the Drug Administration Law and the Food Safety Law have two levels of penalties for prohibited acts, one with a range of fines and compulsory measures for standard violations, and another when the circumstances are serious. NMPA's predecessor agency released Principles on the Discretionary Application of Drug and Medical Device Administrative Penalties (2012). Those Principles set forth circumstances in which the central agency would consider conduct to be severe, warranting enhanced penalties:[22]

- Production of medicine that does not meet statutory requirements and causes "serious consequences," a term which is defined as injuries that are more than mild injuries, i.e., moderate disabilities, general or severe damage to organs and tissues, or other serious harm to human health
- Use of illegal ingredients or alteration of the formulation in such a way that there are unspecified serious consequences
- Obstruction of investigations or concealing evidence
- Failure to inform the public if there is a discovery of a safety issue that could cause harm to human health and life
- Failure to suspend sales and issue a recall when there is a possible safety issue that could cause harm to human health or life[23]

These Principles also set forth other circumstances in which penalties should be heavier, such as when the violation involves controlled substances (narcotic or psychotropic drugs) or medicines for pregnant women or infants, results in injuries, or if the company has engaged in conduct that obstructs investigations or evades detection. NMPA amended these Principles in 2024, and added cosmetics to the scope, among other amendments. The amended version will take effect in August of 2024.

[21] State Food and Drug Administration (2014).

[22] State Food and Drug Administration (2012).

[23] Ibid., Article 12.

Similar situations qualify as severe in the provinces, which have jurisdiction over domestic companies and the regulatory agent enterprises of overseas license holders for drugs, devices, cosmetics, and health foods. Shanghai established administrative penalty rules that state that, unless applicable laws and regulations decree otherwise, local officials can consider a series of aggravating factors when determining whether to impose "severe" penalties. These include:

- Offenses that cause serious illness or death or severe economic harm
- Offenses that take place during a natural disaster or public health or social accident
- Offenses that repeat the same conduct within the same year
- Company personnel who refuse to cooperate or retaliate against government officials
- Company personnel who hide or destroy evidence or property sealed by local enforcement officials
- Company personnel who otherwise conceal or destroy evidence[24]

One new element of administrative penalties for China FDD violations that has emerged is individual-level liability. This means penalizing responsible individuals within the company, particularly when the circumstances are deemed serious, according to the considerations above. Some version of individual liability has been included in the framework laws and regulations for some time, but it became more pronounced with respect to the penalties for serious food safety violations in relation to infant formula and other exported products between 2004 and 2008. The 2008 revision of the Food Safety Law included provisions that permitted individuals to be held accountable, including one that permits individuals responsible for certain prohibited acts, such as selling contaminated meat, to be detained for a certain length of time.

In 2018, NMPA and the Ministry of Public Security released a document that described the overall policy of individual liability. It connects the policy to the Four Most Stringents Policy that generally underlies food and drug regulatory enforcement. Under this policy, Xi Jinping has called for the strictest standards, the strictest supervision, the strictest penalties, and the strictest accountability in the areas of food and drug safety.[25] As the text below shows, the individuals would be held accountable regardless of whether they were acting with the scope of their authority. In other words, both they and the company are accountable:

Provisions on Strengthening Food and Drug Safety Law Enforcement

Deep Understanding of the Meaning

[24] Shanghai Market Supervision Bureau (2022), Article 11.

[25] Jinghua Times (2015).

Implementing the provisions of "punishing the individual" for food and drug violations is an important measure to fully implement the "four most stringents" requirements of the Chinese Communist Party Central Committee on food and drug safety, and to increase law enforcement in the food and drug field. Strengthening the authority of food and drug supervision and law enforcement, and comprehensively improving food and drug safety, is very significant. Food and drug supervision and administration departments and public security organs at all levels must strictly follow the requirements of administration in accordance with the law, and conscientiously implement the provisions of laws, regulations, and rules on "punishing individuals," and they must seriously investigate the legal responsibility of the relevant responsible persons and employ the most stringents supervision and the most severe punishment to ensure food and drug safety.

Clarifying the Scope of Individual Responsibility in Accordance with Law

Individuals who engage in food and drug violations shall be investigated for their legal responsibilities in accordance with the law. Where a [company] engages in illegal acts of food and drugs, in addition to punishing the [entity itself], the directly responsible person in charge of the [company] and other directly responsible personnel shall be held accountable according to the law. The person in charge who is directly responsible is the person in charge who plays the role of deciding, approving, instigating, condoning, or commanding the illegal act carried out by the [company], and is generally the relevant person in charge of the [company]. Other directly responsible persons are those who have specifically committed the illegal act and played a greater role in the [company's] violation of law. They can be either be the production, operation, and management personnel of the unit, or the employees of the [company] . . . .[26]

* * *

Since the issuance of this document, the framework laws and regulations have been amended to include greater responsibility for individuals, including the Drug Administration Law, the Medical Device Supervision and Administration Regulations, and the Cosmetics Supervision and Administration Regulation. Punishments vary between fines that are a portion or multiples of the individual's income, to debarment for years, or life, to administrative and, in extreme cases, criminal detention. Although this policy has been in place for several years now, there is not clear evidence that it is used liberally.

In 2022, China established four rules in each of the regulated spaces discussed in this book (food, drugs, medical devices and cosmetics), detailing the responsibilities for primary responsible individuals in certain managerial roles at drug, device, and

[26] State Food and Drug Administration (2018).

cosmetic registrant (or filer) companies and in food companies holding manufacturing or distribution licenses, including general and special food companies. These rules set forth the responsibilities for the higher level of the enterprise, including the legal representative, which is a position that all companies have for an individual who is generally responsible for the civil activity of the company. For drug and device companies, there is also an "enterprise responsible person" position that overseas general complained with drug and device regimes, and then there are individuals in more specialized leadership roles, such as those responsible for safety, quality and post-market surveillance.[27]

These rules set out not only the qualification requirements for individuals to hold these rules, but they also set forth their job responsibilities and provide guidance on when they can be individually liable for failing to meet those job responsibilities, including for the supervision of compliance with the increasing body of complex food, drug, device, and cosmetic rules. It is increasingly clear that these entities need to be staffed with individuals, who understand corporate responsibility in these specialized spaces and in some cases are expert in important fields like quality.

3.10 Administrative Reconsideration

Once an investigation, inspection, or other administrative procedure is concluded and a decision has been made, a company has various avenues to address objections. Some of the considerations about whether to exercise these rights are discussed below.

Administrative reconsideration and administrative litigation are the two main channels to appeal final administrative decisions, such as fines or product seizures. But there may also be other channels to file objections and engage with the agencies in a dialogue prior to that point. For example, after a site inspection of a manufacturing facility the inspected entity will have a chance to lodge objections to the findings during the final close-out meeting of the inspection, and then 10 days after the final report (see Chap. 5).

The Center for Drug Evaluation (CDE) also has a process for applicants to object to decisions on drug approvals. Under the revised Drug Administration Law and

[27]NMPA (2022) Administrative Supervisory Provisions on the Implementation of Safety and Quality Primary Responsibilities of Drug Marketing Authorization Holders, https://www.nmpa.gov.cn/xxgk/fgwj/xzhgfxwj/20221229195805180.html [Last Visited October 31, 2023]; NMPA (2022) Administrative Supervisory Provisions for the Implementation of the Main Safety and Quality Responsibilities for Medical Devices, https://www.nmpa.gov.cn/xxgk/fgwj/xzhgfxwj/20221229195418188.html [Last Visited October 31, 2023]; NMPA (2022) Administrative Supervisory Provisions for the Implementation of the Main Safety and Quality Responsibilities for Cosmetics, https://www.nmpa.gov.cn/xxgk/fgwj/xzhgfxwj/20221229195623179.html [Last Visited October 31, 2023]; SAMR (2022) Administrative Supervisory Provisions for the Implementation of the Main Safety and Quality Responsibilities for Foods, https://www.gov.cn/zhengce/2022-09/26/content_5726881.htm [Last Visited October 31, 2023].

Drug Registration Regulation, CDE issued a draft rule that allows an applicant to dispute a rejection of a drug application to CDE within 15 days; within 50 business days of the receipt of the decision, it may apply to CDE to establish an expert committee to consider the matter further.[28] Thereafter, the applicant can move on to administrative reconsideration procedures which may be handled by NMPA's dedicated Administrative Reconsideration Office.

Once a given decision is final, the individual or entity that is the subject of the administrative decision, such as a warning, a fine, a seizure of product, an order to cease production or distribution, or a revocation of a license,[29] may apply within 60 days to the same or another administrative agency or designated agency for reconsideration of the decision. There are various rules on the jurisdiction to hear these rules, and administrative agencies and local governments will have their own rules. For example, if the decision was made by an agency, such as SAMR, then the application would go to SAMR. If the decision was made by a lower-level government, then the application would go to the local government at the next highest level of government.[30]

The entity or individual can also apply for review of the provisions underlying the administrative act, but only for certain rules, e.g., those of departments of the State Council and county level or above local governments department rules. Rules by the main administrative agencies handling China FDD regulation, e.g., SAMR, NHC, and the rules of the provincial level governments (e.g., Shanghai, Guangdong), are excluded.[31] During the pendency of administrative reconsideration, the penalty can still be executed, absent court intervention or exercise of discretion by the deciding administrative agency.

Once an application for reconsideration is received, it will be processed by the designated administrative reconsideration authorities within the legal affairs office of the agency. The deciding official may respond to the application in writing and present evidence, which the private applicant can examine, absent concerns about state secrets, confidential commercial information, or individual privacy. The decision should be rendered in 60 days, barring complexity warranting additional time. An administrative reconsideration decision is then appealable through administrative litigation.

3.11 Administrative Litigation

After receiving an administrative decision from an agency that is appointed under the law to make it, an individual or a company can seek either (1) administrative reconsideration or, unless the law provides that administrative reconsideration is a

[28] Center for Drug Evaluation, State Drug Administration (2019).

[29] Standing Committee of the National People's Congress (2017b), Article 6.

[30] Standing Committee of the National People's Congress (2017b), Articles 12–15.

[31] Standing Committee of the National People's Congress (2017b), Article 7.

required first step, (2) to challenge the decision in court through administrative litigation, pursuant to China's Administrative Litigation Law. The company can also pursue administrative litigation after administrative reconsideration if they are unsatisfied with those results.

Various factors can influence the decision to use the courts over agency reconsideration. For example, courts might be considered a fairer venue, being separate from the agency. Or, perhaps the issue is more procedural and legal in nature, such as failure to grant the parties to a decision a chance to be heard on certain issues. Courts may be better positioned to handle these issues as opposed to more technical FDD issues, for which agency expertise might be more useful.

Administrative litigation can be initiated against national or local level agencies. Depending on the level of agency involved, the court of original jurisdiction may change. China generally has four levels of courts, district, intermediate, provincial-level high courts, and the Supreme People's Court. For example, while basic level district courts have jurisdiction over cases filed against local government departments, cases against a local government above the county level would need to be filed in the intermediate level court. Like administrative reconsideration, administrative litigation can be used to challenge a decision by an agency to issue a penalty, such as a fine or a warning, or a decision to reject a license application. It can also be used to challenge an administrative agency's failure to respond to an application within a statutorily designated period or for an agency's refusal to perform certain duties that are required under the law.[32] As with civil cases, there is no real discovery process, but the parties do exchange evidence to support their respective cases, absent considerations (e.g., personal privacy or confidential commercial information) of confidentiality.[33]

As will be discussed below, suing an agency, particularly an agency that has the power to approve marketing licenses on a regular basis, is a significant step for a company to take. Companies often want to preserve their relationships with agencies and may therefore not quickly seek judicial assistance under the Administrative Litigation Law. However, these cases do happen. An important example from a case decided by the Supreme People's Court is the Vellpharm case.

3.11.1 Vellpharm v. State Food and Drug Administration [now NMPA] (2015)

This case took place before the 2017 reforms that drastically changed the system of drug approvals in China (see Chap. 5). At the time of the case, it could take an applicant for a clinical trial between 1 to 2 years to obtain permission from the drug regulator to even commence its study. NMPA's inefficiencies had created a backlog of, by some estimates, over 20,000 of drug and device applications (clinical trial,

[32] Standing Committee of the National People's Congress (2017a), Article 12.

[33] Standing Committee of the National People's Congress (2017a), Article 32.

marketing, renewal, amendment). These delays stood in contrast to the timelines in the regulations that gave NMPA and its Center for Drug Evaluation a certain number of days to review application materials for completeness, organize an expert review, and decide on an application. Pursuant to the 2007 version of the Drug Registration Regulation (DRR), CDE had 90 days to review a drug clinical trial application, which could be extended if there was a request to the applicant for a supplemental submission. Once the review was complete, the State Food and Drug Administration had 20 days to make its final decision. (Now CDE makes decisions on clinical trial applications, and it has 60 business days to object to the materials submitted. If 60 business days pass without a decision, the clinical trial may proceed.)

Vellpharm applied in 2009 for permission to conduct a clinical trial of its antihypertensive drug, felodipine controlled-release tablets. At that time, Vellpharm's active ingredient was manufactured by the Zhejiang Juhua Group in China, but the final product was made in Vietnam. Therefore, the drug would be registered as an imported drug. As will be discussed elsewhere in this book, active pharmaceutical ingredients (API) in China used to receive their own drug registrations. Vellpharm's API did not have such an approval to present to NMPA, nor did it present other evidence of legal sourcing channels. At the time the following requirement in the 2007 DRR applied:

> **Article 95**
>
> For an imported drug preparations application, documents to evidence the legal channels of contact packaging materials or containers of the drug, documents to evidence the legal channels of drug raw material [API] and excipients must be provided. For raw material [API] and excipients of drug that have not been approved by SFDA, standardized information of relevant production processes, quality specification, and inspection methods should be submitted.

NMPA accepted Vellpharm's clinical trial application on April 20, 2009, and transferred the application to CDE for review 10 days later. The review took until February 2010, at which time CDE made a recommendation to NMPA's drug registration division. NMPA sent the case back to CDE for further findings on August 9, 2010 (reasons unknown), and CDE sent back its reply on November 12, 2010. On that same day, NMPA rejected the application on the grounds that Vellpharm had failed to submit either evidence of approval of its active ingredient in China or sufficient documentation to show its origin and conformance with standards in the country of origin of the drug.

At the time, the DRR required that technical reviews be completed in 90 days from the date of acceptance of a complete dossier; following that, NMPA had 20 to 30 days to make its final decision.[34] During the technical review, if materials

[34] State Administration for Market Regulation (2020), Articles 150–151.

were missing, CDE was expected to ask the applicant to supplement its application. The Vellpharm clinical trial application took 1.5 years to complete after acceptance, and CDE did not issue a request for supplemental materials.

Seemingly angered by this substantial delay and ultimate rejection, which at the time was typical, Vellpharm filed suit in the Beijing First Intermediate Court. It argued that NMPA's decision violated the law on three grounds: NMPA should have notified Vellpharm to submit supplemental material as to the API's origins, the approval opinion was not properly sealed, and NMPA/CDE exceeded statutory deadlines in making the approval.

The Intermediate Court found for the agency, stating that the seal was not improper, it was not legally required to request supplementary materials from the applicant, and the timing was not improper because the Administrative License Law and the DRR permit the review time to be extended. Vellpharm appealed to the Beijing High Court. That court was slightly more sympathetic to Vellpharm, noting that the timeline clearly violated the regulations, but that the timing was not material, and the court could not cancel the review because Vellpharm's application was defective.

Vellpharm appealed to the Supreme People's Court, raising similar issues. The Court's reasoning is below.

Supreme People's Court of the People's Republic of China

Administrative judgment

(2014) Xing Ti Zi No. 27

Vietnam Vellpharm Pharmaceutical Liability Co., Ltd. (hereinafter referred to as "Vellpharm company") v. The State Food and Drug Administration of the People's Republic of China (hereinafter referred to as the "State Food and Drug Administration") drug administrative license case, Beijing Higher People's Court in May 2012. The Administrative Judgment (2012) Gao Xing Zhong Zi No. 90 was made on the 18th, which has become legally effective. Vellpharm dissatisfied and applied to this court for a retrial.

.... [The court described the reasoning of lower district court and the Beijing High Court (a provincial level court), which had both rejected the plaintiff's claims and found that NMPA had acted within the rules on timing of its decision and within its discretion when it did not ask for supplemental materials as to the origin of the raw material at issue. The Supreme People's Court then went on to recount the parties' arguments in the case, and to frame the issues for its decision.]

This court believes that the main controversies in this case are: whether the timing of [NMPA's review] exceeded the statutory time limit, whether the sample inspection result was a necessary condition for [NMPA] to take the actions that are the subject of the complaint, and whether [NMPA] should have requested that the appellant submit supplementary materials during the review process.

Regarding the first issue, Article 42 of the "Administrative Licensing Law" provides that: "Except for administrative licensing decisions that can be made on the spot, the administrative agency shall make an administrative licensing decision within twenty days from the date of accepting the application .... If a decision cannot be made within twenty days, it may be extended for ten days with the approval of the person in charge of the administrative agency, and the applicant shall be informed of the reasons for the extension, except if other [applicable] provisions in laws and regulations provide otherwise."

Article 152 of the Drug Registration Regulations states: "The [NMPA] shall make an approval decision within 20 days; if it cannot make a decision within 20 days, the timeline may be extended for 10 days with the approval of the leadership of the Administration, and the applicant shall be notified of the reasons for the extension of the time limit."

In this case, after [NMPA] accepted the drug registration application submitted by the appellant on April 20, 2009, the relevant application materials were transferred to the Center for Drug Evaluation on April 30, 2009. On February 9, 2010, the Center sent its "Comprehensive Evaluation Opinions" to [NMPA]. On August 9, 2010 [NMPA] returned this Opinion to the Center for Drug Evaluation for further review. After the Center reissued its Comprehensive Evaluation Opinion on November 12, 2010, [NMPA] made the decision on that same day not to approve Vellpharm's above-described application for registration (to conduct clinical research).

In this case, the technical review ... of CDE clearly did not comply with the provisions of Article 150, paragraph 1 of the Drug Registration Regulations, and in accordance with the provisions of Article 45 of the "Administrative Licensing Law" after the deduction of the time for technical review, ... the [timeline] also exceeded the legal time limit. The court of first instance's failure to include the time of the [NMPA's] first review of [CDE's] "Comprehensive Review Opinions" between February 9, 2010 and August 9, 2010 in its analysis of [the legality of the timeline], and its determination that the [timeline] did not exceed the statutory deadline were in error.

Regarding the second issue, from the provisions of the DRR one can see that the technical review takes the technical materials submitted by the applicant as its object and evaluates the safety, effectiveness, and quality controllability of drugs. Sample testing uses the physical sample as its object, and evaluates the aforementioned three aspects [safety, effectiveness, and quality controllability] of the drug through [registration] standard confirmation and physical inspection. The objects and methods of [technical review and sample testing] are different, but the purposes overlap. Therefore, there is a possibility of drawing different conclusions on the same matter. Therefore, they should be considered collectively and the final conclusion cannot be reached based on one of them alone.

In this regard, Article 92 of the [DRR] stipulates: "[CDE] shall form a comprehensive opinion based on technical review opinion and sample inspection results and submit that comprehensive opinion to [NMPA] together with relevant materials, and [NMPA] will make an approval decision based on the comprehensive opinion. If the [requirements of the] regulations are met, [NMPA will issue] a Drug Clinical Trial Approval Document; if the requirements of the regulations are not met, [NMPA] will issue an Approval Opinion Notice and explain the reasons." Accordingly, [NMPA] must make an approval decision based on the "comprehensive opinions," which include the sample inspection results, rather than just on consideration of the technical review opinions.

In this case, [notice of decision that is the subject of Vellpharm's complaint] was issued on November 12, 2010, but the China Pharmaceutical and Biological Products Inspection Institute issued its Drug Registration Inspection Report Form on November 17, 2010. This action clearly violated the above-described provision [of the DRR]. The court of first instance and [NMPA] maintain that as long as either the technical review report or drug registration inspection report does not meet legal requirements, NMPA can, in accordance with the provisions of Article 154 of the [DRR], reject the drug application. [This Court] does not support such reasoning.

Regarding the third key issue, Article 91 of the [DRR] provides: "[CDE] shall organize pharmacological, medical, and other technical personnel to review the application materials within the specified timeframe. If necessary, [CDE] may require the applicant to supplement the materials and explain the reasons." This Article gives [NMPA] the discretion as to whether to require applicant [to submit] supplement materials.

However, the exercise of discretionary power should conform to the legislative purpose and help protect the legitimate rights and interests of the stakeholder. Article 3 of the [DRR] states: "Drug registration refers to a procedure through which NMPA decides whether it agrees to approve an application of a drug registration applicant in accordance with legal procedures and based on its review of the safety, effectiveness, and quality of the drug to be marketed." Combining the review object specified in [Article 3] with the relevant content in Article 154 [of the DRR] for refusing approval, it is possible to establish the following presumption: drug registration should revolve around the safety, effectiveness, and quality controllability of the drug, and through a review of the application materials, research items, [drug ingredients], production sites, and samples, among other materials, a determination of whether the circumstances in paragraphs (4) [design and implementation of research items does not support safety, efficacy, or quality], (6) [non-compliance of sourcing of ingredients] and (7) [failure of onsite inspection or sample test] of Article 154 [of the DRR] are present, and after that, based on the existing materials and circumstances, whether it is sufficient to determine that the drug that is the subject of the application cannot meet the requirements of safety, effectiveness and quality. Under those circumstances [NMPA] may make a decision not to approve [the application].

And if the application materials are duplicative, deficient, or inaccurate, and on that basis it is not possible to form a certain belief about whether the drug meets the safety, effectiveness, and quality control [requirements], or rather items (1) [duplicative data used in applications without justification], (2) [application determined to be untrue and applicant cannot remedy] or (5) [failure to submit supplemental information within the required timeframe] of [Article 154 of the DRR] are present, then the applicant should first be required to provide explanations or supplementary materials. Under the circumstances that the applicant cannot reasonably explain or give a supplement, then [NMPA] may make a decision not to approve [the application].

Therefore, although Article 13 of the [DRR] states that: "The applicant shall provide sufficient and reliable research data to prove the safety, effectiveness, and quality controllability of the drug, and be responsible for the authenticity of all the information," this provision does not mean that [NMPA] can make a decision to reject the application only if the information provided by the applicant is insufficient or the authenticity of the application is doubtful. [NMPA] should prudently perform its review responsibility and fulfill its obligation of reminding [the applicant about these deficiencies] in good faith. And, should it not be possible to further evaluate the application by requesting that the applicant explain or supplement relevant materials to determine whether the drug in dispute meets the safety, effectiveness and quality controllability, the applicant's clinical trial application shall be rejected on the grounds that the information provided by the applicant is inadequate, insufficient, and does not meet requirements.

In this case, although the materials that [Vellpharm] submitted about the API manufacturer Zhejiang Juhua Group Company did not contain the relevant approval documents for API production, in accordance with Article 95 of the [DRR] [an applicant like Vellpharm, which does not have an approval of the API from NMPA] shall submit standard research materials related to the production process, quality indicators, and testing methods of the API.

In this regard, although [Vellpharm] did not proactively submit the report, as described above, [NMPA was unable to] determine based on the existing materials whether [the drug] could meet the requirements of safety, effectiveness, and quality controllability. Therefore, [NMPA] should have requested that [Vellpharm] submit the above-mentioned materials instead of directly deciding to reject the application. [NMPA's] method of handling [this application] required [Vellpharm] to pay another registration fee of more than 40,000 [RMB] after supplementing the materials to permit [its]... drug to restart the review process. It did not fully protect the applicant's relevant legitimate rights and interests, and [NMPA's] judgment was improper.

As for the arguments raised by [Vellpharm] that the "Notice of Examination and Approval Opinion" exceeded the authority of the agency, that [NMPA] did not treat a Vietnamese enterprise fairly, [that NMPA] charged registration fees that were higher than those for domestic companies, and that the court of second instance violated legal procedures, there is no factual and legal basis, and this court does not support [those arguments].

In summary, [notice number] 2010104265 issued by [NMPA] is illegal pursuant to items 1, 2, 3, and 5 of Article 54 (2) of the Administrative Litigation Law, and [such notice] shall be revoked. Vellpharm's request to revoke the accused decision is based on law and should be supported. The first-instance judgment rejecting Vellpharm's litigation request, and the second-instance judgment rejecting the appeal and maintaining the first-instance judgment are incorrect and should be corrected. According to Article 61 (2)–(3) of the Administrative Litigation Law and Article 78 of the Supreme People's Court's Judicial Explanation on Several Issues on the Implementation of the Administrative Litigation Law 70, [this Court's] judgment is as follows:

1. Revoke the judgment of the Beijing High People's Court (2012) Gao Xing Zhong Zi No. 90 Administrative Judgment;
2. Withdraw examination and approval opinion notice issued by [NMPA] number 2010104265;
3. Order [NMPA] to re-examine [Vellpharm's] "Application for Clinical Trial of Felodipine Controlled Release Tablets" within the statutory time limit. . . . .

* * *

When interviewed about the case, Vellpharm's Chairman stated that the company had pursued this litigation because it was tired of NMPA's practice of taking excessive time to examine applications and failing to communicate with applicants.[35] As discussed above, this issue of clinical trial applications consuming a year or more or having to be refiled for insignificant reasons, resulting in waste and deepening China's drug lag, was one of the primary issues that the policy under the Opinions on Deepening Reform of the Review and Approval System to Encourage Drug and Medical Device Innovation issued by the General Offices of the State Council and the Central Committee of the Chinese Communist Party, in 2017, was intended to address. Although communication with NMPA can still be challenging, the situation has improved.

Another striking aspect of this case is the effectiveness of a court in resolving an issue with an administrative agency. There are numerous examples of courts overturning agency decisions, but the Vellpharm case took direct issue with NMPA's exercise of its discretion. Although companies are often very reluctant to challenge NMPA for fear of damaging relationships with the agency in ways that

[35] Lu (2016).

could affect future applications—and there is some truth to that view—the Vellpharm case is an example of what can be achieved through formal processes in the right set of circumstances.

3.12 Administrative and Civil Enforcement in Practice

Based on experience with hundreds of administrative and civil cases in China, and with the above-described legal framework in mind, what follows is a summary of the administrative adjudication and enforcement environment that companies face in China.

First, if a company experiences an administrative issue in China in the form of an administrative inquiry, investigation, or litigation, for example, it should immediately consider hiring an administrative lawyer. There are a number of administrative law experts, including former agency officials or judges, who know the system and procedure, understand how to talk to officials, and think proactively about the future consequences of their statements.

Some companies will send their internal personnel (sometimes counsel) to handle these issues in meetings with administrative agencies or in court. Administrative agencies may like that approach, viewing a representative of a party to the litigation as more honest than a lawyer. However, company officials may be flustered during official visits, make inconsistent statements, and agree to sign official documents, referred to as "interview notes" or memoranda (which become evidence in the government's case), when they may not reflect the actual situation. It may be prudent to have an employee or representative of the company at meetings to show officials that the company is taking the issue seriously, but that person should be accompanied by a lawyer or prepared by one.

Another approach is to send public affairs or government affairs personnel to handle government inquiries (typically only if the case is not in court.) The purpose of the government affairs team is to establish and maintain a good relationship with the government, so their negotiating posture will be different than a lawyer. What's effective may depend on the case, including if technical issues of regulatory interpretation are involved, such as in a case about labeling.

3.12.1 How Do Enforcement Issues Begin?

Enforcement against a company for issues with a regulated FDD product can begin in several ways in China. It can happen through regular surveillance, such as inspections of facilities, stores, or hospitals, or testing of product in the market. Indeed, there are robust testing programs at various points for foods, drugs, medical devices, and cosmetics. Some sampling is random; some is conducted according to planned programs, such as for medical devices, as described below.

Enforcement can also begin when a consumer or a company employee raises a report to the government. All agencies, including NMPA and its provincial counterparts, have report (*jubao*) handling systems. There are procedural rules governing the handling of these complaints and requiring that officials investigate all of them to determine whether they have merit. These rules also protect the reporter's identity. Consumer complaints can be legitimate, but they can also be quite spurious. As described earlier, China has "professional consumers" who purchase products simply to bring complaints and collect multiple small amounts of compensation.

Enforcement may also begin when officials in China read about enforcement issues with similar products overseas. There are rules that require reporting of serious safety information about drugs, devices, and cosmetics abroad and actions taken in response, including recalls or warnings. Officials may see these reports and decide to inquire as to whether China is affected.

In addition to more routine testing, there may be campaigns or concerted enforcement efforts. In 2014, for example, NMPA launched a campaign against certain issues in the medical device industry, called the 5 Rectifications Campaign.[36] Officials were told to seek out and investigate certain types of common violations. Now, local officials conduct medical device sampling (i.e., testing of the devices against their registered specifications) on an annual basis. In another example, various antitrust investigations occurred regarding the fairness of prices set by foreign drug makers in China.[37] These types of campaigns or special investigations can be politically charged and may require legal and policy help.

3.12.2 Investigations and Establishment of Cases?

When enforcement begins, companies may receive substantial requests for documentation supporting product content and labels, e.g., testing reports, label reviews, and manufacturing records. These may be requested by local officials through more informal channels: verbally via phone or, in some cases, even via social media tools, such as WeChat. Further, when these materials are requested, there may not be a clear indication as to the allegation of wrongdoing that the agency is investigating. Some officials may refuse to clearly state the content of the complaint, while others may send a photo (perhaps with the complainant's name redacted). Nevertheless, companies will often comply with the documentation and meeting requests because the alternative is a much more formal and less friendly process in which a case established.

Once an agency in China begins to issue written requests for documents and formally "establishes a case" (*li an*) in its system, the investigation takes on a more official character that will almost certainly lead to an administrative penalty. Establishing a case means that the agency has what it believes to be concrete and

[36] Wang (2014).
[37] Lane (2016).

supportable evidence of wrongdoing and feels comfortable that higher-level officials and those in any legal department would support a notice of penalty.

The informal investigation stage can be intense and requires the presentation of sufficient evidence (discussed below) and statements from the company to show that it has complied with the law. Agencies have a fair amount of discretion regarding what is sufficient, but they will want to balance risks to the agency, including the risk that a reporting consumer or whistleblower will sue the agency and allege publicly that the agency and officials did not do their job.

3.12.3 What Evidence Is Needed?

China has no discovery process in the civil or the administrative case context. Administrative agencies can use coercive powers to obtain evidence, including raiding a company and taking away computers and files or simply requesting many documents under the color of their authority. However, for the most part, parties are free to determine what evidence to provide to prove their case. In a civil case, there is no system to demand and compel production of relevant evidence, which makes enforcement via court quite challenging.

There are various types of evidence that administrative agencies will accept in China. Perhaps the best evidence is something from another government body or quasi-government body, e.g., a quasi-governmental industry association, stating that the company's action is compliant. For example, if someone alleges that a medical instrument is an unapproved device, simply provide evidence of NMPA's approval. Or, if the content of an advertisement is deemed to violate the Advertisement Law, provide the advertisement's approval by a local medical products administration. Sometimes, precedent set by many similar cases that result in rulings in nearly identical circumstances is enough for officials to drop the initial inquiry. However, not all cases are so simple.

In cases in which a case in brought in a province other than the company's home province (i.e., where the China affiliate or agent is located), sometimes the home province regulator will be willing to issue a letter as to compliance, particularly if it has investigated this or a similar matter. Sometimes the other province will call upon (i.e., certify) the question to the home province regulators to give an opinion or even assist with the investigation. A statement from the home province asserting compliance would be a significant factor in making a decision.

It is also possible in some cases that the local authorities may certify the matter to a central agency that oversees that area's regulatory policy. For example, provinces may certify questions to NMPA about labeling issues. After severe COVID-19 pandemic lockdowns in 2022, Shanghai's Medical Products Administration certified a question about relaxing the deadlines for companies related to the implementation of various cosmetics rules, which NMPA permitted.[38]

[38] State Food and Drug Administration (2022).

Other forms of evidence that are frequently helpful include label reviews by laboratories that are accredited by the Certification and Accreditation Administration of China (CNCA) and/or the China National Accreditation Service for Conformity Assessment (CNAS). These laboratories are not government agencies, but their accreditation helps. Documents issued by government laboratories (i.e., sometimes the laboratory is under a government agency) are even more persuasive.[39]

In 2017, NMPA also issued a draft rule on evidence in administrative cases. This rule sets forth the different types of evidence that may be acceptable in administrative cases. The universe of evidence was defined by the draft rule to include "documentary evidence, physical evidence, audio-visual materials, electronic data, witness testimony, statements of the parties, inspection reports, accreditation and identification opinions [i.e., expert opinions made by special institutes or professional organizations], and on-site inspection records."[40] The draft includes principles on the collection of evidence, such as when originals are required or copies of documentation may be permitted. In general, some verification of copies of evidence may be needed, such as the company chop (or seal) or an attestation that the two are the same. Under the draft, both originals and copies of NMPA-issued licenses have the same effect.[41] Export or laboratory reports or literature from other countries is generally less persuasive. To be accepted, it needs to be translated into Chinese and may need to go through a legalization process through the foreign affairs agency and Chinese embassy or consulate in the other country. The process can take months, so it is challenging to assemble this evidence on short notice.

3.12.4 When to Seek Reconsideration or Appeal?

If an agency does establish a case after its investigation and determines to issue a penalty, the company will face a question of whether to appeal. As discussed above, there are two paths of appeal: administrative reconsideration, which seeks review by the administrative reconsideration committee of the agency or its supervising agency, or administrative litigation, which seeks court review. Administrative litigation is also available as an appeal of administrative reconsideration. And a company can work its way through the court system from district to intermediate and on to the provincial high court for appeals.

Depending on the type of case (e.g., standard enforcement action versus politically sensitive matter), seeking some higher-level review may be entirely appropriate. This often comes down to (a) whether the company continues to believe it has a strong argument (and the agency is being unreasonable) and (b) whether the appeal will seriously damage the company's relationship with the agency in the future. As

---

[39] Civil Ruling on Retrial of the Product Liability Dispute Between Xianxian Liu and the Changbai Mountain Wine Industry Group (2017).

[40] National Medical Products Administration (2017b).

[41] Ibid, Articles 10–15.

to question (a), another factor to consider is whether the court will be willing to adjudicate technical issues in the case. There are ways to help a court through expert evidence, such as by formally requesting that it to entrust technical issues to expert committees or judicial verification (or "identification") institutes (e.g., those that handle forensic evidence), such as in the case of medical malpractice or vaccine cases. Once the expert institution completes its review it will issue a report on technical issues that the court may rely upon. Even though verification procedures through expert institutes may be available, courts can be reluctant to order them or even accept their conclusions in some cases. Although our focus here is administrative cases, because FDD is primarily an area of administrative law, many of the same considerations also apply in civil cases.

3.12.5 When to Consider Settlement?

Throughout every case, there will be suggestions by authorities that the parties agree to settle their dispute, whether through court mediation or through private settlement. In cases involving agencies, officials may be happy for the company to make the complainant whole as a way of stopping the case. A consideration in this decision is whether settlement in one case will encourage others. For example, professional consumers tend to work in groups; release against one (even if enforceable throughout the country) is not a release against all. Word that a company is willing to settle with one person may encourage claims from others. Further, once a complaint has been lodged with administrative authorities, even settling with the complainant may not end the complaint if the administrative authorities find illegality. Therefore, some may adopt a strategy of trying to discourage consumers from bringing frivolous claims in the first place by settling cases as little as possible.

The thoughts above represent only the most basic summary of what is a much larger and more complex area. Civil and administrative actions are a regular feature of operation for companies in the FDD space in China. As we will continue to see in the chapters to follow, consumers and officials assertively raise questions about the compliance and safety of products, with some companies of larger size potentially facing 10s or over a 100 of inquiries per year. The trend over the last 10 years has only been more enforcement, and those enforcement actions are increasingly addressing technical issues, such as whether the products comply with applicable standards or good manufacturing or distribution practice rules. While China's legal framework for enforcing FDD regulation is complex, developing procedures to handle inquiries—one that includes input from the right people in the company in cooperation with legal and government affairs experts—has helped companies weather them successfully by reaching a reasonable outcome with the officials.

References

Balzano J (2012) China's food safety law: administrative innovation and institutional design in the comparative perspective. Asia-Pac L Policy J 13:2. http://blog.hawaii.edu/aplpj/files/2012/05/APLPJ_13.2_Balzano.pdf

Center for Drug Evaluation, State Drug Administration (2019) Drug Registration Evaluation Conclusion Dispute Resolution Procedures (DRAFT). https://www.cde.org.cn/main/news/viewInfoCommon/f7282e50e8a4daf82bc55a9d3b364fd5.

Chen X (2022) Continue to promote standardization of administrative record-filings. Xinhuanet. http://www.news.cn/politics/2022-11/21/c_1129144960.htm.

China Food and Drug Administration (2013) Legislation procedure rules. http://www.moj.gov.cn/pub/sfbgw/flfggz/flfggzbmgz/201503/t20150306_145444.html.

Civil Ruling on Retrial of the Product Liability Dispute Between Xianxian Liu and the Changbai Mountain Wine Industry Group (2017) Jilin Province High Court, Civil Judgement No. 2363

Jinghua Times (2015, May 31) Xi Jinping: 4 "most stringent" food safety measures. http://finance.sina.com.cn/china/20150531/021922308353.shtml.

Lane EJ (2016) China gears up for antitrust pricing drive against drugmakers. Fierce Pharma. https://www.fiercepharma.com/pharma-asia/china-gearing-up-for-anti-trust-pricing-drive-against-drugmakers.

Lu K (2016) Vietnam drug maker sues CFDA: changing the method for shutting the door on drug approvals. Finance Net. http://www.1cjnews.com/cjxx/xf/99064.html.

Ministry of Justice (2020) Draft guiding opinion on strengthening the protection of trade secrets and confidential business information during the course of compulsory administrative license procedures. http://www.moj.gov.cn/pub/sfbgw/lfyjzj/lflfyjzj/202008/t20200821_150811.html.

National Medical Products Administration (2017a) Drug and medical device evaluation and approval information confidentiality administrative measures. https://www.nmpa.gov.cn/directory/web/nmpa/xxgk/fgwj/gzwj/gzwjzh/20170602180301359.html.

National Medical Products Administration (2017b) Food and Drug Enforcement Evidence Principles (Draft for Comment). Article 10. https://www.nmpa.gov.cn/directory/web/nmpa/xxgk/zhqyj/zhqyjzh/20171207134201164.html.

Shanghai Market Supervision Bureau (2022) Rules for application of administrative penalty discretionary standards. Article 11. https://www.shqp.gov.cn/yingp/ypgg/20220328/917964.html.

Standing Committee of the National People's Congress (2011) Administrative compulsory measures law. http://www.gov.cn/flfg/2011-07/01/content_1897308.htm.

Standing Committee of the National People's Congress (2017a) Administrative litigation law

Standing Committee of the National People's Congress (2017b) Administrative reconsideration law

Standing Committee of the National People's Congress (2019) Administrative licensing law

Standing Committee of the National People's Congress (2021) Administrative penalty law

State Administration for Market Regulation (2020) Drug registration regulation. https://gkml.samr.gov.cn/nsjg/fgs/202003/t20200330_313670.html.

State Council of the People's Republic of China (2013a) Decision of the State Council on canceling or decentralizing certain administrative examination and approval items. http://www.gov.cn/zhengce/content/2013-05/17/content_1306.htm.

State Council of the People's Republic of China (2013b) Strictly controlling new administrative licenses. http://www.gov.cn/zwgk/2013-09/26/content_2495516.htm.

State Council of the People's Republic of China (2014) Decision of the State Council on canceling or decentralizing certain administrative examination and approval items. http://www.gov.cn/zwgk/2014-02/15/content_2602146.htm.

State Council of the People's Republic of China (2017) Procedural regulations on enacting administrative regulations. http://www.gov.cn/zhengce/2020-12/26/content_5574507.htm.

State Council of the People's Republic of China (2018) Notice on promoting the reform of "separate on of licenses" across the country. http://www.gov.cn/zhengce/content/2018-10/10/content_5329182.htm.

State Council of the People's Republic of China (2019a) Decision of the State Council on canceling or decentralizing certain administrative examination and approval items. http://www.gov.cn/zhengce/content/2019-03/06/content_5371253.htm.

State Council of the People's Republic of China (2019b) Open government information regulations. http://www.gov.cn/zhengce/content/2019-04/15/content_5382991.htm.

State Council of the People's Republic of China (2020) Decision of the State Council on canceling or decentralizing certain administrative examination and approval items. http://www.gov.cn/zhengce/content/2020-09/21/content_5545345.htm

State Food and Drug Administration (2012) Principles of drug and device penalty discretionary application. https://www.nmpa.gov.cn/xxgk/zhqyj/zhqyjzh/20120919163701336.html.

State Food and Drug Administration (2014) Provisions on food and drug administrative penalties. https://www.nmpa.gov.cn/directory/web/nmpa/xxgk/fgwj/bmgzh/20140428120001181_6.html.

State Food and Drug Administration (2018) The Ministry of Public Security of the General Administration of Food and Drug Administration issued a notice on increasing food and drug safety law enforcement and strictly implementing the regulations on punishing people for food and drug violations. https://www.nmpa.gov.cn/xxgk/fgwj/gzwj/gzwjzh/20180125183401913.html.

State Food and Drug Administration (2022) Reply letter of NMPA related to registration and filing matters for cosmetics during the prevention and control of the novel coronavirus pandemic. https://www.nmpa.gov.cn/directory/web/nmpa/xxgk/fgwj/gzwj/gzwjhzhp/20220427173951109.html.

Wang K (2014) China tightens supervision and enforcement in the medical device sector. https://www.ropesgray.com/en/newsroom/alerts/2014/march/china-tightens-supervision-and-enforcement-in-the-medical-device-sector.

4 Common General Principles of China Food and Drug Regulation

4.1 Separation of Imported and Domestic Pathways

One of the hallmark features of China's food and drug regulatory system is the separation between imported and domestically made finished products. This often rigidly-applied distinction has been the basis for two separate regulatory pathways, which has various implications for regulatory policy, corporate arrangements, and supply chains.

4.1.1 Imported Product Pathway

The imported pathway established permission for products that are made and finished (except for certain re-packaging steps) outside of China. Typically, the National Medical Products Administration (NMPA) or the State Administration for Market Regulation (SAMR), in the case of health foods, licenses these products for importation and marketing. The license or regulatory permission holder typically must be an ex-China (i.e., located outside of China) manufacturing company. The foreign manufacturer/license applicant must appoint an entity in China to act as its agent on regulatory and other matters.

With the imported pathway, the approved version of the product abroad serves as a predicate for China's approval. In other words, there is an expectation that China's approval matches the overseas approval not only in terms of the holder, but also in terms of the indications and other aspects of the labeling (e.g., dosage form, route of administration). This principle is imperfectly applied in practice and, as discussed below, China's labeling is not always an exact translation of the foreign labeling. While NMPA allows applications for imported products in the drug and device spaces that have not yet been approved in any jurisdiction, it expects that proof of approval of this foreign predicate will later be submitted. This system places constraints on adjustments to the license that are tailored to specific jurisdictions;

J. C. Balzano, *China Food and Drug Law*, Law for Professionals,
https://doi.org/10.1007/978-3-031-61901-4_4

China may have different requirements, particularly under its Pharmacopeia, for example.

In more recent years, this policy for overseas license holders has not always proven well-suited to China's goals of encouraging a robust domestic FDD industry, particularly with respect to drugs and medical devices. Biotech companies in China seeking to grow and acquire new technology partner with foreign companies and acquire the rights to engage in research and development and to commercialize their technology in China, with an increase in these deals over 2016–2022. Agreements often call for the product to begin as an imported product (produced outside of China) and transfer manufacturing into China after some years. For drugs, once the product was domestically produced in China it would be considered generic.

This arrangement and China's rules force the ex-China entities to hold the marketing authorization and be liable for the product, along (to varying extents dependent on the space) with a China based company, which serves as the marketing-authorization holder agent (see, e.g., Drug Chap. 5). For a small ex-China entity that, for example, wants to license its products to a partner in the China market, subjecting themselves directly to China's vast and complex (rules are very fluid, ambiguous, and mostly only in Chinese) regulatory system is daunting, to say the least. But a quick technology transfer presents its own substantial practical burdens and risks. This results in substantial burdens on ex-China entities to understand how to comply with an onerous regulatory framework and a disincentive for local companies to become independent of their overseas partners.

4.1.2 Domestic Pathway

The domestic pathway grants permission to products that are manufactured, for the most part, entirely in China. There is no official definition of "domestically manufactured," but there are various practices and nuances, which the chapters below discuss. Not every space treats the boundary between domestic and imported manufacturing the same. Until recently, these licenses were granted specifically to facilities that actually made these products, but they are now attached to the products themselves (which are also tied to typically one, but increasingly more than one, separately licensed manufacturing facility). Domestic marketing-authorization holders must either have their own production facility or attach themselves to licensed manufacturers and in some cases licensed distributors.

4.1.3 Differences Between the Imported and Domestic Pathway

The requirements for approval under the imported and domestic product pathways should be substantively the same in terms of types of data and the overall standard for approval. In other words, all drug and device manufacturers must prove that their investigational products are safe and effective for intended uses by submitting non-clinical and clinical data or its equivalent and complying with mandatory standards (discussed below). Cosmetic and health-food manufacturers must prove that their products are safe and, for some products, applicants must prove they are efficacious with respect to certain functional claims.

Administratively and practically speaking there are differences between imported and domestic products. For example, until recently no imported drug or device could be approved until it obtained approval in the country of origin or manufacture, and NMPA's policies tended to favor products that were first approved in the U.S., Europe, and Japan. Domestically produced products, on the other hand, could be approved without waiting for approval abroad.

Also, to demonstrate approval abroad, drug and device manufacturers have had to submit proof of overseas approval, such as in the form of a certificate of pharmaceutical product (CPP), certificate of free sale (CFS), or certificate of foreign government. The holder of that proof must be the holder of the license in the country of origin and the products may be required to patch each other. Thus, applicants that make their products outside of China were and, to some extent, still are constrained with respect to the timing of launch in China, the license holder, and the degree of variability between foreign and China-marketed products. Also, imported product makers must prove that they comply with the good manufacturing practice (GMP) requirements in the country of origin and certify that they will comply with China GMP, which now exist for drugs, devices, and cosmetics.

There are also regulatory measures designed to encourage manufacturing in China. For example, in 2016 NMPA developed a priority review category for drug products that transferred manufacturing from abroad into China. Similarly, it instituted another priority review category for drugs produced from operations in China that also supply drugs to the EU and the U.S. and have been inspected by the European Medicines Agency or the United States Food and Drug Administration. Domestic companies are also eligible for government subsidies and tax benefits if, for example, they engage in research and development or technical innovation.

China also designed a priority review pathway for devices that requires domestically held intellectual property. To qualify for the innovative medical device priority review category there must be IP held in China, i.e., patents or a license arrangement registered with the government.

China and NMPA also have tended to permit only domestic companies to participate in the drafting of mandatory standards that form the basis for product specifications. The failure to expressly permit both domestic and foreign firms to participate in China's standard-setting process, pursuant to a general, national-level law called the Standardization Law—which governs standards in all spaces—was criticized by the U.S. as an unfair trade practice in 2018.[1]

Animal testing on cosmetics is another example. Since 2013, certain domestically made cosmetics have been permitted to avoid animal-testing requirements. Imported cosmetics were not (until 2021). The imported products had to undergo animal testing at government laboratories in China, which causes problems for companies

[1] Office of the United States Trade Representative (2018a).

entering China that want to claim cruelty-free status. The only way around animal testing is to use an on-shore manufacturer to complete the last steps in the manufacturing process, so that the company could qualify for a domestic approval. Similarly, in 2023, the government began to allow domestic companies to use their own internal laboratories to conduct the other testing needed for cosmetics registration and filing applications, while currently the makers of imported products must wait for space for testing in local government-designated laboratories in China.

Until recently, one core benefit to the imported pathway was less onerous inspections, because China did not implement its manufacturing and distribution license system in other countries. In China, nearly all makers of drugs, devices and cosmetics must obtain some type of review and approval for their facility, and for drugs, foods, and higher risk devices a distribution license or notification may also be required.

This distinction is less meaningful now, however, because there are programs for inspecting foreign manufacturing facilities that make products for importation into China. As discussed below, there are now regulations regarding inspections of foreign drug and device clinical-study sites and manufacturing facilities. The use of overseas onsite and virtual inspections has generally been increasing, with some slow-down during the pandemic.

For certain types of food, such as infant formula, dairy products, and raw meats, there are also regular, pre-qualification foreign-facility inspections. NMPA has even proposed an overseas inspection regime for cosmetics, which is historically a less regulated area.

The state hospital procurement system favors domestically produced goods Schonberg (2021). The Government Procurement Law requires that government procurement, i.e., the purchasing of goods, projects, and services with public funds, must rely primarily on domestically produced goods. Imported goods can only be the object of government procurement, if certain goods, projects, or services cannot be obtained in China or cannot be obtained under reasonable conditions, the procurement is for use outside of China, or law or regulations provide that imported goods may be procured. A proposed amendment of the Procurement Law would maintain a similar framework.[2] This means that state-owned hospitals, where most individuals obtain their health care, or the center for disease control institutions that buy vaccines must procure domestic products, unless they meet an exception. This should not apply to the pool of private hospitals.

In 2014, China overtly stated it intended to promote the wider use of domestically produced medical devices in efforts to reduce costs in hospitals.[3] Procurement rules now require percentages of certain devices procured for use in the public health system to be domestically produced. Domestic companies can also be eligible for additional subsidies (or tax benefits, such as a 75–100% deduction for research and

[2]National People's Congress Standing Committee (2014), Articles 2, 10.; Ministry of Finance (2022).

[3]Reuters (2014).

development or value added tax exemptions for technology transfer) and grants that overseas marketing-authorization holders may not be able to obtain.

In 2021, the press reported that several ministries released the Imported Product Government Procurement Examination Standard, which set forth an index of products that needed to be 100%, 75%, 50%, or 25% domestic. This document—which is not public—provides that 100% of certain medical devices must be domestically produced.[4]

Another system called "volume-based procurement" also tends to favor domestic companies by giving preference, in the form of a certain percentage of sales in a locality, to companies that win local bids, usually by lowering their prices. These companies can produce goods at lower costs, win bids, and thereby obtain a dominate market share.

However, domestic companies can also face challenges. They are subject to closer jurisdiction by local authorities and to the planning of the central government in China, which can put them in challenging positions. These companies are also subject to foreign-exchange restrictions that add administrative burdens to transferring money outside of China to make investments in foreign technology.

4.2 Evolution of the System of Pre-Approvals

Perhaps NMPA's greatest power is its ability to issue, amend, and renew (usually every 5 years) pre-market approvals in the form of product registrations, i.e., its power to issue and cancel administrative licenses. As discussed, it is otherwise a small agency with few resources that must rely heavily on other central agencies and organizations, as well as its local counterparts in the provinces and lower levels of local governments, to surveil for violations and enforce any compulsory administrative penalties (e.g., fines, seizures, recalls).

4.2.1 Reduction of Pre-Approvals/Licenses

The general trend with respect to pre-approvals is, as described in Chap. 3 on administrative law concepts, removing unnecessary or unlawful pre-approvals, moving from licenses to less onerous record filings or implicit approvals (which operate like notifications), and shifting or building administrative resources in the form of additional personnel and technology (e.g., web platforms) to facilitate faster, more efficient approval processes. An important example of capacity building is the addition, starting in 2015, of hundreds of personnel to the centers that review drug

[4] A. Shalal (2021) China Quietly Sets New 'Buy Chinese' Targets for State Companies – U.-S. Sources, Reuters. https://www.reuters.com/business/aerospace-defense/china-quietly-sets-new-buy-chinese-targets-state-companies-us-sources-2021-08-02/. [Last Visited July 4, 2023].

and device applications to expedite drug-marketing approvals (described in Chap. 2).

At the same time, clinical trial approvals for drugs and certain medical devices went from approvals to "implicit" approvals: If there is no objection by NMPA after 60 business days, the drug clinical trial or high-risk Class III device trial may proceed based on the clinical trial protocol that the applicant submitted to NMPA. Another example is that, as of 2018, all non-special-use cosmetics may be marketed after submitting a complete record-filing in the form of applicant and product information and laboratory testing results.

4.3 Creation of True Product Licenses

The other important trend separates product and facility licenses. This trend is important because it provides significant flexibility for smaller drug, device, and even health food companies to pursue market entry without having to invest in building a manufacturing facility. They can contract out the responsibilities for manufacturing and distribution without losing control over the product.

4.3.1 History of Non-Product Specific Licenses

Until recently, China FDD regulation has generally held to the notion that the holder of a permission to market a product should have internal capacity to manufacture it and not merely to supervise a contract manufacturing organization. For example, before the Drug Administration Law was revised on December 1, 2019, drugs made in China did not receive individual product licenses. Rather, as discussed in Chap. 5 on drug regulation, NMPA issued a manufacturing license to the facility. NMPA evaluated the safety and effectiveness of products and then ultimately issued an approval number for the drug that was bundled to the facility license. The situation was similar with certain domestically manufactured cosmetic products.

Also, although China permitted contract manufacturing for drugs, the developer of a drug or device still had to have its own manufacturing license and therefore its own facility, personnel, and quality system. It could not simply rely on a contract manufacturer alone. This requirement made it difficult to be a small research and development company and still hold a product license.

Pre-2019, registration holders of imported drugs or devices were, on paper, referred to as manufacturers but, in practice, the system worked differently than for domestic products. The licenses were viewed as licenses issued to the foreign company (the marketing-authorization holder abroad) for the importation of a specific product.

Over time, the imported drug, device, and cosmetic licenses came to function in certain ways like marketing authorizations, even with respect to contract manufacturing, transfer, and liability. If the foreign license holder had a contract-manufacturing arrangement abroad, NMPA generally would not question that the manufacturing site was different from the address of the license holder. Also, although there was no pathway to transfer imported drug licenses (IDL) between companies, NMPA and its predecessor agencies permitted transfer via a process to amend the name of the IDL holder. On certain forms for registrations or supplements (e.g., for license transfers), the IDL, device, or cosmetic license holder and the manufacturer were listed separately, thus acknowledging the possible separation.

One other feature of the system had significant impact. Because China tied eligibility for product approvals to entities that possessed facilities to manufacture or distribute products, they would only permit entities that possessed those facility licenses to engage in other product-related activities, such as advertising or, in the case of food, applications to establish enterprise standards (i.e., product standards that were specific to a given company and thus could differ in certain respects from national standards). In other words, a manufacturing or distribution license or permission was required to apply for permission for an advertisement (all drug, device, and special food advertisements must be approved). As imported products only received permission to import, their local affiliates had to obtain distribution licenses (not always achievable) or partner with entities in China that held the requisite license, in order to engage in those activities.

4.3.2 Evolution of the Product License System

This system of tying the license for the product to one facility has evolved. As stated above, imported licenses gradually began to function like product licenses. Reforms also began to change the system to encourage smaller companies to come into the market.

In 2014, the Medical Device Supervision and Administration Regulation (the device framework rule) reform made various changes that moved devices towards a marketing-authorization holder system. First, for domestic products a manufacturing license was no longer necessary to apply for a product license, thereby distancing the two processes. And for certain devices that qualified as "innovative devices," which is a priority pathway described in the device chapter, the license holder was permitted to contract out manufacturing without holding its own manufacturing license.

A breakthrough for drugs came in 2015 when the National People's Congress granted special permission for the State Council to enact a pilot marketing-authorization holder program. As will be described below, this pilot allowed individuals and research entities to apply for drug-product licenses for domestic products and contract out manufacturing and distribution, while remaining responsible for the safety, effectiveness, and quality of the product. The program was initially available in ten provinces.

In 2017, China announced a plan to expand the marketing-authorization holder (MAH) concept nationwide for drugs and devices, although amendments to the existing legislation have taken a long time. The program was not officially expanded to the food and cosmetic spaces, but the flexibility it created has begun to creep into those areas, too.

NMPA also began a pilot program—the medical device "registrant" pilot program—that allowed individuals and research entities in 10 provinces to contract out manufacturing to one or more facilities with the appropriate manufacturing license. Like the drug plan, the entities that hold the product's marketing permissions must meet conditions related to their ability to supervise quality in the manufacturing and distribution environments. The registrants (i.e., MAHs) must fulfill a variety of monitoring and reporting responsibilities related to product quality, safety, and effectiveness. The device MAH pilot was expanded nationwide in 2021. This MAH policy will be described in the chapters devoted to individual regulatory spaces herein.

4.3.3 Restriction of Supply Chain Through the Product Life Cycle

Another feature of China FDD regulation has been a general requirement that the product be developed, made, and refined from the same production line at the same facility that will later produce the final product.[5] This strict rule has placed significant limits on research and development and supply chain flexibility. This rule is coupled with a general reluctance to permit segmented manufacturing, including restrictions on contract manufacturing (as discussed above), but particularly across borders.[6] It also has made it difficult to start making a product abroad and later transfer manufacturing to China without a lengthy and costly technology transfer process.

Consider this scenario. A health-food product is developed in the United States and the manufacturer wants to bring it to China and manufacture the product there, which should make the Chinese government very happy. To do so, the manufacturer can't simply submit an application, which would require product testing anyway. Rather, because the NMPA requires testing of market-ready products that will feed production under the license, the manufacturer would have to wait until it has constructed or purchased facilities in China before it can even begin to contemplate submitting its application.

This system may be evolving. Draft rules are now considering permitting R&D to take place in one facility in China or abroad and allow multiple facilities to later be qualified to produce the drug. There are still special considerations for certain products, such as biologics, blood products, and controlled substances such as narcotic and psychotropic drugs but, overall, the MAH system should permit substantially more supply chain flexibility if it continues its current track. The

[5] See, e.g., National Medical Products Administration (2022a), Article 22.

[6] National Medical Products Administration (2022a), Article 69. The Draft shows that segmented manufacturing might be considered in the future but only with approval.

system for changing manufacturing sites has also evolved to permit this change to occur more efficiently.

4.3.4 Specialized Facility Licenses

Unlike the United States, but like Europe and Japan, China has a separate system of manufacturing and distribution licenses in nearly every space. The exception is cosmetics, which require manufacturing licenses but do not require distribution licenses (although there are now a set of concrete legal requirements for distributors). Marketing authorization holders must ensure that domestically produced products are made in manufacturing facilities that are licensed and qualified as compliant with good manufacturing practice, which includes a site inspection. They must also ensure that licensed and qualified distribution facilities handle the product downstream in the market before they reach user entities or retail stores, depending on the type of product. Even with cosmetics the marketing-authorization holder is charged with ensuring that the distributors meet legal requirements.

The facility license system is more efficient than it was before. Simple entity business licenses can be obtained in advance of the facility licenses. Manufacturers of drugs, for example, no longer have to obtain both a manufacturing license and a good manufacturing practice certification as they did before. Those processes are now merged into one license to reduce complexity.

China no longer rigidly requires a one-size-fits-all license for every manufacturer or distributor in certain areas. For example, Class I device makers only need a notification as opposed to a device manufacturing license, and Class II device distributors, which distribute everyday products to consumers (e.g., condoms or thermometers) only need a record-filing, as opposed to a license. China has recently developed an exemption catalogue to remove the requirement of any permission (whether filing or license) for certain lower-risk Class II devices. Food vendors that distribute pre-packaged foods in certain places also may be exempt from a license requirement.

There is no system of foreign-facility registration, except for infant formula and certain foods (e.g., dairy) and other animal products (e.g., raw meat and fish products). As discussed above, in place of registering and licensing foreign facilities that export products to China, NMPA requires evidence that the manufacturing facility is operating in accordance with its home country's manufacturing rules and is conducting overseas inspections. NMPA enforces China good manufacturing practice regulations and other standards on foreign facilities. Although increasingly frequent, these inspections do not occur at regular intervals, and overseas entities may go several years between inspections, unless there is a significant safety event or non-compliance that results in a for-cause inspection.

4.4 Requirement for Local Testing and Clinical Data

China's default requirement for data to support product marketing is laboratory tests and clinical studies run in China on Chinese patients. Another feature of the food and drug regulatory landscape is the heavy reliance on government-run laboratories and hospital sites to support approval processes and enforcement. This means that applicants for product approvals are required to use only certain labs accredited by the government in China, which are usually run or accredited by the government. Similarly, when it comes to enforcement, reports from domestic laboratories and experts are most persuasive, even if the reports come from well-qualified foreign laboratories. For example, a food-label review from a government-accredited China Inspection Body and Laboratory Mandatory Approval (CMA) laboratory helps ensure distributor confidence and rebut claims of non-compliance by consumers. China also requires a certain number of tests on cosmetics and some for medical devices take place in CMA laboratories.

Human studies must be done at hospitals that the government accredits. This means that studies or clinical trials for devices or drugs are most often conducted at state-run hospitals. This is also the case for human-meal studies (i.e., testing on humans) needed to support functional claims for health foods. Although China has established rules to accept foreign clinical data and test results, that system is still evolving, and it is not clear how foreign laboratory results will be assessed.

4.5 Reliance on Foreign Approvals

As noted above, China requires imported products to show proof of approval and marketing abroad. One basis for this policy may be that China does not want to serve (or be exploited) as an experimental base for new products. Also, China's approval process for new products was not always as rigorous as it now is, and it still does not conduct pre-market inspections for overseas manufactured products. Reliance on approvals from overseas health authorities has remained a centerpiece of imported product applications.

Even as China's regime as evolved and grown, it has retained this requirement. All products, including drugs, medical devices, cosmetics, and health foods, require a certificate of free sale (CFS) or other document issued by a foreign authority, officially evidencing approval in the country of origin. That approval must be held in the name of the applicant for marketing approval in China. In some circumstances, this may be relaxed somewhat, such as in the case of cosmetics and health foods, wherein China has begun, in theory, to accept market data evidencing the safety of the product outside of China in lieu of a CFS.

Although officially there are no limits under the regulations as to where the foreign approval can come from, as discussed above, NMPA has in certain programs previously favored approvals from the United States, the European Union, and Japan. This reliance on approvals and guidance from these three jurisdictions is a theme in NMPA regulation. For example, in 2018 the Center for Drug Evaluation introduced a priority program specifically for drugs that filled an urgent clinical need and had been approved in the last 10 years in the U.S., EU, and/or Japan.

4.6 Use of Technology

NMPA has gradually employed technology to make application systems more efficient. This includes various forms of online databases, reference tools, and portals for product and facility license applicants, as well as methods for more electronic communications between reviewers and applicants. More recently this has included downloadable software, such as e-stamping software that applies an electronic version of the applicant or its agent's chop (every company in China has a corporate seal that binds the company).

4.6.1 Websites and Online Databases

The Chinese websites of NMPA, the National Health Commission (NHC), and like provincial agencies have become increasingly sophisticated, comprehensive, and well updated. NMPA's website, for example, boasts thousands of documents dating back several years, a strong search function, and many valuable subpages through which stakeholders can find application forms and guidance documents and track recalls and enforcement issues. The Center for Drug Evaluation (CDE) and the Center for Medical Device Evaluation (CMDE) also have increasingly informative websites, which are updated daily.

The websites post notices of new laws, regulations, rules, and documents, and drafts for comment as they emerge. They also post guidelines, "explanations," and "question and answer" documents to explain those pieces of legislation. They post announcements of expedited program predeterminations (i.e., for stakeholder comment objection) and announcements of product approvals. They post examples of cases in which products were penalized for legal violations, recall notices, warnings concerning product safety issues, cancellations of product licenses (for varying reasons), and orders for drug and device manufacturers to revise their labels.

NMPA and its provincial counterparts have put product, facility, and clinical trial information online in searchable databases. Each of the product spaces NMPA regulates has databases of approved licenses or successful filings. These databases include limited information, such as the name and address of the license holder, manufacturer, and regulatory agent for imported products. For drugs and devices, the scope of use (or indications) may be in the public database entry. The databases also include a validity period.

CDE has a site allows tracking of the stage of review of marketing, clinical trial, supplement, and renewal applications. Light bulbs will be not in place (review not started), dark (review completed), lit (in line for review), or flashing (review in progress) depending on the stage. This can allow companies (including third parties that may be seeking to reply on an innovative product to make a generic) to track application progress.

NMPA and local authorities publish information on manufacturing and distribution licenses and filings for the entities that produce and distribute different regulated products. The existence of a facility license is also described on general business licenses for an entity, which are available online. This information serves a valuable supervision function and allows for companies and individuals to conduct diligence.

NMPA and the provinces have developed other electronic platforms, including one for adverse reaction and event reporting, which requires a login. This has gradually come to replace a system of paper reports. Under the new Drug Administration Law, there is now a system for registering data underlying the approvals for drug ingredients (active ingredients and excipients) and packaging materials, cosmetic ingredients, and medical device raw materials and components similar to a master-file system in the United States or Europe, to permit information about those materials, once approved, to be referenced by other applications. Similar ingredient or component registration platforms also exist for cosmetics and medical devices.

Individuals may also make reports of suspected regulatory violations, wrongdoing, or counterfeit products via online platforms.

4.6.2 Online Submission Systems

NMPA has also moved towards electronic application filing systems. For example, there is an electronic system for filing Class I medical device applications, and non-special use cosmetic applications for both imported and domestic products may be filed online at the local levels. A similar online filing system was developed for certain imported and domestic health foods. Most drug and device applications are now submitted entirely through electronic means, CD-Roms *and/or* an upload of documents to the web platforms of CDE or CMDE.[7] CDE has its own e-stamping software, and in 2024 it plans to launch software to assist applicants with compiling their applications and software to assist CDE in conducting its formality review before the substantive technical review begins (see Chap. 5).[8]

The use of online application submission systems, particularly the approach that NMPA takes to them, is significant. First, a great deal can be accomplished through these systems in terms of how the permissions of the application are managed. For example, a specific clinical trial may be linked to a given entity of the sponsor (even when there is not an official application pathway to change the sponsor). Second, often only domestic entities can access to these accounts. This leaves overseas entities at a disadvantage with respect to communication with NMPA, submission of confidential materials, and guidelines that are not publicly posted on China government websites.

---

[7] National Medical Products Administration (2023).

[8] Center for Drug Evaluation (2023).

Restricting these accounts to China-based entities prohibits individuals abroad at drug companies from engaging in various aspects of the regulatory process. They are not able to obtain certain application information nor submit information remotely and must rely on the transfer of information through affiliates or third parties. This arrangement can make it more difficult to protect any sensitive confidential information associated with these applications.

4.7 Reliance on Conformity Standards Versus Safety and Effectiveness Data

Rather than standards acting as a reference or voluntary guidelines (as they might be in other places), China has mandatory standards for all of these spaces. These standards greatly shape and limit the options for products in all of these spaces in different ways, and there are rarely exceptions from applicable standards.

There is nuance in the different spaces, but applicants for drugs, devices, cosmetics, and health foods and marketers of general foods must comply with these standards for ingredients, testing, product specifications, and labeling, among other areas. Drug, device and cosmetic, products are also registered under a specific product standard—a set of specifications that forms a product-control strategy (i.e., quality standards/performance indicators and relevant tests)—that references various mandatory standards throughout for tests or ingredient specifications, for example.

These standards fit within a national standardization regime, which is governed by a framework law called the National Standardization Law. The National People's Congress approved an amendment to the Standardization Law in 2017. Like its predecessor, the revised Standardization Law acknowledged a structure in which there are national mandatory standards (GB), a body of recommended national standards (GB/T), industrial standards made by China's various central government agencies (e.g., YY = medical standards), local standards made by local governments and their agencies, enterprise standards made by individual companies setting some higher standard in some respect, and collective standards made by a group of companies through a quasi-governmental industry association.

National standards may be written by agencies like NMPA or its affiliated centers or other entities, but they are officially enacted by the Standardization Administration of China, which is now within SAMR. There are also specific regulations or proposed regulations governing the standards for different spaces. For example, NMPA has issued measures on medical device standards, national and industrial. There are regulations governing the enactment of food-safety standards, which both SAMR and NHC play a role in approving. And, in 2017, NMPA proposed measures on drug standards, which governed both the standards incorporated into the China Pharmacopeia and product standards negotiated as part for the registration process.

The significance of the government standard regime is that in many ways applicants must go through a lengthy process to design specifications that account for these mandatory standards without forcing them to re-design products that may already be on the market abroad. Very recently, the drug and device rules allowed that some deviation may be possible, if adequate data is presented, but the practical significance of these provisions is still emerging.[9] If applicants for product approval cannot comply with these applicable mandatory standards (or seek an exemption), they may be forced to make changes that they would not otherwise make, based on a sound design process, just to be able to register the product and speak truthfully about its compliance with regulators.

China's technical standards tend to be very granular and prescriptive. Compliance with these prescriptive standards is much easier to evaluate, likely a consequence of the government's distrust of official discretion and capacity to evaluate complex FDD issues.

China sees these standards serving a basic safety function, guarding against the unsafe design of products and establishing uniform criteria against which products may be tested. The National Standardization Law and related regulations prioritize the mandatory nature of standards that govern areas that affect human health. Standards are often prescriptive, limiting the ability to design new products or processes, and the amendment process can leave limited opportunities for public comment on draft standards are narrow, leaving only 7 or 10 days for comment on a fairly complex document.

4.8 Use of Experts

Recently policy documents have also called for increased use of experts and efforts to recruit those with science degrees into agencies and to further professionalize the team of inspectors that inspect manufacturing or distribution facilities of regulated products or sites of clinical research. NMPA has established rules for the use of experts during the registration application review process, including medical key opinion leaders and professors. It released a set of general procedural rules in 2014,[10] and then two more specific sets of rules, one covering drug registration application review and one covering devices. These rules regulate the compilation of a database of experts, each appointed for a 5-year term (subject to review and renewal), according to a set of qualification requirements, and they set forth procedures on the use of those experts during application review.

CDE and CMDE may consult these experts on a one-off basis, such as through phone communications or emails for simple common problems, or they can convene an expert meeting, which may be open to the applicant or the public in certain cases. Experts must commit to keep information about their consultation confidential. The experts must also certify that they have no conflicts of interest. CDE and CMDE may publish general lists of the experts online. For example, CDE published a list of 626 doctors, professors, and researchers, who could serve as experts in 2020 when

[9] See, e.g., State Administration for Market Regulation (2021), Article 19.

[10] National Medical Products Administration (2014).

the 2020 drug evaluation and review expert selection rules were issued.[11] The experts on the list appear to be from China, but the 2014 rules contain limited provisions on the use of experts from other countries and jurisdictions.

4.9 Internationalization

Internationalization as it is used in this section refers to harmonization with standards issued by international bodies. China FDD policy documents and regulations often have references to China's desire to adopt international best practices and standards. In the case of food, new regulations contain provisions permitting importation of novel foods—at least in principle—if the exporting company can demonstrate that international standards cover it.

In 2022, the National Medical Products Administration reaffirmed its commitment generally to adopt international standards and participate in international organizations. China committed to:

> Strengthen exchanges and cooperation in the international regulation of drugs, deeply participate in the formulation of international regulatory rules, actively transform and implement international technical standards and guidelines, and promote the modernization and internationalization of supervision. Actively participate in the formulation and revision of international rules, deeply participate in the work of relevant international organizations, such as the World Health Organization (WHO), the International Drug Regulatory Authority Alliance (ICMRA), the International Medical Device Regulatory Authority Forum (IMDRF), and the International Cosmetic Regulatory Alliance (ICCR), in order to promote transformation and implementation of relevant guiding principles.[12]

There are various other examples of this phenomenon. As discussed above, NMPA joined the International Council for Harmonisation of Technical Requirements for Pharmaceuticals for Human Use (ICH) as a permanent member in 2017. NMPA has adopted the Electronic Common Technical Document as a basis for the drug marketing application process and has been translating, adopting, and training personnel on numerous other ICH guidance documents. This is an effort to harmonize its standards with those in other jurisdictions. Various standards from the International Standards Organization (ISO) have been adopted into China's standards system that affects all of the regulated products discussed in this book.

NMPA is considering joining other international organizations and programs, such as the Pharmaceutical Inspection Cooperation Scheme (PIC/S), which is an organization of regulatory authorities focused on harmonization of good manufacturing practice standards. It is a member of the International Medical Devices Regulators Forum (an association of device regulators that build common rules), and an increasingly active participant in the World Health Organization.

---

[11] Center for Drug Evaluation (2020); Center for Medical Device Evaluation (2020).

[12] National Medical Products Administration (2022b).

NMPA or provincial regulators are also eager to know what other countries are doing and how to review international materials. Over the years, I have attended and spoken at conferences organized by Chinese administrative agencies to obtain insight into "international experience." U.S.-, EU-, and Japan-based companies are sometimes invited to closed-door conferences in which they discuss their methods of compliance, and in the early years of NMPA's overseas inspection program, it was at times used as a way for inspectors to study the operations of these facilities.

This desire to internationalize has a limit, however. China will often (a) make changes (sometimes very significant changes) to international standards to adopt them to China's circumstances and (b) will often not adopt international standards or practices if it would mean structural change to NMPA's regulatory practices. And the implementation of these standards and schemes may fall short in practice. For example, China still asks for many more documents and data beyond what would be required under international standards at certain stages of drug development and continues to apply its own rules.

In practice, internationalization also arguably has another aspect. The world comes to China, but China also comes to the world. Earlier China FDD regulations expected that companies would comply with the manufacturing rules in their countries of origin. But as the rules have progressed, China now expects that companies incorporate Chinese laws and regulations into their standard operating procedures and manufacturing practices and, as will be discussed in the China drug and device chapters, that they will be inspected by China authorities for compliance.

China is a party to various international treaties that may have some, but still limited formal impact, in the FDD area. Some of these treaties are either not yet or only partially implemented and some are part of the significant effort of China to increase its participation in the system of international trade earlier this century. Although this book is not about international trade, some discussion here and in other chapters is warranted.

As part of China's entry into the World Trade Organization (WTO), China is part of the Agreement on Sanitary and Phytosanitary Measures, which encourages members to take a balanced approach to food safety and plant and animal regulation by protecting the public health without imposing unnecessarily protectionist restrictions on trade. China is required to notify measures that may affect trade to the WTO Secretariat and allow member comment. In practice, China does engage in safety-based regulation by employing expert advice during the rule-making process, but as the chapters that follow will show, its rules often place significant restrictions processes in the form of approvals, filings, and registrations, which give officials broad discretion. China notifies proposed rules, as applicable, to the WTO for stakeholder comment, but it sometimes does so after there have been significant opportunities for comment in Chinese domestically and the rules are either finalized or unlikely to change.

China is also a party to the WTO Agreement on Trade-Related Aspects of Intellectual Property Rights (TRIPS). Pursuant to TRIPS, China is required to give drugs containing "new chemical entities" data protection (i.e., protect the data from reliance by generic or other follow-on competitors).[13] As discussed in Chap. 5, although this protection has existed in China's law for some time, it is not implemented in practice. At best, China provides some data protection for drugs for which the marketing authorization holder is in a country with which China has a bilateral agreement that binds it to grant data protection.

China is a party to other conventions that are relevant to its biosecurity regime that is discussed in Chap. 7 and that regulate plant and animal resources. For example, it is a party to the United Nations Convention on Biological Diversity and the Nagoya Protocol on Access to Genetic Resources and the Fair and Equitable Sharing of Benefits Arising from Their Utilization.[14] This Protocol provides a legal framework for the benefit sharing related to utilization of genetic materials of plant, animal, or microbes. China issued a draft rule on implementing the Nagoya Protocol including the requirements for approving sharing arrangements in 2017, but never finalized it. The Biosecurity Law (2021) contains a high-level provision related to protections for biological resources that has not yet been expanded on by implementing regulations. However, China has developed an analogous regime for human resources that is implemented quite stringently through Human Genetic Resource Regulations (see Chap. 7).

Related to plant and animal protection, China has adopted and implemented the Convention on International Trade in Endangered Species of Wild Fauna and Flora (CITES). This treaty is implemented through a system of permits for the import and export of plant and animal materials that are on a list issued by the National Forestry and Grassland Administration,[15] the local counter-parts of which also issue the permits. It should be noted, however, that the list of plants and animals protected by CITES is not the only time such a permit or a certificate is required for the import or export of plant or animal materials. For example, if the material is on another protected plants list an export permit could also be required.[16]

In addition to international agreements, China also has bilateral agreements with different nations that have content related to the FDD area. For example, in 2020 the US and China signed the Economic and Trade Agreement Between the Government of the United States of America and the Government of the People's Republic of China (Phase I Agreement). This agreement called for patent linkage and patent term extension related to aspects of pharmaceutical products. Chapter 5 briefly discusses how both systems have been implemented to date. The China Switzerland Free Trade Agreement provides for terms of regulatory data protection for chemical drugs and biological products. As noted above, although it has not expressly stated as much, NMPA has arguably honored this treaty in practice. This will also be discussed in Chap. 5.

---

[13] WTO (2017), Article 39.

[14] Convention on Biological Diversity (2021).

[15] National Forestry and Grassland Administration (2023).

[16] National Forestry and Grassland Administration (2021).

4.10 Intellectual Property

Concerns about the protection of intellectual property can certainly influence the regulatory decisions that stakeholders make in the FDD area, including whether to approach the whether to approach the market and how early in a product life cycle. There is often criticism of China on multiple fronts and by countries around the world with respect to policies that unfairly require or encourage intellectual property transfer to Chinese firms or the government and with respect to outright theft. For example, there are allegations that China pressures foreign companies (sometimes indirectly) to partner with Chinese firms, due to their better market access, and to transfer technology to those firms.[17]

In 2018, the Trump Administration issued findings of its investigation into "China's Acts, Policies, and Practices Related to Technology Transfer, Intellectual Property, and Innovation...." This document found that China uses "vaguely worded provisions," verbal or in-person interactions with officials, and seemingly ad hoc expert panel reviews to encourage technology transfer to China and further its goal of becoming a leader in the area of pharmaceuticals and advanced medical devices set forth in the document, Made in China 2025. It also alleged that because China-based entities or partners act as the applicant in administrative processes, those partners are able to use purported government requirements to obtain greater access to a partner's technology.[18]

And there are other allegations: In 2022, FBI Director Christopher Wray stated that "China has stolen more data than every other nation combined."[19] Shortly before that the member states of the European Commission denounced cyberattacks for purposes of industrial theft and espionage emanating from within China.[20] And, Japan has stepped up efforts to combat export of high-level scientific research to China, amid fears of its misuse.[21]

Other programs or efforts have also raised concerns. For example, China established a government-funded Thousand Talents Program that gives grants and funding to professors, researchers and innovators overseas,[22] but participation in the program was alleged to be a means to target information or sectors that China deems important for its own development. Some individuals involved did not disclose participation to corporate or university employers in the United States.[23]

Each of these issues touches the FDD area. China's rules do favor partnerships with local companies on various issues, including clinical data transfer and access to the procurement system. China has also given subsidies and grants to biotech

[17] Bateman (2022), Chapter 9.

[18] Office of the United States Trade Representative (2018b).

[19] Folmar (2022).

[20] Council of the European Union (2021).

[21] Kakuchi (2020).

[22] Working Group on Chinese Influence Activities in the United States (2018).

[23] Lewis (2023).

companies that partner with foreign companies to develop and commercialize drugs and devices, and in the process may obtain valuable know-how related to the making of the products. Or, certain China based consultants can use generally worded regulations to request information regarding the formulation of products and production processes prior to an application submission without making an attempt to tailor the requests or distinguishing between "must haves" and "nice to haves" to protect against unlawful disclosure.

There is also a concern because NMPA and provincial regulators ask for a substantial amount of confidential information during application and inspection processes (more than other countries with significant innovative drug markets, in some cases, but not unlike requests from regulators in other markets). For example, additional questions about manufacturing are asked as part of follow up conversations or requests for "supplemental documents," which are deficiency letters received after the technical review of the application begins. The level of information expressly required in the rules of the various spaces has gradually increased. For example, NMPA now wants to know, in detail, how cosmetic ingredients and medical device raw materials are made and has rules that require that they see executed batch records that show each step of manufacturing of drugs. Further, NMPA guidelines now call for sets of individual records (although coded to remove names) from clinical trials in certain areas. Regulators requiring *access* to such records (if needed) is not unusual, but defaulting to a full set of records is perhaps more than is needed.

The requests for information may not be not consistent across applications, as reviewers may take different positions on the level and form of information, and it is not clear whether NMPA's follow-up questions are the result of misunderstanding about the content of the application, inexperience as to what level of information is needed to evaluate safety, effectiveness and quality for marketing (no one should dispute NMPA's right to assess these aspects of products), or the belief that it needs to obtain more information about technology, which may be novel. The reasons for these requests are simply not very clear.

The risk posed by the requests for such a level of information are compounded by the fact that—despite some notable improvements in IP protections in the FDD area (see, e.g., Chap. 5)—China does not provide consistently strong protections that other markets do, such as strong, predictable regulatory data protection/market exclusivity for innovative drugs. A system of invention and method of use patents is available for pharmaceuticals, but recent statistics indicate that the invalidation rate is fairly high, at 51% (all claims invalidated),[24] with 16% of other cases resulting in partial invalidation.[25] Therefore, the original holder of the information may not be able to protect its intellectual property rights using legal channels. Nor is there a high level of clarity and predictability about when certain information will be released to the public or applicants for "open information" (China's version of the Freedom of

[24] Wang (2021).

[25] Xu (2021).

Information Act), which provides competitors with a channel for seeking information in a way that could later shield it from legal challenges.

All of these issues raise practical concerns: that information that is not needed to evaluate the product might be accidentally disclosed, mistakenly transferred, or stolen by partners or their employees (not a risk, unique to China, but a risk nonetheless) during the preparation or submission processes.

Whether or not every story or allegation in this area is true may be up for debate—we cannot resolve the debate here—but concerns about IP disclosure, coupled with weak or non-existent regulatory exclusivities, concerns about patent-invalidation, and efforts to move manufacturing to China by encouraging domestic manufacturing or conditioning certain benefits on IP ownership in China, can significantly impact decision-making in this area.

In the following chapters, we will see many of these core principles unfold in greater detail in each of the individual spaces. Although there are certainly differences amongst the spaces on many of these issues, these common principles and themes will be evident throughout the coming chapters.

References

Bateman J (2022) U.S.-China technological "decoupling." https://carnegieendowment.org/2022/04/25/countering-unfair-chinese-economic-practices-and-intellectual-property-theft-pub-86925.

Center for Drug Evaluation (2020) Notice of the Center for Drug Evaluation of the National Medical Products Administration on the Release of the "Administrative Measures for External Experts of the Center for Drug Evaluation" and the List of the First Batch of External Experts of the Center for Drug Evaluation. https://www.cde.org.cn/main/news/viewInfoCommon/d2da7eb076e7c7f97f22b8be2dffcdac.

Center for Drug Evaluation (2023) Notice on updating "Technical Requirements for Electronic CD-Rom Application Materials" and other documents. https://www.cde.org.cn/main/news/viewInfoCommon/2969c293179bd697dbb64c454926dd80.

Center for Medical Device Evaluation (2020) Administrative measures for external experts. https://www.cmde.org.cn//xwdt/zxyw/20200220154033835.html.

Convention on Biological Diversity (2021) Nagoya protocol on access and benefit sharing. https://www.cbd.int/abs/about/.

Council of the European Union (2021, July 19) China: declaration by the high representative on behalf of the European Union urging Chinese authorities to take action against malicious cyber activities undertaken from its territory. https://www.consilium.europa.eu/en/press/press-releases/2021/07/19/declaration-by-the-high-representative-on-behalf-of-the-eu-urging-china-to-take-action-against-malicious-cyber-activities-undertaken-from-its-territory/.

Folmar C (2022, November 15) FBI head: China has 'stolen more' US data 'than every other nation combined.' https://thehill.com/policy/cybersecurity/3737251-fbi-head-china-has-stolen-more-us-data-than-every-other-nation-combined/.

Kakuchi S (2020, December 2) Japan steps up protection of research from espionage. https://www.universityworldnews.com/post.php?story=20201202220022645.

Lewis D (2023, January 5) China's thousand talents plan to entice researchers home boosted their output. https://www.nature.com/articles/d41586-023-00012-5.

Ministry of Finance (2022) Draft of the government procurement law. http://gks.mof.gov.cn/gongzuodongtai/202207/t20220715_3827392.htm.

National Forestry and Grassland Administration (2021) National key protected wild plant list. http://www.forestry.gov.cn/c/www/gkbmgz/300099.jhtml.

National Forestry and Grassland Administration (2023) Appendix to the Convention on International Trade in Endangered Specials of Wild Fauna and Flora. http://www.forestry.gov.cn/c/www/gsgg/368208.jhtml.

National Medical Products Administration (2014) Administrative rules on the use of outside experts (trial implementation). https://www.nmpa.gov.cn/zwgk/rshxx/rshzhc/20141224120001354.html?type=pc&m=.

National Medical Products Administration (2022a) Drug administration law implementing regulation draft for comment. https://www.nmpa.gov.cn/directory/web/nmpa/xxgk/zhqyj/zhqyjyp/20220509222233134.html.

National Medical Products Administration (2022b) Notice of the General Affairs Department of the State Food and Drug Administration on further strengthening the service work of foreign-funded enterprises. https://www.nmpa.gov.cn/xxgk/fgwj/gzwj/gzwjzh/20220627171348198.html.

National Medical Products Administration (2023) Notice on the implementation of electronic drug registration application reporting. https://www.nmpa.gov.cn/xxgk/ggtg/ypggtg/ypqtggtg/20221130190751164.html.

National People's Congress Standing Committee (2014) Government procurement law. http://czj.panzhihua.gov.cn/zwgk/zcfg/zcwj/1684856.shtml.

Office of the United States Trade Representative (2018a) 2018 Special 301 Report. https://ustr.gov/sites/default/files/files/Press/Reports/2018%20Special%20301.pdf.

Office of the United States Trade Representative (2018b) Findings of the investigation into China's acts, policies, and practices related to technology transfer, intellectual property, and innovation under Section 301 of the Trade Act of 1974. https://ustr.gov/sites/default/files/Section%20301%20FINAL.PDF.

Reuters (2014, August 18) China calls for greater use of homegrown medical devices. https://www.reuters.com/article/china-medicaldevices/china-calls-for-greater-use-of-homegrown-medical-devices-idUSL4N0QO34020140818.

Schonberg A (2021) Government procurement and sales to state-owned enterprises in China. https://www.uschina.org/reports/government-procurement-and-sales-state-owned-enterprises-china#:~:text=Track%20Travel%20FAQs-,Government%20Procurement%20and%20Sales%20to%20State%2DOwned%20Enterprises%20in%20China,improvements%20over%20the%20last%20decade.

State Administration for Market Regulation (2021) Medical device filing and registration administrative measures. Article 19

Wang J (2021) Overview of patent invalidity in China's pharmaceutical field. IPR Daily. https://mp.weixin.qq.com/s/W3oMTP6iBR3W6ru9im16Tg.

Working Group on Chinese Influence Activities in the United States (2018) Chinese influence and American interests: promoting constructive vigilance. https://www.hoover.org/research/chinas-influence-american-interests-promoting-constructive-vigilance.

World Trade Organization (2017) Agreement on trade-related aspects of intellectual property. Article 39. https://www.wto.org/english/docs_e/legal_e/31bis_trips_04d_e.htm#7.

Xu J (2021) Interesting statistics for pharmaceutical patent invalidations. https://www.linkedin.com/pulse/interesting-statistics-pharmaceutical-patent-invalidations-jian-xu/.

National Forestry and Grassland Administration (2021) National key protected wild plant list. http://www.forestry.gov.cn/ [illegible]

National Forestry and Grassland Administration (2023) Appendix to the Convention on International Trade in Endangered Species [illegible]

National Medical Products Administration (2022) [illegible]

National Medical Products Administration [illegible]

National Medical Products Administration [illegible] Department of State Food and Drug Administration on further strengthening the supervision of [illegible] funded enterprises [illegible]

[illegible]

5 Pharmaceutical Regulation in China

5.1 Introduction

The amount, complexity, and compatibility with international norms of drug regulation in China has increased substantially in the twenty-first century, beginning with China's entry into the World Trade Organization. While many factors have driven this regulatory reform, two stand out as important. The first is the growing domestic demand that better quality, affordable health care be made available for a wider sector of the population. This includes important clinical needs in the therapeutic areas of cardiovascular treatment, diabetes, infectious disease, oncology drugs, and many others. The second is an overall policy to advance innovation and technology in the biopharmaceutical sector, which became a stronger force around the time of the 13th and 14th Five Year Plans in 2016 and 2020, respectively, and the Made in China 2025 Policy. It is also responsible for important legislative amendments and regulatory reform programs. Indeed, with these and other forces at work, China has become the second largest healthcare market by some measures,[1] although its innovative drug market still remains a growth area.[2]

China's drug market has traditionally been dominated by generics, active pharmaceutical ingredients and their component chemicals, and Chinese medicine. Some multinational innovative drug makers have been bringing their drugs to China for several years—some of the earliest ventures date back to the 1980s—but usually those drugs came to China later than other markets, perhaps ten or more years after being approved for marketing in the rest of the world. This "drug lag" was in part a product of the structure of China's drug regulatory system and the long 5–10 year delays that those structural features (e.g., long, cumbersome clinical-trial approvals that could only be sought for imported drugs after foreign approval) caused.

[1] International Trade Administration (2023).

[2] Yeoh (2021).

J. C. Balzano, *China Food and Drug Law*, Law for Professionals,
https://doi.org/10.1007/978-3-031-61901-4_5

In 2015, China's government put in place more regulatory resources (e.g., more funds and drug application reviewers) and began to encourage and approve innovative drugs at rapid pace, one that has continued in the years since. China has shown particular interest in approving drugs that are developed, manufactured, and/or brought to market first in China, and which meet designated clinical needs (e.g., oncology) and have "independent intellectual property," or IP, owned domestically in China or not encumbered by restrictive licenses from foreign companies. In 2021, China approved 47 innovative drugs,[3] and in 2022, it approved 50.[4]

In terms of quality, China continues to confront how to manage drug safety and quality in the context of its emerging industrial infrastructure. This includes ensuring that the businesses that produce active pharmaceutical ingredients and precursor chemicals make them in accordance with specialized drug manufacturing rules, and that China and its drug industry establish facilities that can handle the manufacturing of more sophisticated pharmaceutical products, such as biologics. The latter firms may be making these products both as their own developed products and some are engaging in contract manufacturing for foreign companies that are selling abroad. Either way, adherence to strict good manufacturing practice rules has been a goal, one with challenges, for China.

More recently, a new area of biosecurity (see Chap. 7) has developed. This area addresses not just how to develop safe and effective medicines but how to do so without threatening China's national security by misusing biotechnology. The rules in this area treat the data generated during development as a national resource that must not be exploited. To avoid such exploitation, China restricts the use of human biospecimens and related data by foreign or foreign-invested companies and the exportation of those samples and data.

5.2 Content of China's Drug Market

As discussed, China's drug market consists of traditional Chinese medicine (beyond the scope of this book) and what was at one point considered western medicine. The majority of the innovative drug market in China consists of small molecule drugs (chemical drugs), and more and more of these drugs are approved each year. But biological product development and approvals are increasing rapidly in various areas, including advances in certain areas of regenerative medicine, such as cell and gene therapies. There are number of approved gene-therapy development efforts by Chinese companies and Chinese-foreign joint ventures, some resulting in gene-therapy products approved for marketing.

For many years, China's drug market consisted primarily of generic products or copies of medicines sold abroad, and the state insurance plans primarily reimbursed

[3] Sun and Gao (2022) Four "fast lanes" including the Breakthrough Therapy Program facilitate drug R&D and marketing, People's Daily Online.

[4] MedSci (2022).

generics. As will be discussed below, many of these were either drugs formulated, made, and tested according to existing standards (i.e., recipes) or copies of drugs marketed abroad. It was not until recently that a generic maker had to select an innovative or other well-established product against which to demonstrate that its drug was bioequivalent or that there was no significant difference in the rate and extent of absorption of the generic drug versus the reference-listed drug. A similar conceptual shift has occurred for biologics, moving from biologic applications that needed less data because they were made according to an existing standard to the demonstration of a follow-on biologics' biosimilarity to a reference-listed innovative biologic.

From a regulatory perspective, China's drug market roughly has four groups of drug makers. Many of the policies, laws, and regulations have been built around encouraging certain groups of stakeholders and types of activities.

First, a substantial amount of activity is still from homegrown generic/biosimilar makers, and the system still supports those drug makers in certain of ways, including the system for reimbursement and government procurement by public hospitals (described below), which pits generics against innovators from an early stage of marketing and asks those two groups to compete for the lowest price. As an incentive, the winning bidder in the procurement processes gets a certain bulk part of the market share in a particular province. Various features of the system are favorable to generics: Innovative drugs can be copied fairly quickly usually with no data or market exclusivity to protect the innovator (only Chinese medicine and generics have a formal type of exclusivity). China's generic makers include state-owned firms and smaller hybrid private/public firms engaged in biosimilar development. In 2018, generic and active pharmaceutical ingredients (API) manufacturing accounted for almost twice the share of the market held by patented small molecule drug and biologics manufacturing.[5]

Second, large overseas companies import their drugs to China, typically after launch in other markets (with some notable exceptions). Due to complexities in China's system, they are often made offshore, although some companies have transferred manufacturing to China and have their local affiliates serve as the marketing-authorization holder.

Third, China's emerging group of domestic innovative drug companies have been growing rapidly since 2015 or 2017, when certain restrictions on manufacturing and the timing of clinical trials in China were lifted. These companies often develop their drugs for China and/or license the rights to drugs from overseas- based biotech firms. In these deals, the China-based innovator might take control of development and commercialization of the drug, perhaps with the goal of moving manufacturing to China in the future, at least for those drugs sold in China.

Fourth, perhaps relatively unknown, are the overseas biotech companies from which China-based companies license drugs. These smaller biotech companies must still be the marketing-authorization holder if the drug is made outside of China

[5]Daxue Consulting (July 19, 2022).

(as explained below) and as a result take varying degrees of interest in expanding and entrenching their own interests in the China market. They are different from large companies in that they do not necessarily have the resources to be directly involved on the ground in the China market and, in some cases, grant another company rights in their product via a contract, allowing a China-based company to obtain approval and market the drug in China.

5.3 China's Drug Paradigm and Policy Shift

As discussed, China's strong push toward innovation in the biopharmaceutical industry began in 2015 and 2016, with Made in China 2025 and the 13th Five Year Plan, respectively. These policies encouraged the building of a strong innovative pharmaceutical industry in a market dominated by generics and beset by delay. Two core documents specific to drugs have shaped the approach to encouraging drug and device innovation in China since that time.

The first is the State Council's Opinion on Reform of the Drug and Device Approval System ("Document 44"), issued in August 2015. Document 44 came from the State Council itself, which means it enjoyed the support of high-level leaders.[6] This Opinion was not an express plan innovation but established a number of important policies intended to encourage innovation often by increasing administrative efficiency. One of the main purposes of Document 44 was to clear away a substantial application backlog that had collected for different types of drug and device applications (registration/marketing, amendment, and renewal). China committed to devote new resources to reviewing and approving these applications and, in the future, to devote resources to priority or innovative applications. Indeed, after this document, the National Medical Products Administration (NMPA) released several regulatory documents establishing new categories of drugs for priority review on the basis of various criteria, including place of manufacturing and time and place of approval.

Of the policies set in motion by Document 44, arguably one of the more influential is the definition of "new drug" as new to the world. For years, China had a registration category (explained below) for small molecule drugs and biologics that had not been approved anywhere in the world. But Document 44 stated that "new drug" is a drug that has never been approved anywhere; shortly thereafter, NMPA published a set of "registration categories" that indicated that the definition of an "innovative drug" would be tied to this newness. This was intended to reduce drug lag and bring medicines to China first. In the years following this change, NMPA has tied a number of benefits and proposed benefits, including those relating to priority review and intellectual property, to this distinction.

Document 44 put in place other important mechanisms. It aimed to increase China's role in international development, encouraging participation of medical

[6] State Council (2015).

institutions in international multicenter trials, to upgrade the quality and professionalism in technical reviews of drugs and devices, and to professionalize inspections, among other goals. It set the stage for important reforms that would find their way into later policy documents and into the newly revised Drug Administration Law (DAL), in 2019.

Document 44 was followed in 2017 by an opinion issued by the General Offices of the State Council and the Communist Party, "Deepening of the Reform of the Evaluation and Approval System and Encouraging Innovation for Drugs and Medical Devices," referred to by some as "the Innovation Opinion." The general offices that issued this Opinion serve coordination functions, including coordinating and advising on inter-agency relations. Still, issuance by these important offices of both the main executive organ of the government and the CCP showed the commitment to this push toward innovation.[7]

The Innovation Opinion took arguably bolder steps to encourage innovation. It expanded resources for clinical trials and increased the speed of trial approval and commencement, coordinating NMPA, ethics committee, and human genetic resource (HGR) approval. It called for the continued building of expedited programs/green channels to accelerate the review and approval of drugs and devices serving urgent clinical needs, including orphan (rare disease) indications. It established policies to implement IP protections for innovative drugs, including linking patent protection to the generic review process (patent linkage), extending patent terms to compensate for long marketing application reviews (patent term restoration) and protecting the valuable clinical data of innovative drugs (regulatory data protection).

Like Document 44, the Innovation Opinion included several other reforms that would serve as the basis for revisions to practice and legislation for both drugs and devices (e.g., review teams and confidentiality protections). These reforms will be discussed in both this and medical device chapters of this book (see this chapter and Chap. 6). Since that time, documents policy documents have reinforced the focus on innovation and capacity building of regulatory agencies to review applications and effectively inspect manufacturing facilities. Those documents have suggested other reforms, such as the 14th Five Year Plan for National Drug Safety and Promoting High-Quality Development which also includes sections on using digital surveillance methods and new technologies for drug testing, as well as regulatory reforms to support new important product areas, such as cell and gene therapies (which have become a focal point of China's push for innovation in the drug space).[8]

Also, as noted in Chap. 1, in 2023 following a long period of delay and arguably retrenchment of reform, the State Council issued another Opinion on "Further Optimizing the Conditions for Foreign Investment and Further Endeavoring to Attract Foreign Investment." This Opinion promised further reform (without

[7] State Council General Office, Central Committee General Office (2017).

[8] National Medical Products Administration et al. (2021d) The 14th Five Year Plan for National Drug Safety and Promoting High-Quality Development.

specifics) to improve the procedures for obtaining marketing authorizations for drugs that are approved abroad and that transfer manufacturing to China.

While China has made great strides to increase its level of innovation and research and development, it has not established the strong IP protections for that market that exist in other innovative markets, such as market and data exclusivities, and there are still challenges with providing transparent and reliable access to reimbursement through the state insurance plans, which cover the majority of the population, or to the locally run procurement processes by which public hospitals buy drugs. There have been definite improvements in adding innovative drugs to the national health care reimbursement list, referred to as the National Reimbursement Drug List (NRDL), which is now updated annually—prior to 2017 it was rarely updated. Also, China has made more procurement rules available. However, the decision-making surrounding these processes is still opaque, and companies that are included on the NRDL have to reduce their prices sometimes by up to 60% or more.[9] Similarly, the volume-based procurement system rewards a lower price with a high-market share in a given locality--, whichever company wins the bid gets a large percentage of the market. The drugs that win in this system have to compete with generics (once available), although some originator drugs with lower prices have won bids. This chapter is not about reimbursement and procurement, but it the issues in that area are noted for purposes of completeness.[10]

The remainder of this chapter primarily discusses the processes for development, evaluation, and monitoring of drugs to ensure they meet standards for safety, effectiveness, and quality.

5.4 Drug Definitions and Categorization in China

China's drug regulatory system does not rely heavily on the wording of definitions in laws or regulations. As discussed herein, the laws and regulations contain these definitions, but determinations are made case by case between the agency and the applicant, and guidance/interpretations that give more certainty or concrete methods of determination are not frequently published. Still, it is possible to understand the organization of different products into categories by looking at the definitions and other guidance.

China generally classifies drugs by looking at three aspects: (1) general type (small molecule, biologics, Chinese medicine, *in vitro* diagnostic regulated as a drug), (2) registration category (based on various factors, including status of approval in the world and in China, generic versus innovative, produced overseas and imported in finished form or produced domestically in China) and (3) predominantly for biologics, intended use/nature of the medicine (e.g., therapeutic versus preventive). Registration category is the most critical for determining the pathway

---

[9] Lew (2019).

[10] ChemLinked (July 25, 2022).

and priority that a candidate medicine will receive. These aspects will be discussed in detail below.

5.4.1 Definitions

The term "drug" was defined as follows in the DAL in 2001:

> Pharmaceutical drugs refers to substances that are used to prevent, treat, and diagnose human disease and intentionally regulate human physiology and that have set indications or primary treatment functions, use methods, and dosages, including Chinese herbal medicines, Chinese sliced medicines, finished Chinese medicines, chemical raw materials and their preparations, antibiotics, biochemical, radiopharmaceuticals, blood serums, vaccines, blood products, and diagnostics.

On August 26, 2019, the Standing Committee of the National People's Congress passed the first full revision of the DAL since 2001. That revision did little to expand the definitions in the law, but the definition of drug was simplified. The first part remained nearly exactly the same:

> Pharmaceutical drug refers to substances that are used to prevent, treat, and diagnose human disease and intentionally regulate human physiology and that have set indications or primary treatment functions, use methods, and dosages.

The DAL indicated that drugs could be small molecule, biological, or traditional Chinese medicine. The Law did not contain a definition of biologics, but biological product is defined in a guideline:

> Biologics are preparations using microorganisms, cells, animal or human derived tissues or bodily fluids, and other such materials made using biological technology that are used for the prevention treatment or diagnosis of disease. For the purpose of standardizing the registration application and management of biological products, these products are classified as preventing biologics, therapeutic biologics, and in vitro diagnostic reagents managed as biologics.[11]

The same guideline also defines therapeutic and preventive biologics. In terms of two popular areas of development, cell and gene therapy products are therapeutics (as one might guess) and vaccines are "preventive biologics"—used for the inoculation of humans to prevent and control the occurrence and prevalence of diseases. These include vaccines in the National Immunization Program and those that are not in the National Immunization Program.[12]

Therapeutic biologics refer to those used for the treatment of human diseases, such as proteins, polypeptides, and their derivatives produced by engineered cells

[11] National Medical Products Administration (2020d) Biological Product Registration Categories and Application Material Requirements.

[12] Ibid.

(such as bacteria, yeast, insect, plant, and mammalian cells) using different expression systems; cell and gene therapy products; allergen products; micro ecological products; biologically active products extracted from human and animal tissues or bodily fluids or prepared by other processes, such as fermentation.[13]

IVDs regulated as drugs are those used for blood-source screening and those that use radionuclide tracers. The DAL also contains no distinction between innovative drugs and generic drugs. From various guidelines, it is possible to understand NMPA's definition of generic drug:

Generic drugs are those that imitate a marketed originator drug. They are divided into two categories [of domestically produced generics] one is those that imitate originator drugs marketed abroad but not in China [Category 3, see below] and the second is those that imitate originator drugs that are marketed in China [Category 4, see below]. These generic drugs must be equivalent to the originator drug [the reference preparation.] in terms of quality and therapeutic efficacy.[14] The generic drugs in these two categories must have the same active ingredient, dosage form, strength, indications, route of administration, use method and dosing regimen as the reference product.[15] A third category of generics produced abroad and applying for marketing in China, Category 5.2, see below, has the same requirements as Categories 3 and 4 above.

In the case of Category 3, if the applicant presents complete research data proving reasonability of a variation may the strength, use method and dosing regimen differ from that of the reference preparation.

Guidelines contain a definition of both generics biosimilars.

> A biosimilar refers to therapeutic biologics similar to a registered originator product in terms of quality, safety and effectiveness. The amino acid sequence of a biosimilar candidate should be the same as the original product. . . .[16]

Later guidelines defined the degree of similarity and some product-specific guidelines further elaborated on the methods for demonstrating similarity. Unlike generic drugs, biosimilars must use originator reference products that are approved and registered in China.[17] Of the 30 biosimilars approved between 2019 and this writing in 2023, all or nearly all appear to have relied upon originator products approved for import.[18] However, in 2018 NMPA released a notice that allowed the

---

[13] Ibid.

[14] China Food and Drug Administration (2016b) Explanation of the Small Molecule Drug Registration Category Reform Plan.

[15] National Medical Products Administration (2020f) Chemical Drug Registration Categories and Application Material Requirements.

[16] China Food and Drug Administration (2015d) Technical Guiding Principles for the Research and Development and Evaluation of Biosimilars (For Trial Implementation).

[17] Center for Drug Evaluation (2021d) Technical Guidelines for Similar Evaluation and Indication Extrapolation of Biosimilars.

[18] BigMoleculeWatch (2023).

importation of biological products for studying biosimilars, provided that those originator biologics had at least received clinical trial approvals in China. Given the relatively broad exemption that China allows from patent protection for studying follow-on drugs for purposes of administrative approval, studies can begin many years in advance of patent protection.[19] In any event, as will be shown below, there is no registration category for domestically produced biologics that reference products are only approved overseas.

5.4.2 Registration Categories

Central to the reformed registration categories was the concept of a "new drug" as redefined in Document 44, in 2015 (discussed above). This definition was not included in the revised DAL—although it is stated in lower level NMPA guidelines—but it has been a central qualifying criterion for various benefits for drugs under Chinese law, from patent term extension to certain expedited programs.

The definition in the DAL provides some broad guidance on what constitutes a drug, i.e., that it has a therapeutic intended use based on the claims made in its labeling, its intentional regulation of physiology, and establishment of indications and specific use methods.

A discussed above, the DAL and its implementing regulations divide drugs into three general groups: chemically synthesized drugs, biological drugs, and Chinese medicine. A fourth group (not too often mentioned) is in vitro diagnostics that are regulated as drugs. Biological products are divided into therapeutic and preventive biologics (i.e., vaccines). The next level of organization is a division into different registration categories. The registration categories determine the documents that need to be submitted with the clinical trial and marketing applications and, more recently, the applicability of expedited review programs and patent and other intellectual property protections.

Previously, there were fourteen categories of biologics and twelve categories of chemical drugs. But these categories caused a substantial amount of confusion because they could easily overlap. For example, for biologics, a drug could be both a drug not yet marketed either in China or overseas (prior Category 1) and a monoclonal antibody (Category 2). These categories resided in appendices to the Drug Registration Regulation (DRR), which was enacted in 2002[20] and amended in 2005,[21] 2007, and, most recently, in 2020.[22]

Once Document 44 emerged, NMPA revised the registration categories for small molecule drugs in 2016, in a document separate from the DRR. The new categories

[19]National Medical Products Administration (2018f) Matters Concerning the One-time Import of Reference Drugs for Biological Products Used in Clinical Trials.

[20]State Drug Administration (2007) Drug Registration Regulation (Trial Implementation).

[21]State Food and Drug Administration (2005a).

[22]State Food and Drug Administration (2007b) Drug Registration Regulation.

focused more clearly on a distinction based on the level of innovation, which was tied to the "newness" of the drug to the world, and to its place of manufacture. Therefore, an innovative drug was a drug with a new chemical entity (Category 1), meaning that it had not yet been approved for marketing domestically or overseas, and an improved new drug (Category 2) was one with an improvement (e.g., new route of administration or indication) that had not been approved anywhere in the world. There were special generic categories, which had previously not existed, divided based on whether the reference drug that the generic demonstrated equivalence to was marketed overseas (Category 3) or in China (Category 4). Drugs that had already received approval for marketing in other countries and were imported into China were in Category 5, with 5.1 being for originator drugs approved abroad and 5.2 for generics.[23]

These revised categories did not create a place for a domestically produced, innovative drug if there is already a drug marketed in China. That option only exists for imported drugs. In other words, once a drug is marketed inside China, if a maker in China wanted to fully develop its own version—even if it generated all new data—it would have to go through the generic pathways, or perhaps develop a new-to-the-world innovation that pushed it into another category, such as Category 2 for improved new drugs. When companies use a technology transfer pathway to apply for domestic marketing authorization, on the basis of an existing imported drug registration that they or their overseas affiliate or partner hold, the domestic application must be under a generic category. This relegates some drugs that may have valuable innovations (e.g., improvements in that do not qualify them for Category 2), to generic status even though they may be supported by full safety and effectiveness data.

Ultimately, the categories were further revised after the 2019 DAL. They were removed from the DRR and included in a series of separate registration guidelines covering small molecule drugs, therapeutic biologics, and preventive biologics. The Tables 5.1 and 5.2 compare the categories as they were in 2007 DRR and in the 2020 DRR.

A similar re-categorization of preventive and therapeutic biological products took place in 2020. Unlike small molecule drugs, NMPA did not issue a reform plan for biologics in 2016. As discussed above, the growth of biologics in China has been slower than small molecule drugs but has been increasing as China encourages growth in advanced areas, such as regenerative medicine therapies (including cell, gene, and tissue engineered therapies).

There is a separate set of registration categories for preventive biologics, i.e., vaccines. It is structured similarly to the categories for therapeutic biologics. The same structure exists with innovation defined entirely by newness, so that additional next-in-class innovative products produced in China can only really come on the market as biosimilars.

---

[23] China Food and Drug Administration (2016c) Notice on the Work Plan to Reform the Registration Categories for Chemical Drugs.

Table 5.1 Small molecule drugs (i.e., chemical drugs)

2007 Registration categories	2020 Registration categories
Category 1: New Chemical Entities Not Marketed in Anywhere in the World	Category 1: Innovative Drugs that are not marketed in China or overseas: This category refers to drugs containing new compounds that have a clear structure and pharmacological utility and have clinical value.
(i) Drug substance and its preparations made by synthesis or semi-synthesis	Category 2: Improved new drugs not marketed in China or overseas. This means drugs for which there is an improvement on the basis of a known active ingredient to aspects of the drug such as the structure, dosage form, formulation and manufacturing process, route of administration, or indications.
(ii) Chemical monomers (including drug substance and preparations) extracted from natural sources or by fermentation	2.1: Optical isomers containing known active ingredients prepared by resolution or synthesis methods, esters of known active ingredients, salts to known active ingredients (including salts containing hydrogen bonds or coordination bonds), changes to the acid radicals, bases or metal elements of known salt active ingredients, or forms of other non-covalent bond derivatives (such as complexes, chelates or inclusion compounds), which also have obvious clinical advantages.
(iii) Optical isomers (including substances and preparations) obtained by chiral separation or synthesis	2.2: Drugs that have known active ingredients that have new dosage forms (including new drug delivery systems), new formulation techniques, or new routes of administration, and that have obvious clinical advantages.
(iv) Drugs with fewer components that are derived from multi-component drugs already on the market	2.3: A new compound preparation that contains known active ingredients and that has obvious clinical advantages.
(v) New combination products	2.4: A new indication of a drug with a known active ingredient.
(vi) A preparation marketed in China but with a new indication that is not approved anywhere else in the world	**Category 3:** Drugs for which a domestic applicant seeks to make a generic copy of an originator drug that is marketed abroad but not marketed in China. This type of drug should establish quality and therapeutic equivalence [consistency] to the reference product.
Category 2: Drug preparations with a new route of administration not on the market anywhere in the world	**Category 4:** Drugs for which a domestic applicant seeks to make a generic copy of an originator drug that is marketed in China. This type of drug should establish quality and therapeutic equivalence [consistency] to the reference product.
Category 3: Drugs Marketed Outside of China	**Category 5:** Drugs marketed abroad that apply for marketing in China.
	5.1: Originator and Improved Drugs Marketed Abroad that are applying for marketing in

(continued)

Table 5.1 (continued)

2007 Registration categories	2020 Registration categories
(i) Drug substances and preparations with or without a change in dosage form, but no change of route of administration	China. Improved drugs must have a clear clinical advantage.
(ii) Combination products with or without a change in dosage form, but no change in route of administration	5.2: Generic drugs marketed abroad that are applying for marketing in China.
(iii) Preparations with a new route of administration that are marketed outside of China	
(iv) Preparation marketed in China with a new indication already approved for marketing outside of China	
Category 4: Drug substances/preparations with a change in acid or alkaline radical (or metallic element) but without any pharmacological change, and the original drug entity [active ingredient] is already approved for marketing in China.	
Category 5: Drug preparations that are approved in China with a new dosage form, but no change in route of administration.	
Category 6: Drug substances/preparations that comply with an existing national [drug] standard.** [Analogous to a generic].	

5.5 Generic Small Molecule Drugs

As discussed above, generic drugs are going through a similar evolution in China as they did in other countries—how to best ensure the appropriate degree of sameness between a generic drug product and a reference product that was approved based on full data.

The prior system allowed generics to come to the market based on lesser or no data and no standardized comparison to a chosen reference product; rather, the products needed to demonstrate compliance to an existing drug standard. There were not rigorous requirements for judging whether generic candidates met the standards and could demonstrate sufficient equivalence with an originator drug whose approval was supported by a full dossier of safety and effectiveness data. Also, the method of selecting that standard/reference, some of which may not have been representative of an originator product and the quality of the standard itself were problems in that system. This system began to change in 2007 when the Drug Registration Regulation evolved to state that generics should be similar to existing drugs, but concerns about standards to evaluating this equivalence remained and

Table 5.2 Therapeutic biological products

2007 Categories	2020 Categories
Category 1: Biological products not marketed anywhere in the world	Category 1: Innovative Biological Products: Biological products not marketed inside or outside of China.
Category 2: Monoclonal antibodies	Category 2: Improved Biological Products: Improvements to safety, effectiveness, and quality control of biologics marketed inside or outside of China, and have a clear advantage.
Category 3: Gene and somatic cell therapies and their preparations	2.1: Improvements to dosage form or route of administration for marketed products.
Category 4: Allergen products	2.2: New indications that have not been approved in China or abroad and/or new populations.
Category 5: Multi-component products that are bioactive and extracted through fermentation from human and/or animal tissues or bodily fluids	2.3: New compound products of biological products with similar products on the market.
Category 6: New combination products made from biological products that are already on the market	2.4: Major technical improvements, such as recombinant technology to replace biological tissue extraction technology, changes in amino acid sites or expression systems or host cells that have obvious clinical advantages.
Category 7: Products that are marketed overseas, but are not marketed in China	Category 3: Biological Products Marketed Inside or Outside of China
Category 8: Certain strains used for preparing microecological products that are not yet approved	Category 3.1: Products produced and marketed abroad applying for marketing inside China.
Category 9: Products that do not have an identical structure with marketed products and not yet marketed anywhere in the world (including amino acid locus mutation or absence, modification caused by a different expression system, deletion, changed interpretation, and chemical modifications of the product).	Category 3.2: Products marketed abroad but not inside of China and applying [for approval] to market and product in China.
Category 10: Products with a preparation method that is different from the marketed version, such as different expression systems or host cells.	Category 3.3: Biosimilars
Category 11: DNA recombinant technology products, such as those that use recombination technology to replace synthesis, extraction, or fermentation technologies.	Category 3.4: Other biologics
Category 12: Products converted from non-injectable to injectable forms or from topical to systemic use, and not yet marketed inside or outside China.	

(continued)

Table 5.2 (continued)

2007 Categories	2020 Categories
Category 13: Marketed products with a change in dosage form but not in route of administration.	
Category 14: Products with a change in route of administration (exclude category 12).	
Category 15: Biological products approved according to existing national standards.	

some proposals might allow the applicant to select the reference drug without significant constraint.

A 2015 decision of the Standing Committee of the National People's Congress authorized reforms that required that generic makers demonstrate that their product met the quality and effectiveness of an originator drug, and to conduct a "consistency evaluation." In other words, the concept of a generic as a drug based on a reference product that had been previously approved on the basis of a full safety and effectiveness dossier (e.g., after randomized, controlled clinical trials) was officially introduced.

The Standing Committee of the National People's Congress:

Authorization of the State Council Launching the Drug Marketing-authorization Holder System Pilot Program and Decisions on Related Issues

(Adopted at the 17th meeting of the Standing Committee of the 12th National People's Congress on November 4, 2015)[24]

In order to promote reform of the drug review and approval system, encourage drug innovation, improve drug quality, and provide practical experience for further reforming and perfecting the drug regulatory system, the 17th Meeting of the Standing Committee of the 12th National People's Congress decides as follows:

The State Council is authorized to launch a pilot drug marketing license holder system in ten provinces and directly-administered municipalities, Beijing, Tianjin, Hebei, Shanghai, Jiangsu, Zhejiang, Fujian, Shandong, Guangdong, and Sichuan, and to allow drug research and development institutions and

[24]Xinhua News Agency (November 11, 2014).

researchers to obtain drug approval documents and numbers and to bear responsibility for the quality of those drugs accordingly.

It is agreed that the State Council may organize reform of drug registration classification, improve drug quality, and promote the transformation and upgrading of our Country's drug industry. To this end, in accordance with the relevant provisions of the Drug Administration Law of the People's Republic of China, the production of drugs for which there is an existing national drug standard shall comply shall comply with national drug standards and meet the quality and efficacy of an originator drug. For drugs approved for production that are already on the market overseas but are not yet marketed in China, if there is no national drug standard, the drug should meet the quality and efficacy of the originator drug. The China Food and Drug Administration shall formulate and revise relevant national drug standards in a timely manner in accordance with the aforementioned requirements.

The trial period authorized by this Decision is three years, starting from the date of implementation of this Decision. The China Food and Drug Administration shall formulate specific pilot plans, which shall be reported to the Standing Committee of the National People's Congress for the record after approval by the State Council. During the pilot program period, the State Council shall strengthen the organization, guidance, supervision, and inspection of the pilot work to ensure the quality and safety of drugs. After the pilot program period expires, if practice proves [this approach to be] feasible, the "Drug Administration Law of the People's Republic of China" will be revised and perfected. If practice proves that an adjustment is unsuitable, the implementation of the [original] provisions of the "Drug Administration Law of the People's Republic of China" will be restored. The drug approval number obtained during the pilot program period will continue to be valid after the pilot program period expires. Before the expiration of the pilot program period, the State Council will submit a report on the implementation of this Decision to the Standing Committee of the National People's Congress.

This decision will take effect on November 5, 2015.

In October 2018, the NPC Standing Committee extended the marketing-authorization holder pilot program for another year to permit a longer debate regarding the provisions in the draft DAL.[25] The revised DAL, which was ultimately

[25] Pu (October 24, 2018).

passed in August 2019 and effective December 1, 2019, formalized the marketing-authorization holder system nationwide.

NMPA further developed the reference product approach to generics through a new set of registration categories (as indicated in the definition and decision above). NMPA also released a set of requirements for conducting "quality and equivalence evaluations" between reference products that were either what would be considered innovative drugs in other countries (i.e., approved on the basis of full data) or "internationally known" generics.

Finally, NMPA, with the help of expert committees, developed a series of reference product lists stating which innovative drugs could serve reference products. NMPA developed rules and registration categories, allowing reference products that, among other things, permit a drug that is only approved in another country to be a reference product in China. (The same approach applies to biosimilars.)

The implementation of the consistency evaluation was not simply prospective. Generics on the market were required to go back and demonstrate therapeutic and quality consistency to an originator drug or an "internationally known generic." This was a phased-in program, in which generic drugs that were on the Essential Drug List (i.e., those that were fully reimbursable under the state reimbursement plans) were required to complete the evaluation by 2018 if they were approved prior to 2007 and were small molecule drugs with oral solid dosage forms; among those, if the drug required a clinical trial to prove efficacy or there are other special circumstances the deadline was extended through 2021.[26] Those that could not locate a reference product were required to conduct a clinical trial to stay on the market.

Based on the 2015 change, NMPA has been developing a growing body of guidance for both generic drugs and biosimilars, including on the consistency evaluation itself, the selection of an appropriate reference product, and labeling of generic drugs.

Overall, the process of constructing a generic application based on a reference product is still evolving. In terms of selecting the reference product, a generic maker can either select one from the Reference Product Lists, which regularly compiled in installments by a group of experts organized by the Center for Drug Evaluation, or the generic maker can apply for another reference product, selecting from those marketed abroad or in China.

In 2019, CDE issued a set of procedures for selecting a reference product, with an order of preference for reference products. The preference is for originator drugs, defined in this rule as drugs "approved for the first time globally and supported by full and complete safety and effectiveness data." The order of preference is:

[26] State Council (2016b) Opinion on Conducting the Quality and Therapeutic Equivalency Evaluation for Generic Drugs.

1. Originator drugs marketed inside of China.
2. Drugs made inside of China after a confirmatory review or after an approved technology transfer.
3. Originator drugs not imported into China (i.e., products on the market abroad).[27]

If an originator drug is not available because it is no longer in production or there are quality issues with that drug, then an internationally recognized drug approved in the U.S., E.U., or Japan or one that is approved to be made in China or transferred there after a technology transfer may be used. If none of these reference products exist, then NMPA has discretion to approve another drug as a reference product based on its evaluation of its safety, effectiveness, and quality.

5.6 Biosimilars

Chinese law did not have a regulatory structure for developing and obtaining marketing approval for true biosimilars until recently. Rather, like small molecule generic drugs, an applicant could design a biologic against the existing standard for one on the market, which could therefore potentially be approved based on less data. This was the structure that existed prior to the establishment of regulatory guidance for comparative testing of a follow-on biological product against a reference product to establish a sufficient degree of similarity to permit the follow-on product to enter the market based on that data alone.

The original Guidelines on the Development and Evaluation of Biosimilars (Trial Version) emerged in 2015 and defined a biosimilar as a "therapeutic biologics that is similar to a registered originator product in terms of quality, safety, and effectiveness."[28] That Guideline required that the biosimilar candidate demonstrate similarity to a reference product, an originator biologic approved by NMPA based on full safety and effectiveness data, i.e., not a product approved through an abbreviated pathway with lesser data requirements. Unlike small molecule generics, this Guideline did not state that foreign drugs could be reference products.

To develop biosimilars, China requires comparability studies against the reference product at every stage of development. If earlier studies, e.g., on manufacturing, demonstrate similarity between the biosimilar and the reference product, and that similarity can be verified by further non-clinical studies, then subsequent clinical studies may be limited. The remainder of the Guideline lays out guidance for evaluating the quality of the biosimilar and its similarity, in terms of safety and effectiveness, through non-clinical and clinical studies.

Once the Guideline was issued, NMPA issued additional product-specific guidelines for the development of certain types of biosimilars. These include

[27] Center for Drug Evaluation (2019b).

[28] China Food and Drug Administration (2015d) Technical Guidelines for the Research and Development and Evaluation of Biosimilars (Trial Implementation).

guidelines for basal insulin,[29] cetuximab[30] (oncology), pertuzumab[31] and trastuzumab[32] (oncology), tocilizumab[33] (arthritis), adalimumab[34] (arthritis, Crohn's Disease, and other various indications), liraglutide injection[35] (diabetes), and rituximab[36] (oncology and autoimmune diseases), among others. Importantly in 2020, the agency developed a biosimilar pathway as part of the therapeutic biologic registration categories, which will be discussed below, and later in 2021 it issued a further guideline on demonstrating and extrapolating indications from existing data.

NMPA approved China's first biosimilar in February 2019, with its approval of the marketing registration application for rituximab injection (Hanlikang), which was developed by Shanghai Fuhong Hanlin Biopharmaceutical Company. At the time NMPA claimed that China had the most biosimilars under development of any nation, with nearly 200 applications approved for development, many of which had already entered Phase III.[37] Reports are that the biosimilars market could quadruple in China over by 2025, and those participating are a mix of multinationals, home-grown biotech innovators, and more traditional China follow-on companies.[38] Since the first biosimilar, NMPA has approved approximately 30 more.[39]

5.7 Standards of Approval

Like other countries, China requires that applicants for drugs demonstrate that their drugs are safe and effective when used under intended conditions and that they can manufacture them to applicable quality standards. The evidence of safety and effectiveness must come from clinical studies of some sort and, as the next few

[29] Center for Drug Evaluation (2022c) Guidelines for clinical trial design of once-daily basal insulin biosimilars.

[30] Center for Drug Evaluation (2022b) Guidelines for clinical trial design of cetuximab injection biosimilars (Trial).

[31] Center for Drug Evaluation (2021a) Guidelines for clinical trial design of Pertuzumab Injection Biosimilars.

[32] Center for Drug Evaluation (2020d) Guidelines for clinical trials of trastuzumab biosimilars for injection.

[33] Center for Drug Evaluation (2021b) Guidelines for clinical trials of Tocilizumab Injection Biosimilars.

[34] Center for Drug Evaluation (2020b) Guidelines for clinical trials of Adalimumab Injection Biosimilars.

[35] Center for Drug Evaluation (2020e) Guidelines for the Design of Clinical Trials of Liraglutide Injection Biosimilars.

[36] Center for Drug Evaluation (2020c) Guidelines for clinical trials of Rituximab Injection Biosimilars.

[37] National Medical Products Administration (2019a) The First, Biosimilar Drug Rituximab Approved for Marketing.

[38] Xi (March 2, 2020).

[39] BigMoleculeWatch (2023).

sections will show, China usually requires at least some clinical evidence from patients in China. In some cases, substantial data can come from outside of China, but those are usually rare drugs or drugs that are deemed to be urgently needed, in which case NMPA may require data in China to be generated by the applicant post-marketing as a condition of staying on the market.

For imported drugs, NMPA will also want to understand the evaluation of other regulators outside of China have said about the drug. In some cases, they may request that the applicant provide more information that was submitted to foreign regulators or NMPA/CDE officials may inquire during meetings with applicants about the approach taken by other agencies. However, NMPA does not recognize the determinations of foreign regulators as conclusive and makes its own decision after reviewing the evidence.

The drug law and regulations do not articulate a definition of drug safety, but NMPA and CDE have issued various guidelines as to how to collect and evaluate safety information in increasingly sophisticated ways, such as through periodic reporting and analysis during clinical trials. Some evaluation may be conducted by a data-safety monitoring committee, which is a group of Chinese and potentially international experts appointed by the sponsor that periodically evaluate data during the trial.[40] At the time of this writing, NMPA was also seeking comments on the adoption of ICH E19 on "a selective approach to safety-data collection in specific, late state pre-approval and post-approval clinical trials."[41] This guideline covers justifications for a reduced safety data collection approach during clinical studies.

NMPA and CDE will consider non-clinical safety evaluation materials, reported serious adverse reactions, particularly suspected unexpected serious adverse drug reactions (SUSAR) that occur during development or safety profiles abroad for drugs that are already on the market. As we will see, sponsors in China must work with the clinical trial investigators to submit reportable events, including SUSARS as individual events reports and periodic development safety update reports.

Effectiveness is evaluated by whether a clinical trial meets its endpoints (i.e., clinical outcomes) under its protocol, which may be evaluated by NMPA at pre-development meetings (discussed below) and will be evaluated during the clinical trial application "implicit" approval process (also discussed below). In general, NMPA will accept evidence that a drug is superior to a comparator or a placebo. It will also accept non-inferiority evidence, on which it released a guideline in 2020. Non-inferiority requires the demonstration that, although a drug is not superior to a comparator, drug data demonstrate that its effects are within a clinically acceptable range.[42]

---

[40] Center for Drug Evaluation (2023a) Guidelines on the Summary Analysis and Reporting of Safety Information During Drug Clinical Trials.

[41] Center for Drug Evaluation (2023b) Notice regarding the public solicitation of ICH "E19: Selective collection of safety data in specific pre-market or post-market clinical trials" implementation suggestions and Chinese version.

[42] Center for Drug Evaluation (2020f) Guiding Principles on the Non-Inferiority Design of Drug Clinical Trials.

In 2021, NMPA released a draft guideline that suggested that oncology drugs might have to prove superiority to other drugs in their class to be approved, perhaps signaling the end of the approaches described above in this area. Ultimately, however, a closer of read of the guideline by experts arguably showed that, instead, it was about patient-centered outcomes and well-designed clinical trials that maximize safety.

Quality control is also a factor in approval under the DAL. NMPA evaluates quality through various metrics. First, it approves the quality control standards used in manufacturing, requiring that they come from the China Pharmacopeia or that the applicant demonstrate through data that some other quality standard is acceptable. Quality is also assured through the requirement for registration testing before the marketing application and sometimes before the clinical trial application, and new pre-marketing good manufacturing practice (GMP) "verification" inspections for domestic companies (and possibly in the future for producers of imported drugs). These and other methods will be explained further in the sections below.

Although not always expressly stated, it is apparent that NMPA also evaluates other factors in its review. One core one is clinical need. As we will discuss below, many of NMPA's expedited review programs are designed around a concept of clinical need, i.e., approving drugs for which there is no other comparator on the market in China. Clinical need for a drug is determined by an expert review of whether comparable therapies exist in China. China is not alone in expediting urgently needed drugs, but NMPA can also make clinical need a factor in approval itself. In other words, it may not approve or, more likely, substantially delay drugs it determines China does not need, even if they meet standards for safety and effectiveness. A need for a higher quality product may be a factor. NMPA has stated that it will not approve drugs for which it has concerns about quality and the marketing-authorization holder's capacity to ensure quality.

There are other soft factors that NMPA may consider, some more evident than others. Whether a product is imported or domestic has not expressly been a criterion for approval, although sometimes there are indicators that NMPA looks more favorably on certain products, such as vaccines, if they are made in China. There are announcements of newly approved products on NMPA's website that appear to tout that a drug is home-grown. And some of those announcements also emphasize that the drug represents "independent intellectual property." Independent intellectual property is a term of art used in Ministry of Science and Technology (MOST) documents as a preferred type of arrangement, under which China-based companies own or have long-term intellectual property rights in technology.

More specifically, independent intellectual property refers to a Chinese person or entity that owns the rights to the intellectual property such that it can freely exercise those rights with no or minimal restrictions. The Chinese party must at least have the rights to continually develop the intellectual property and to use the improved product. The Chinese party can receive the intellectual property by its own research and development, transfer, contractual or otherwise, including donations, and through corporate acquisitions or mergers, among other such ways. If the Chinese party obtains the rights via contract it should have them for at least 5 years.[43]

[43] State Council General Office (2008).

The following announcement is an example:

> On December 17, [2018 NMPA] approved the first domestically produced PD-1 monoclonal antibody-Teriprizumab injection (trade name: Tuoyi) for marketing. This drug is innovative biological product drug independently researched and developed by Chinese enterprises with completely independent intellectual property rights. It is used to treat local progress or metastatic melanoma after the failure of previous standard treatments. . . .
>
> As the first domestically-produced monoclonal antibody drug targeting PD-1 that was approved for marketing in [China], Teriplizumab has been supported by a major national science and technology project. This product has been in clinical research and development since the beginning of 2016, and more than 20 clinical trials are currently in progress, including clinical trials in the United States. In March 2018, the [NMPA] officially accepted the application for marketing registration of this product, and included it in [priority review and conditional approval programs]. [CDE, CFDI, and NIFDC] and other relevant units worked together to actively communicate and guide the applicant, solve the technical problems encountered in the review in a timely manner, and prioritize the technical review, on-site inspection, and laboratory testing. . . .[44]

This announcement gives insight into how China, through NMPA and other agencies such as MOST, guides domestic development and how these economic policies are then reflected in drugs' approvals. The concepts of independent intellectual property and a science and technology grant (not unusual for other countries), which primarily go to domestic products, are on display. This does not make these factors criteria for approval, but these are one type of drugs, i.e., domestically produced innovative medicines with domestically held intellectual property rights, that NMPA likes to see approved. But, as the expedited program section below will show, other foreign-made innovative drugs can also be a priority, particularly when they are approved in the U.S., E.U., or Japan and when the applicants can demonstrate a significant clinical advantage in terms of safety, effectiveness, and/or quality.

5.8 Overview of the Drug Development and Marketing-Authorization Process

The following sections will focus on the procedures governing the development and marketing of drugs. The pathways to bring a drug into China became increasingly varied over the course of the 2010s. While China earlier was principally a generic market that did not see innovative drugs or their copies until after they had been on the market elsewhere for several years, China has recently created a pathway for innovators to bring their drugs first to China, or simultaneously with development in the rest of the world. As we will see, the increased trend in biosecurity and preference for domestic products may be hampering that pathway and threatening its success. However, progress may be seen in the removal of significant roadblocks,

[44]National Medical Products Administration (2018d) First Domestic PD-1 Antibody Drug Toripalimab Injection Approved for Marketing.

such as the requirement for imported drugs to first be approved abroad prior to commencing clinical trials in China.

What we will examine here is the paradigmatic pathway to bring an innovative medicine to China. This often starts with some type of clinical development in China, whether through a local study or a China-based site in an international multicenter clinical trial (i.e., a global trial). Although theoretically there are now ways to obtain approval to market a drug in China without conducting a clinical trial there, most often some sort of local clinical development to demonstrate clinical experience in Chinese patients is thought to be necessary. Some preclinical laboratory testing may be done in China but, generally, the innovative candidates brought into the country by overseas companies rely on ex-China preclinical data and the applicant will certify that such preclinical work has been done in accordance with good laboratory practices; China also has its own rules and system for certifying local laboratories.

One of the first steps in drug development for China is to determine whether the drug will be made inside China or overseas. The distinction is meaningful, because (as discussed in Chap. 4) China adheres to rigidly defined imported versus domestically produced pathways. This choice must be made prior to development; once an investigational drug is considered imported the NMPA expects it to be an imported product that is finished overseas once it comes on the market. Manufacturing locations cannot be changed from overseas to domestic during development. Imported versus domestic is also an all-or-nothing proposition. Although it may be possible particularly for small molecule drugs to import some of the raw materials, the processing of drug substance must take place either in China or abroad. This is particularly the case for biologics, for which NMPA may expect that early materials, such as cell lines, must made inside of China for the product to be considered domestically produced. More about these distinctions will be discussed below.

The imported versus domestic distinction also determines the location of the sponsor of the trial and marketing-authorization holder (MAH). If the drug is imported, the sponsor/MAH must be an overseas company and if it is domestic, then it must be a company in mainland China. Companies that make drugs Hong Kong, Macao, and Taiwan also fall under the imported drug pathway, because those jurisdictions are each governed by separate sets of drug laws. Also, an imported drug sponsor *may need to be* and an imported drug marketing-authorization holder *must* be a company that holds an overseas approval (e.g., from the U.S. Food and Drug Administration, the European Medicines Agency, the Ministry of Health, Labor and Welfare in Japan, or other health authority) in the country of manufacture or origin in its own name. This is what is referred to as the "certificate of pharmaceutical requirement" (CPP requirement). Once tied to a specific overseas approval, the drug in China must match in terms of labeling and indications, strength, dosage form, and route of administration.

The CPP requirement used to be so strict that development could not even begin in China until the drug was approved abroad, because the CPP in the sponsor's name had to be presented to obtain approval for the clinical trial from NMPA. In 2017, around the time of the Innovation Opinion, NMPA removed the CPP requirement for drugs that had not been approved anywhere in the world.[45] Those drugs did not have to have a CPP in the sponsor's or marketing-authorization (MA) applicant's name at the time of filing. However, the 2017 notice did not entirely remove this constraint. Once the drug is approved overseas, a CPP in the holder's name must be presented with a subsequent application. Therefore, for example, if an MA applicant has a drug that has not been approved anywhere it can file its application in China without a CPP, but if, 5 years later, the drug has been approved in the European Union when MAH needs to renew the drug's authorization in China, the MAH must present a CPP in its name in order to keep the license active.

For these reasons, the imported versus domestic distinction and the CPP requirements create substantial burdens. The MAH of an imported drug must plan how to ensure it meets the CPP requirement for the next renewal and match the imported drug MA with that in China. And the all-or-nothing distinction creates challenges in diversifying supply chains, because a domestic drug producer cannot segment, for example, part of its manufacturing in China and part in Singapore to use resources efficiently. We will see how these requirements play out in other ways in the sections below.

Approval of most drugs requires some clinical research in China. Clinical trials in China require a series of approvals and steps to establish one or more clinical trial sites within China's network of accredited hospitals (more on this below). An ex-China sponsor will often engage one or more clinical research organizations (CROs) to assist with mapping out the available sites and initiating contact regarding the clinical trial, including negotiating the clinical trial agreement that frames various aspects of the site-sponsor relationship. Clinical trial agreements between the sponsor and each site are required by China's Good Clinical Practice (GCP) regulations and by the human genetic resources (HGR) regulation.

While preparing for site engagement, the sponsor of the trial will also work with an agent entity in China (every application involving a foreign sponsor for an investigational drug produced overseas must appoint such an agent) to submit a clinical trial approval application to NMPA's CDE for approval of all or one of the phases of the clinical trial. The clinical trial application approval process by NMPA's predecessor agencies used to take nearly a year or more (sometimes 2 years for biological products), but now the rules provide that clinical trial applications will be deemed approved in 60 business days from payment of the fee and acceptance of the application, if there are not objections or a rejection from CDE.

This clinical trial approval is then coupled with ethics committee approvals from the individual sites, and finally a human genetic resources (HGR) approval (see Chap. 7), to allow the trial to begin. We will discuss each of these approvals below. After HGR approval is in place, one or more phases of the trial are run, potentially taking several years. During the clinical trial, the sponsor will need to ensure that it is

[45] China Food and Drug Administration (2017b) The Decision Concerning Matters on the Adjustment of Imported Drug Registration.

appropriately monitored and run according to Chinese law and other applicable anticorruption and health care laws. If there are changes to the protocol or other features of the trial, amendments to the clinical trial and HGR approvals must be filed.

When the trial is complete, the sponsor files a marketing application, which undergoes a technical review by CDE, including review by committees of experts. CDE makes a recommendation to NMPA to approve or reject, and NMPA then makes the final decision to issue the marketing authorization. Once issued, the marketing authorization is usually valid for 5 years. The approval letter may set forth conditions for renewal, such as additional studies that must be run and completed within a designated period of time.

Once approved, post-market processes, i.e., pharmacovigilance, annual reporting, and periodic GxP (manufacturing practices or distribution practices) begin. Post-market requirements have become increasingly robust in China over the last 10 years and include post-market individual adverse drug reaction and group adverse event reporting, periodic safety reporting, risk management plans, reports and implementation of recalls, labeling changes (e.g., new warnings), drug annual reporting, and drug re-evaluation (i.e., rethinking of the safety and effectiveness bases for approval). Drug marketing-authorization holders must also monitor their manufacturers and distributors (if separate) to ensure that they are following quality rules, which are known as good manufacturing and good supply practice.

Third parties will also have their own obligations. This includes the third parties involved in clinical trials, such as contract research organizations and site management organizations, and third-party manufacturers and distributors, such as contract development and manufacturing organizations.

5.9 Non-Clinical Development

China requires non-clinical research and laboratory testing to establish safety prior to human (i.e., clinical) testing, primarily in the form of randomized, controlled clinical trials.

No prior approval is necessary to conduct non-clinical testing, but if it is done in China, the DAL and NMPA's implementing rules require that the testing be done in institutions that are certified as compliant with China's good laboratory practices (GLP).[46] The GLP rules include requirements for personnel, facilities, equipment, procedures, testing, quality control, and recordkeeping of the institutions that conduct non-clinical safety evaluations, which are defined as "tests for the purpose of evaluating drug safety, including safety pharmacology tests, single-dose toxicity tests, repeated-dose toxicity tests, reproductive toxicity tests, genotoxicity tests, carcinogenicity tests, local toxicity tests, immunogenicity tests, dependence tests, toxicokinetic tests and other such tests related to the evaluation of drug safety."[47]

[46] China Food and Drug Administration (2017d) Good Laboratory Practices.

[47] *Ibid.* Article 4.

Not all companies conduct this testing in China; some simply submit this data with their clinical trial application, along with certifications for any foreign laboratories. Unlike in other areas where China has localized development requirements, pre-clinical testing on drugs does not necessarily need to be done in China and, indeed, in 2018 NMPA largely abolished the need for batches of investigational product to be tested at laboratories in China prior to the commencement of the trial; an applicant can submit its own manufacturing and test reports. Such testing in China may still be required at the discretion of the CDE reviewer or routinely in the case of some drug products, such as vaccines.

5.10 Registration Applications and Fees

NMPA approves registration applications for drugs. These registration applications include clinical trial applications, marketing applications, market-authorization renewal applications, and supplements (e.g., to add strength or make changes to the approved manufacturing process, among other changes). NMPA's local counterparts in the provinces approve facility licenses for companies that engage in drug manufacturing and distribution, which will be discussed below.

For each of these applications, there is a registration fee. Prior to 2015, registration fees were minimal, and NMPA had not adjusted them since 1995. In 2015, the State Council established a policy to restructure the system and regularly adjust the fees to facilitate reduction of the application backlog and more efficiently approve innovative drugs and devices (National Development and Reform Commission 2015). The policy was established in Document 44, which stated:

> Adjust fee charging policy. Integrate and merge drug and medical device registration [licensure], approval, and registration [*dengji*] fee items. In accordance with the principle of generally balancing revenue and expenditure, the registration fees for pharmaceuticals and medical devices shall be increased and adjusted every five years. Appropriate preferential treatment will be given to small enterprises applying for registration of innovative drugs and medical devices. Fee income is included in the fiscal budget, and separate management of revenue and expenditure is implemented. The funds required for the review and approval work shall be arranged through the financial budget.[48]

The fee schedule from 1995 was thus adjusted in 2015 and then again in 2020. Table 5.3 compares the two systems. Each drug application is tied to a fee and payment required for acceptance of the application. An applicant adding a strength (i.e., a new specification for the drug), increases the fee by 20%. Each strength per dose of the drug (e.g., 5 mg versus 15 mg) receives its own registration number.

The fees under the 1995 schedule were minimal, particularly compared with user fees in the United States, which, in some cases, can be over a million dollars. Further, the 1995 fees no longer matched the demands of the drug approval process. The

[48] State Council (2015).

Table 5.3 Registration Fees (1995)[a]

	Clinical research/ Human observational study		Manufacturing review and approval		Pre-production transfer for official manufacturing	Tech transfer/ Generics	
Project amount review and approval [*converted to USD*]	Initial Approval	Renewal	Initial	Renewal	Production Approval Fee	Initial	Renewal
Category I	390 USD	544 USD	670 USD	3900 USD	1560 USD		
Category II	390 USD	544 USD	670 USD	3900 USD	1560 USD		
Category III	390 USD	544 USD	544 USD	3110 USD		235 USD	235 USD
Category IV	310 USD	------	544 USD	3110 USD			
Category V	310 USD	------	544 USD	1560 USD			

[a]State Drug Administration (1995) Notice on Adjusting Drug Approval and Inspection Fee Standards

Table 5.4 Registration Fees (2020)[a]

Type of project		Domestic production	Foreign production
New Drug Registration Fee	Clinical Trial	$30,070.95	$58,888.94
	Marketing Authorization	$67,659.63	$93,016.33
Generic Drug Registration Fee	Marketing Authorization with No Clinical Trial	$28,755.34	$56,502.06
	Marketing Authorization with Clinical Trial	$48,677.35	$78,623.00
Supplemental Registration Fee	No Technical Evaluation Needed	$1,503.55	$1,503.55
	Technical Evaluation	$15,599.30	$43,909.85
Drug Re-Registration Fee (Once in Five Years)		Determined by the provincial pricing and finance department	$35,583.95

The new fees reflect the evolving regulatory structure, which is focused on marketing authorizations for products (as opposed to licenses tied to manufacturing facilities) and the increased need for resources to support NMPA's goals of approving innovative technology. Fees for supplements only apply if the supplement requires approval, but not if it is subject to a record-filing or annual report requirement. Waivers are available for small companies. Drugs made in Hong Kong, Taiwan, and Macao are considered overseas/imported drugs. Fees for drugs approved for emergency use are not reflected here and are subject to a separate fee schedule

[a]National Medical Products Administration (2020o) Republishing the Fee Standards for Drug Registration

1995 fees reflected a time when clinical research was structured differently, generics were the predominate class of drugs, the registration categories were tailored differently, and approval for manufacturing and the product itself were bundled together.

The 2015 fee schedule increased these fees dramatically, although still not as high as those of regulators overseas and, pursuant to Document 44, the fees were revised 2 years later in 2020 (Table 5.4)

5.11 Clinical Development

Over the past 10 years, China has increased its status to be one of the more common destinations for clinical trials. By some metrics it is among the top five most popular sites for clinical trials in the world,[49] and the largest producer of scientific articles as of 2018.[50]

[49]Hawkins (April 8, 2022).

[50]Tollefson (January 18, 2018).

China's regulation of clinical trials is informed by several concerns. On one hand, there is concern over the treatment of subjects in those trials, and inadequate supervision of the trial by institutional ethics committees. China has routinely been concerned about exploitation of the large numbers of subjects; China does not want to be a testing ground for experimental treatments or drugs that have not yet been adequately vetted in the rest of the world. Similarly, it has been concerned about poor treatment of subjects by doctors and domestic companies. The problems have ranged from inadequate explanation of the risks to subjects or improper vetting of new risks associated with innovative therapies (e.g., stem cell therapies) to inadequate informed consent[51] and poor handling of investigational product and related storage methods.[52]

More recently, it has been concerned with the competence of its clinical trial infrastructure to support goals of developing innovative technologies. As a result, data integrity, including record-keeping and analysis techniques, have become the focus of regulation. Finally, especially since 2015, biosecurity and concerns about the misuse of data generated from tests on Chinese subjects has become a separate area of regulation of clinical studies, under the Humana Genetic Resource Regulation (discussed in Chap. 7).

In the late twentieth and early twenty-first centuries, China hosted smaller studies that were used to study the equivalence of generic drugs or to bridge findings for drugs that already had been marketed abroad for the Chinese population. But it now plays host to gene therapy and oncology trials that have produced drugs that were later studied for marketing in the United States. China produced important early pre-clinical and clinical data during the COVID-19 pandemic (including quickly sequencing the virus) that assisted with research abroad into therapeutics for COVID-19, which were later tested in the U.S. and Europe.

As discussed above, policies and regulations have also made clinical development challenging. These include an early regime that restricted development operations of large drug makers, delaying approvals of trials until drugs were on the market elsewhere in the world and even then failing to act on clinical trial applications (CTA) for months or years at a time. Another structural factor causing delay has been the progress of China's health care system and the need for a larger group of skilled physicians and other health care professionals that understand the basics of drug development, including clinical trial design and maintenance of data integrity, and supportive hospitals to provide the necessary resources and infrastructure for trials. There were instances of data fraud or poorly kept records that rendered trial results questionable.

While the other two challenges have been improving substantially, biosecurity has created delays. Each trial must be approved by the Office of Human Genetic Resources (OHGRA) (now the National Health Commission, see Ch. 7), and this has led to months, in some cases years, of additional delay after all other necessary

[51] Jiang et al. (2022).

[52] Poli (August 3, 2018).

approvals have been obtained. The HGR regime also has restricted the transfer of samples and to laboratories and analytics firms. Indeed, it is now virtually impossible to obtain approval for overseas sample transfer.

Therefore, clinical development is complex with challenges and rewards on a number of fronts. One the one hand, China seeks to become a hub of innovation and science, encouraging drug makers to develop and ultimately produce their most cutting-edge products there. Conducting clinical trials in China may be cheaper than in other parts of the world and presents greater recruitment opportunities. The quality of the data is better than 5 years ago, with strong concerns about data integrity diminishing to the point that data from Chinese trials are now more helpful in supporting approvals before other drug regulators. For example, in 2019 the FDA approved Brukinsa, which marked the first time that the agency had approved a drug with efficacy data that was primarily generated in clinical trials in China.[53] Still, data integrity concerns remain.[54]

What is required to start a clinical trial in China? A clinical trial requires (1) an approval to use the unapproved, investigational drug on a trial basis from NMPA and other approvals and (2) an infrastructure for the trial itself, including sites (hospitals) and typically third parties to assist the sponsor(s) with running the trial. These third parties typically include clinical research organizations (CROs), site management organizations (SMOs), laboratories, and recruitment organizations. Each of these elements will be discussed in turn below.

5.11.1 The Clinical Trial Application

For a foreign drug, this substantial undertaking typically takes three to four months to prepare, because of the need for translation of various materials. The application follows International Counsel on the Harmonisation of Requirements for Registration of Pharmaceuticals for Human Use (ICH) guidelines in many respects, but it must also follow special China guidelines for small molecule drugs and biological products. These “Registration Guidelines” and “Application Acceptance Guidelines” are critical in this respect and call for additional items and potentially special formatting.

Prior to 2019, NMPA required roughly the same set of application materials for both clinical trial and marketing-authorization applications. But that has changed and the clinical trial application process is now more streamlined. Currently, NMPA has adopted the ICH M4 Common Technical Document Modules for application materials, which has been translated into Chinese and adopted for China’s requirements.

An important issue has been the amount of manufacturing information required at this early stage. For example, instead of master batch records, NMPA requires

[53] Mathias and Maddipatla (November 14, 2019).

[54] Loftus (February 9, 2022).

executed batch records for biological products, which provide a near recipe for the drug. NMPA may also inquire further into the manufacturing process through meetings and supplemental requests for information. Such a significant amount of information to provide early in the development process creates a risk of disclosure if the submission process is not secure.

5.11.2 NMPA/CDE Meetings

Despite increased transparency related to World Trade Organization accession, the Chinese government is not always known for having an open-door policy. Agencies can be opaque institutions that are difficult to engage with and that issue high-level directives. NMPA has established different channels for stakeholders to seek consultation. But it does not have the same open-door policy, as other health authorities do, to allow for stakeholders to see formal meetings under most circumstances.

Stakeholders might pursue one or more of three options: informal contacts and meetings, open-window/blinded-call consultations, or formal meetings. The first type allows the company or its personnel to interact with NMPA or CDE personnel because of existing connections or through chance interactions at conferences. Stakeholders might reach out through WeChat or other more informal platforms and ask general questions. A number of large-scale plans are based on the general missives from WeChat conversations. Some of these informal meetings and communications can contain very meaningful information.

"Open window" consultations occur weekly at a designated time for a designated area, such as drugs. Companies can call or email questions to NMPA. Some will visit in person and talk to an official assigned to answer questions. This method does not allow an informed conversation to take place, as the official is there to answer general questions and will likely not have had the opportunity to review materials prepared by the company in advance. The same is true of blinded calls, in which a company calls up and asks questions of an NMPA or CDE official. The questions are general and off-the-cuff rather than a meaningful discussion of a specific issue. That said, blinded calls and scheduled consultations can be a helpful data point. The applicant can also ask general procedural questions or address minor technical issues if it has a login (for domestic entities or through its agent) to the "applicant window." CDE will reply for general non-technical questions within 10 days and technical questions within 15 days.[55]

Formal meetings on "key" technical issues and points of development are permitted under specific circumstances. These meetings include face-to-face meetings, teleconferences, videoconferences, and written responses from CDE to company questions. CDE has issued a rule dividing these meetings into three categories:

Type One Meetings: These meetings may be requested when an applicant encounters "major safety issues" during the course of a clinical trial or "major

[55] Center for Drug Evaluation. Office Services.

technical issues" during the development of a breakthrough therapy, or other prescribed circumstances (see below). These meetings should be scheduled within 30 days after an application.

Type Two Meetings: These are meetings that are held at certain important stages in the research and development process, such as prior to the first submission of a clinical trial applcation to discuss issues such as whether there is a sound research basis for the drug's development or whether safety risks to subjects are controllable. These meetings may also be held at the end of phase II, prior to the start of phase III pivotal clinical trial (to discuss major technical issues), and prior to the submission of a marketing application, as well as prior to the issuance of a marketing authorization (to discuss post-marketing risk control).

CDE rules set forth further circumstances in which applicants may apply for these Type II meetings, such as:

- Prior to the submission of a conditional approval or other priority review application.
- Prior to a new clinical trial to ensure that the materials are complete and that the trial is feasible. For applications to open a China site of an international multi-center trial or where clinical trials have already been approved in countries with developed regulatory systems such a meeting is not necessary. This is a time-saver associated with international multi-center trials.
- Prior to the submission of a marketing application of therapeutic and preventive biologics.

Type II meetings should be scheduled 60 days after an application for a meeting.

Type Three Meetings: These meetings encompass all other meetings that an applicant requests. CDE's rule sets forth the following circumstances when such a meeting must be requested:

- the applicant wants to add an indication or a new combination drug to the trial (when necessary, based on existing data)
- core technical issues during the development of drugs with an urgent clinical need or drugs that treat an orphan indication
- serious questions during the development of complex generics, equivalency evaluations or re-evaluations (e.g., choice of a reference product or selection of a bioequivalence standard)
- design of important non-clinical research (e.g., related to cancer)
- disagreements during the review process about requests for supplementary materials or the results of evaluations
- questions about drug research in cutting edge technical areas
- questions about post-market changes that are not covered by the regulations (particularly for biological products or Chinese medicines) related to categorization of the change, supporting materials, or post-change management plan
- problems with safety/risk assessment during a clinical trial
- other questions, such as questions about design of post-market clinical research

These meetings are supposed to be scheduled within 75 days of an application for a meeting.

The applicant must prepare materials for the meeting, such as the protocol and data. Materials are submitted through the applicant window. The meeting will be approximately 60–90 min and produce an agreement (or, if not, articulate the differences between the parties) in a set of minutes either finalized on the spot or shortly thereafter. While these meetings have been a vast improvement on the prior system, in which meetings were only granted less frequently, there can still be difficulties obtaining a meeting or obtaining an outcome that guides development going forward.[56]

5.11.3 Clinical Trial Sites

Clinical trials must be conducted at accredited sites that have the appropriate expertise for a given study. Previously, sites had to be inspected and approved by health care authorities and NMPA, but since roughly 2019 they can become sites through a record-filing process. NMPA has a database of these sites that lists their accreditation status and areas of expertise. These sites are largely China's state-owned hospitals, but some are privately owned.

While a detailed discussion of China's health care system is beyond our scope, some introduction into the categorization of hospitals and other health care units in China is required. China's NHC and its local counterparts issue licenses to a broad class of institutions called health care institutions (*yiliaojigou*). There are various types of health care institutions, including hospitals (both public and private), basic health care institutions, and other institutions. Some of these are very basic clinics, but clinical trials are typically conducted at the hospitals.

Hospitals are divided into three classes, Class I, II and III. Tertiary hospitals, of which there are 2749 as of 2019,[57] and 3275 as of 2021,[58] are hospitals that provide high-level and specialized medical services to individuals in different regions. They also perform medical education and conduct scientific research. Each class of hospitals is divided into three standard subclasses—A, B, and C—and tier III also contains one specialized subclass of the best hospitals, called "special hospitals." Tier A of Class III is the highest tier and contains 1651 hospitals. Tiers II and I are lower-level institutions that provide more basic medical care. For example, Tier II hospitals are those that meet hospital-grade standards in China, provide medical services to multiple communities, and undertake some teaching and research. There are approximately 10,848 such Tier II hospitals.[59]

---

[56] Center for Drug Evaluation (2020a) Administrative Measures for Communication and Exchange Regarding Drug Research and Development and Technical Review.

[57] National Health Commission (2020a) 2019 China Health Industry Development Statistical Report.

[58] National Health Commission (2022) 2021 China Health Industry Development Statistical Report.

Tier I hospitals are hospitals that provide preventive treatment and rehabilitation services to specific communities and are approved by prefectural local governments. There are approximately 11,264 such hospitals.

Of these hospitals, as of this writing, NMPA and NHC have accredited 1311 for drug clinical trials, including four sites in Hong Kong,[60] which, pursuant to NMPA rules, must be Tier II A and above.[61] Competition for special hospitals to conduct clinical trials can be fierce. These hospitals are subject to their own pressures and may be anxious to bring new trials online and close them out quickly and efficiently in order to generate a regular revenue stream, which is important because they may face funding issues.

One of the first steps in conducting a clinical trial in China is to come up with a list of tentative sites, lead site, and lead principal investigator (PI). The PI may be an employee of the lead site or elsewhere, or contracted with for the study without leading a specific site. A negotiation with a hospital for an arrangement (which is ultimately memorialized in an agreement) to conduct a clinical trial has become an increasingly complex endeavor. This is because hospitals in China are governed by a set of government and internal rules and practices that may clash with the goals of sponsors at times.

Hospitals are often unrepresented by counsel and negotiate through clinical affairs offices, which often have turnover in personnel with varying experience with clinical trials. These hospitals often at least initially reject long, complex clinical trial agreements, although many have become more accustomed to accepting such formats. Others insist on the use of their own rather sparse template contracts, which do not include provisions that sponsors consider critical to protect their rights, such as those covering anticorruption, privacy, and intellectual property. Hospitals also face their own legal pressure. Beginning in 2013 and 2014, the government began to focus on corruption in health care, leading to pressure to ensure that any research funds are flowing to the institution (not investigators) and investigators are not engaging in questionable behavior. Currently, there is increased pressure to comply with human genetic resource rules and biosecurity regulations. As such, hospitals may seek to shift all liability for such issues to sponsors or CROs and ask them to sign certificates attesting to compliance in this area, in addition to clinical trial agreements.

Technology has also supported a positive transformation of the trial process in certain respects. Over the last 3 to 4 years, more clinical trials have employed the digital tools used in other countries. For example, sites and sponsors embraced electronic data capture technology,[62] which has assisted with some of the data integrity issues that manual data collection methods may have caused leading up to the 2015 self-inspection.

[59] *Ibid.*

[60] Center for Food and Drug Inspections. Clinical Trial Institution Filing Database.

[61] National Health Commission, National Medical Products Administration (2019).

[62] Sarpong et al. (2018).

5.11.4 Good Clinical Practice Reform in China

China enacted its first set of good clinical practice rules in 2003 (2003 GCP).[63] They are a simple set of principles concerning the roles of the sponsor, investigator and institution, and monitor, as well as documentation requirements and retention requirements. Some guidelines (e.g., on informed consent and ethics committee composition and function) emerged from that point, but there was little in the way of interpretative guidance of the sparsely worded 2003 GCP.

In 2015, the government began a significant effort to ensure that clinical trials in China produced better quality data and were more compliant overall with GCP. One of the first measures was a requirement that various sponsors conduct a self-verification. On July 22, 2015, the government issued a notice to that effect, and the poor results from this self-verification ushered in a long-term enforcement effort to control and penalize data issues:

> Notice on Conducting Drug Clinical Trial Data Self-Inspection and Verification Work
>
> Enacting authority: State Food and Drug Administration
>
> Document No.: No. 117, 2015
>
> Time of promulgation: 07/22/2015
>
> Implementation time: 07/22/2015
>
> Effective status: Effective
>
> In order to implement the requirements of the Party Central Committee and the State Council to ensure the safety and effectiveness of drugs from the source and the requirements of [The Four Most Stringents Policy, i.e.,] "the most rigorous standards, the strictest supervision, the most severe punishment, and the most serious accountability, and ensuring the safety of the people's diet and medication," [CFDA] has decided to carry out drug clinical trial data verification for the pending [imported and domestic] drug registration applications listed in the [attachment to this Notice]. We declare the relevant points as follows:
>
> 1. From the date of promulgation of this announcement, all applicants for drug registration that have been declared and are pending with [CFDA] must conduct a self-investigation of the circumstances of the clinical trials for pending imported and domestic drug registration applications in accordance with the relevant requirements of the [Drug Good Clinical Practice rules] and the clinical trial

[63] State Food and Drug Administration (2003).

protocol [submitted with their applications] to ensure that the clinical trial data is true and reliable, and the relevant evidence is kept intact.

2. the content of the self-examination includes: (1) checking the consistency of the locked database with the source data, statistical analysis and consistency of the summary report data with source records and the [trial] database; after the lock, [determining] whether the data has been modified and an explanation of the modification. (2) operation and maintenance of the main test equipment such as biological sample analysis and test instruments (e.g., HPLC, LC-MS/MS), the installation and operation of the data management software audit module (Audit trail). (3) Circumstances surrounding screening, enrollment and rejection of subjects in each clinical trial institution, compliance with subject inclusion and exclusion criteria, and random inspections to verify the subjects' participation in the clinical trial. (4) Key data such as the number of cases of protocol deviations, the number of rejection cases, and the number of serious adverse events; the information of the subjects in the information system of the hospital HIS system, such as that for treatment, medication, and the clinical process of the testing assays. (5) Production, purchase, inspection, transportation, preservation, return and destruction of investigational and control drugs, and related bills, records, and sample retention. (6) Collection process of biological samples and records of transportation and handover and preservation; confirmation of biological sample analysis methods, records related to biological sample analysis process and sample retention. (7) implementation of the main responsibilities of the relevant parties in the clinical trial project and regulatory compliance circumstances.

3. Before August 25, 2015, the applicant shall submit an electronic version of the self-investigation report, a scan of the clinical trial contract, and materials related to the main personnel of the research team to the Center for Food and Drug Inspections [CFDI] of [CFDA] for the varieties of drugs listed in the attached document, and attest to the authenticity of the materials through the signature of the legal representative [of the applicant]. The specific requirements of the self-inspection report can be found on [CFDI's website]: www.cfdi.org.cn (Tel: 87559031).

4. If the applicant discovers issues, such as that the clinical trial data is untrue or incomplete, it may file an application for withdrawal of registration application CFDA before August 25, 2015.

5. The food and drug supervision and administration departments of all provinces (autonomous regions and municipalities) are responsible for supervising the self-inspection work of applicants in their [respective] administrative regions. According to the needs of supervision work, they may organize investigations of clinical trials, and the subjects of these investigations should include relevant institutions, such as clinical trial institutions and contract research organizations. The evidence found in the investigation should be promptly locked into evidence, and [the parties involved] shall set forth opinions on the handling [of these

issues]. Relevant circumstances will be reported to CFDA prior to September 8, 2015.

6. CFDA will organize experts to conduct data analysis on the registration applicants' self-investigation materials and conduct [unannounced "flight"] inspections as appropriate. If the clinical trial data is found to be fraudulent or the clinical trial data is incomplete and untrue, [the authorities] will pursue liability of the applicants, clinical trial institutions, and contract research organizations based on the relevant provisions of Article 78 of the Drug Administration Law of the People's Republic of China and Article 166 of Drug Registration Regulation and publicly disclose [the identities] of the of applicants, clinical trial institutions, contract research organizations and their legal representatives and related responsible personnel.

7. During the process of drug evaluation, if the applicant is found to have one of the following circumstances, the application for registration will be rejected in accordance with the relevant provisions of Article 154 of the Drug Registration Regulations: (1) the applicant refuses, evades, or obstructs the inspection or [engages in the] destruction of evidence; (2) clinical trial data cannot be traced to the source or are incomplete; (3) the authenticity is doubtful without reasonable explanation and evidence; and (4) [the applicant] does not submit the self-inspection report.

Applicants who are discovered to have authenticity problems with their clinical trial data during the verification will not have their applications accepted for 3 years. Drug clinical trial institutions that have falsification issues, shall have their qualifications revoked; and the [names of] the personnel, who have violated [applicable] rules during a clinical trial shall be notified to the relevant departments for investigation and handling according to law. The names of applicants, clinical trial institutions, contract research organizations, and their responsible personnel [involved in] falsification [activities] shall be placed on a blacklist.

Special announcement.

Attachment: List of self-checking and verification products for drug clinical trial data

China Food and Drug Administration

July 22, 2015

***

The release of the report on the 2015 self-inspection was significant because many applicants could not demonstrate compliance with the GCP. The results were a mixture of CTA-withdrawals and rejections of the applications; of 1622 applications, 1165 were voluntarily withdrawn, 30 were rejected, and 192 were granted clinical trial waivers.[64]

NMPA responded with a more vigorous program of data inspections, including ordering additional self-inspections and conducting inspections of sites itself, either to close out clinical trials[65] or because of concern about the operation of the trial. NMPA also issued various documents and guidelines to assist sponsors with better data and recordkeeping practices.[66]

NMPA also sought stronger penalties associated with the submission of false data. The DAL and NMPA's rules required accurate submissions, but there was not a clearly applicable provision of the Criminal Code for false statements to the government that covered clinical trial data falsification.

In 2017, the Supreme People's Court remedied that issue with a judicial interpretation that made Article 229 of the Criminal Code on false statements applicable to the intentional provision of false non-clinical and clinical data to drug regulatory authorities. The interpretation applies to a broad range of actors, including personnel from drug and medical device applicants, non-clinical testing institutions, clinical trial sites, and contract research organizations. It makes another article of the criminal code potentially applicable to officials that abuse their position or neglect their duties, resulting in the registration of drugs or devices based on false materials and the occurrence of adverse events.[67]

Despite this innovation, cases citing this interpretation have not been apparent, and the 2017 judicial interpretation was canceled in 2022, and it was replaced with a new judicial interpretation applicable to drugs that made deliberate destruction of non-clinical or clinical data affecting the safety, effectiveness, and quality of drugs an act that seriously endangers human health. The new interpretation was issued pursuant to Article 142-1 of the Criminal Code, which covers drug administration acts sufficient to cause harm to human health. Violations of this provision can potentially result in a prison sentence of 3 years in non-serious cases and 7 years in serious cases. A provision in the 2022 interpretation also made submission of false data to obtain a drug registration a violation of Article 142-1 of the Criminal Law.[68]

After many drafts, NMPA revised the GCPs in 2020. The revision added substantially more content in various areas, including the responsibilities of the investigator, ethics committee, and sponsor, as well as more content on data and safety reporting and record-keeping, which were expanded upon by several other guidelines that were established or revised around the same time.

[64] Liu and Schmitt (February 7, 2017).

[65] China Food and Drug Administration (2015a) Drug Clinical Trial Data Site Inspection Main Points.

[66] National Medical Products Administration (2020i) Drug Records and Data Management Requirements (Trial Implementation).

[67] Supreme People's Court, Supreme People's Procuratorate (2017).

[68] Supreme People's Court, Supreme People's Procuratorate (2022a) Interpretation on Several Questions on the Law Applicable to the Handling of Criminal Drug Safety Injury Cases.

5.11.5 Human Subject Protections and Informed Consent

China requires that sponsors of clinical trials adhere to the Declaration of Helsinki, an international set of general principles governing the research involving human subjects and human materials issued by the World Medical Association. There are number of versions of the Declaration, and the 2000 revision was attached to the 2003 GCP.

Like most GCP elsewhere and the Declaration, China's GCP requires that the subjects be informed of the risks and benefits of the clinical trial, that their participation is voluntary, and that they can opt-out any time without penalty. China's GCP differs from some in that the sponsor is also required to purchase insurance so it can indemnify the site and to compensate the subjects for any physical and economic harm (e.g., loss of wages) "related to the clinical trial." This standard is fairly broad and can be very subject-friendly depending on what is in the clinical trial agreement.

China has also made strides in encouraging more patient-friendly clinical development. For example, in 2022, CDE released a guideline encouraging drug makers to use interviews and surveys to focus more heavily on patient needs.[69] This and other patient-friendly trends, including incorporating a place for real-world evidence into NMPA's guidelines, have created avenues for patient voices. And, despite China's strict controls on the establishment of social groups, some patient advocacy groups (loosely organized) have emerged, while others are state-sponsored.

Revisions to the DAL also made it possible to establish expanded access programs. Although implementation has been slow, China has been developing an official avenue for drug makers to expand access to trials for patients who need the drug but do not qualify as subjects. This could potentially close a difficult hole in the rules that has created legal and regulatory risk for sponsors who want to assist patients but lack a legal basis for giving them the investigational drug.

In practice, there can be problematic dynamics in the treatment environment. Physicians in China face difficulties with patient violence, with doctors and hospitals subject to violence and threats from frustrated patients and their families. A Wikipedia page in Chinese lists various harrowing incidents between doctors and patients, in which patients and their friends have severely beaten, permanently scarred, or held hostage doctors and other hospital personnel, either over treatment or compensation disputes. These incidents do not appear to be limited to a specific geographic region. Hospitals in big cities, like Beijing and Shanghai have had issues. The last one listed occurred on January 20, 2020, when a doctor and other medical staff were murdered after a medical incident that occurred with a patient in the ophthalmology ward.[70]

These incidents and fear about patient retaliation can compete with incentives for compliance with applicable rules, including those requiring that doctors advise

[69] Center for Drug Evaluation (2022d) Guiding Principles on Organizing Patients to Participate in Research and Development.

[70] Wikipedia. List of Medical Injuries in China.

patients of risks and new safety information. Fear of patient retaliation through complaints or violence can also influence evaluations of injuries for compensation (i.e., only injuries that are related to clinical trials are eligible for compensation), potentially arguing there is a connection when none exists. Courts handling injury cases may be similar sympathetic, even setting aside expert determinations in favor of awarding compensation.

5.11.6 Ethics Committees

Another core part of trial preparation is approval of the clinical trial by the ethics committees at the various clinical trial sites. Hospitals that conduct biomedical research on human beings must have an ethics committee, and NMPA rules require that the ethics committee approve the clinical trial. The PI is responsible for applying for the ethics committee approval.

NHC rules, last revised in 2016,[71] govern ethics committee review for biomedical research in medical institutions. In 2023, NHC issued Measures for Ethical Review for Human Life Sciences and Medical Research, which expands the scope of the earlier measures to cover universities and research institutes in addition to medical institutions.[72] China Food and Drug Administration (2017e) These rules are based in concerns about ethical boundaries of human subject research and are similar in form to the regimes in other countries.[73] More recently, the Ministry of Science and Technology released and finalized a draft on ethics review of scientific and technical research activities, adding a further degree of overlap. The MOST draft provides that the NHC rules would apply to the extent their standards are not "lower" than the MOST draft.[74] The GCP first issued in 2003, amended in 2020, also contain rules on the composition, procedures, and legal obligations of ethics committee.

Pursuant to these various rules, ethics committees have at least seven members from the fields of biomedical sciences, ethics, law, sociology, and "social figures," and the committees must have diversity of gender and ethnic minority representation. Ethics committee members serve five-year terms. Each committee elects a chairperson and various vice chairpersons. When reviewing a project, the ethics committee will focus on various criteria, including reasonableness of the risks to the subjects, adequacy of clinical trial materials, including the informed consent form, in informing the subjects of those risks, and the reasonableness of the recruitment

[71] National Development and Reform Commission (2016b) Measures on the Ethical Review of Human Biological Research.

[72] National Health Commission (2023a) Interpretation of Measures on Ethical Review in Human Life Sciences and Medical Research.

[73] National Health Commission (2023b) Measures on Ethical Review in Human Life Sciences and Medical Research.

[74] Ministry of Science and Technology (2023).

criteria and materials, as well as whether the investigator is taking sufficient measures to keep subject information confidential.

NMPA's guidelines on ethics committees espouse similar criteria and require the following materials for approval: the clinical trial protocol, the informed consent form, recruitment materials, the investigator brochure (i.e., information on the drug itself), a blank case report form, NMPA's approval of the clinical trial, and testing of the drug to show compliance with quality standards.[75] This latter category has changed as NMPA no longer regularly requires testing in accredited laboratories prior to initiating a clinical trial in China.

5.11.7 Management of Investigational Product and Quality Considerations

NMPA's expectation is that the investigation drug is coming from a facility that is drug GMP compliant. For domestic drugs, the facility at the time of the clinical trial would potentially still be under evaluation for licensure as a drug manufacturer, with the final GMP and verification inspections occurring prior to approval for marketing. Therefore, there generally has not been flexibility in switching facilities between the time of the clinical trial and manufacture for market, as NMPA expected that would take place through a technology transfer, as described below. (Some recent proposals may allow some new flexibility.) Also, there has been not a pathway for transferring a drug from an imported drug to a domestically made drug at the investigational stage.

If the drug is manufactured overseas, a clinical trial approval serves as the authorization to bring investigational drug into China. As part of the application process, the applicant entity required to attest that the product is made according to good manufacturing practice rules. Like drugs destined for market, certain investigational drugs that come into China must undergo an inspection to ensure that they meet applicable requirements. The applicability of this requirement and the inspection can vary based on the drug and whether it is a small molecule drug or a biologic, or whether it is a controlled substance. A licensed drug distributor (see discussion below) must apply for the applicable inspection at customs and pass the drug on to an appropriately qualified consignee with the necessary drug distribution credentials to store and transport the drug.

In 2018, NMPA released a draft GMP for Investigational Drugs, which was finalized as an appendix to the general GMP in 2022. It requires compliance with the China Drug GMP but sets forth specialized requirements for quality, release of product from product facilities, personnel, labeling, recall, and active and excipient ingredients, among other areas. The sponsor is required to evaluate any contract manufacturing organizations used to make the drug and ensure supervision. These rules also provided labeling requirements for investigational drug and comparator

[75] State Food and Drug Administration (2010a) Drug Clinical Trial Ethical Review Work Guidelines.

products used in the trial under a blinded design. The appendix states that investigational drugs should be made using independent lines and equipment to the extent possible and, if they are manufactured with marketed drugs, special risk assessments and anti-contamination measures should be taken.[76]

The Drug GCP (2020 revision) also stipulate that investigational drugs must comply with quality requirements in China, and sponsors are required to have a plan for ensuring the quality of the investigational supply. Sponsors and investigators must ensure that the investigational drug is only used for the trial and is not diverted. Once the trial is complete, the drug must be returned or destroyed. In some cases, a new trial might commence to follow the original subjects and provide them continued drug supply (depending on the severity of the illness); left-over supply might be used for that purpose.

5.11.8 Clinical Trial Monitoring and Pharmacovigilance

During a clinical trial in China, the GCP requires the sponsor to monitor for safety and to partner with the investigator to communicate issues it the ethics committee and the subjects Issues can arise that require audits during the clinical trials. For example, although data and recordkeeping compliance has improved, there are still issues with under-reporting of adverse events or deviations from the protocol during the clinical trial. However, China's hospitals have made progress in ensuring that data is collected in the proper format and not recorded without the use of forms. Electronic data capture (EDC) systems, which allow for faster input of data, have become more popular, which should assist with data integrity issues presented by paper records.[77] Until recently, paper records were the default for many sites (and in some cases they still are).

The GCP require that sites report collect, analyze, and report adverse drug reactions in accordance with the protocol, including to the sponsor and the ethics committee. Here, China has moved toward harmony with ICH guidelines, including ICH E2B (R3) on transmission of individual case study reports (ICSR) and ICH E2F on Development Safety Update Report for which China has adopted its own guideline. The sponsor must report certain adverse drug reactions received from any source to the investigator, ethics committee, and drug regulatory and health authorities.

CDE also established guidance that requires that suspected unexpected serious adverse drug reactions (SUSAR) are reported directly to CDE via an electronic platform within 7 days for reactions that cause death or threaten life, and within 15 days for other SUSARs. The reports may use ICSR format.[78] Development safety

[76]National Medical Products Administration (2023) Clinical Trial Use Drug Good Manufacturing Practice.

[77]Sarpong et al. (2018).

[78]Center for Drug Evaluation (2018) Standards and Procedures for Rapid Reporting of Safety Data During Drug Clinical Trials.

update reports are required based on the Development International Birthday Date (DIBD) (an ICH concept) or the month and day that the trial was first authorized in China or globally. The first development safety update report in China is due within two months of the first DIBD after the clinical trial is approved in Mainland China.[79]

Another policy-making innovation is the formalization of the role of the data-safety monitoring board in guidelines. These boards, containing experts both inside and outside of China, formed on an ad hoc basis as clinical trials in China became more popular. The concept is similar to that in other countries, in that a DSMB periodically reviews study data to monitor patient safety and the overall conduct and progress of the study. If the DSMB determines that there is an issue with patient safety it may make recommendations to the sponsor concerning the continuation, amendment, or termination of the study. CDE released a guideline on the composition and operation of DSMBs in 2020 further formalizing their role in the clinical development process.[80]

Once the trial is complete and the database is locked, the site will begin close out, and the sponsor will prepare the clinical study report. Under HGR rules, hospital sites and sponsors must fully share all data and records associated with the trial and submit a final report to MOST, including discussing the outcomes and benefits of the trial. Therefore, it has increasingly been the case that some data that may not have been generated at the sites is transferred back to them to meet HGR sharing rules. The last step in the clinical trial process is the data verification inspection, which is conducted for certain types of drugs (e.g., new to the world small molecule drugs, first generics and biological products) if the data will be used to support a marketing application.

5.11.9 Verifications

Verifications apply to more than just clinical trials, but some background here on the emerging verification system overall seems appropriate. From 2020 to 2022, NMPA put in place various rules and guidelines regarding pre-market verification inspections. These can be for clinical trial data or manufacturing sites and can occur at the time of an original marketing application or a supplement. The Center for Drug Evaluation determines during the technical review whether a pre-market verification is required to verify the points in the registration application and conditions for commercial-scale production. In determining whether such a verification is necessary, CDE will consider risk factors related to the product, such as the type of drug, its dosage form, and its registration category and risk factors related to the process environment, including research and development and manufacturing

[79] Center for Drug Evaluation (2020h) Regulations for Safety Update Reports During Research and Development (Interim Version).

[80] National Medical Products Administration (2020g) Clinical Trial Data Safety Monitoring Committee Guideline (Trial Implementation).

entity compliance. CDE rules divide risk factors into high, medium, and low. Category 1 and Category 2 small molecule drugs (i.e., new to the world in some respect) and biologics are risk factors that require pre-market verifications. The risk factors also determine the scope of the verification, i.e., small molecule drugs with active ingredients that have not yet been evaluated should require an inspection of both the preparation and the active ingredient. The CDE rules determining whether to commence a verification are a bit complex as rules go and require some study.[81]

The Center for Food and Drug Inspections organizes the verification. The verification can be onsite or via a paper review of documentation. A verification is not necessarily a good manufacturing practice inspection, but when the drug is produced domestically, it can occur in coordination/conjunction with the final pre-market good manufacturing practice (GMP) compliance inspection. Drugs produced overseas do not require a pre-market GMP inspection, but they can be selected for a pre-market verification.[82]

5.12 Additional Aspects of Clinical Development

Additional potential pieces of clinical development in China are international multi-center trials, investigator-initiated trials ("IITs"), pilot medical tourism/named patient zones, and related to the latter, the development of the concept of real-world evidence.

5.12.1 International Multi-Center Trials

Prior to several innovations in 2015 and 2017, China-sites of International Multi-Center Trials (IMCT) faced several regulatory challenges. These IMCT were restricted to "overseas" applicants under the DRR in effect at that time, did not meet the clinical trial requirement for drug development (absent a separate approval of a clinical trial waiver), and they could not begin in China until the drug was either registered in the rest of the world or the IMCT had entered phase II elsewhere. Furthermore, an applicant could be asked to repeat phase I in China prior to commencing the study.

Despite these disadvantages, IMCT had the distinct advantage of shorter wait times for approval of the trial at a time when it could take one to 2 years to approve a trial (as opposed to the current 60 business days). For this reason, companies would create small multi-center trials in Asia (for drugs that were registered abroad), seek approval of the trial, and then submit a waiver application for the local trial with their

---

[81] Center for Drug Evaluation (2021e) Procedures for the Commencing a Drug Registration Verification and Inspection.

[82] Center for Food and Drug Inspections (2021) Procedures for Drug Registration Verification and Inspection (Trial Implementation) and Five Documents.

marketing application. In 2013, NMPA curbed this practice by requiring that the waiver application be submitted and approved prior to submission of the marketing application, thus adding months to the timelines of overseas companies. As of 2017, approval for IMCT are handled in the same way as other trials, leaving these earlier problems behind.

In 2015, NMPA released guidelines for IMCT trials that required a more robust design and coordination between the China and international sites.[83] These guidelines stated that if more than 20% of the patients were in China, then China experts should be on the independent data-safety monitoring committee. There are references in practice, in some instances, to an unwritten requirement that at least 15% of the patients must be in China, but the regulatory basis for this requirement is unclear.[84]

Document 44 also established a policy to encourage domestic applicants to run their own IMCT. And, in 2017, the NMPA removed the requirement that China sites of IMCTs start later than the rest of the world for small molecule drugs and therapeutic biologics. IMCT thus became a more viable route for drug development in China, and with the establishment of a 60-business day limit on CTA approval, these trials could be run in tandem with local trials if needed to support approval. Indeed, connecting China to an IMCT of an overseas partner and running a local study in parallel has become a common development strategy for a robust data package.

5.12.2 Investigator-Initiated Clinical Trials

Investigator-initiated clinical trials (IITs) have occupied an ambiguous place under Chinese law. These are trials on a drug or device in which an investigator designs and runs the study as a sponsor. An investigator may apply for support from the MAH of the drug in the form of expertise (e.g., assistance with the protocol), funds, and/or supply of drug product.

NMPA has never issued a rule or regulation to expressly permit IITs, and arguably its rules prohibit the study of the drug by individuals or entities that are not either the MAH or hold the rights to develop the drug. As a result, drug companies have approached these clinical trials with caution, concerned that they could be viewed as the sponsor of the clinical trial by NMPA because its rules would not allow any other context in which a drug is supplied to a hospital for study in China. In addition, there were and are anti-bribery concerns, because without a clear reason and use for the study, it could be viewed as the company bestowing an

[83] China Food and Drug Administration (2015b) Interim Guidelines on International Multi-Center Clinical Trials.

[84] China Food and Drug Administration (2015c) International Multi-Center Drug Clinical Trial Guideline.

improper advantage on the health care professional, who is also considered a government official.

An IIT in which an investigator studies records of patients who have used the drug might be permitted, if all hospital-level approvals are obtained (e.g., clinical research committee, ethics committee, and patient consent). In some cases, an observational or interventional study might be permitted if the drug is used on label. Existing studies indicate that drug and device makers are the primary sources of funding for IITs, but hospitals, quasi-governmental industry associations, or contract research organizations are other sources.

This latter group may have been less structured in its approach to IITs. Institutions may view themselves as product developers in their own right because of the now largely obsolete "medical technology" pathway to approve and launch novel medical techniques or methods; these technologies are defined broadly (with potential for overlap with other regulated products) as professional measures "for the purpose of diagnosis, treatment, elimination, alleviation of diseases and pain relief, improvement of bodily functions, prolong life, and assist with recovery."[85] Prior to 2015, there was debate about whether cell and gene therapies should be regulated as medical technologies or biological products. Hospitals were harvesting and engineering these technologies, sometimes with serious safety issues, for a number of years. There were and are hundreds of IITs related to cell and gene therapies. Ultimately, the government decided that NMPA and the NHC would have joint responsibility for these products and that cell and gene therapies should be approved as therapeutic biological products before use, as they are in other countries. Even today, some companies pursue regenerative medicine development in this manner, without a CTA from NMPA, to lower the costs of their operation.

IITs have been popular in other areas. There was a rush of IITs during the early stages of COVID-19 in China, when there were large numbers of infected patients. Doctors somehow obtained drugs that were not approved in China and tried them on patients with varying degrees of success. The State Council and government agencies encouraged these IITs through an official policy document.[86] Investigators later faced difficulties transferring data from these studies to others abroad due to China's HGR regime (see Chap. 7).

Interestingly, despite apparent contradictions with the DAL and related rules, practice has seen NMPA sometimes accept IITs to support approval in China. For example, during COVID-19, the drug Favilavir, an antiviral originally used to treat influenza, was conditionally approved for a COVID indication based on an investigator-initiated clinical trial of 70 patients. Another approval for the antiviral remdesivir was to be based on an investigator-run study at the China-Japan Friendship hospital until it was ultimately suspended due to lack of enrollment.[87]

[85] National Health Commission (2018b).

[86] Ministry of Science and Technology (2020).

[87] Wang et al. (2020).

Beginning in 2020, HGR added another layer of regulation to IITs. Specifically, OHGRA became increasingly reluctant to approve IITs when either the investigators or associated companies applied. In the summer of 2020, OHGRA indicated that if a company received either data or intellectual property from an IIT, it would be required to obtain an international collaboration approval with the hospital and investigator. This policy from OHGRA presented companies with a choice to either purely fund the research but receive no results, or receive results and go through the cost and administrative burden of an HGR approval.

In December 2020 NHC issued a draft rule on IITs, which as of fall 2021 is applied as a final rule on a pilot basis in certain provinces and provincial-level municipalities, Beijing, Shanghai, Guangdong, and Hainan. This was the first such rule that expressly legitimized the use of IITs to study drugs and medical devices. It was issued by NHC without apparent assistance from NMPA, but addressed the issue of NMPA approval for a test article, and answered other important questions about IITs.

First, it defined an IIT as a study conducted by a medical institution on individuals or their health care information. It included both interventional and observational IITs but stated that an IIT would be, by definition, not for registration purposes. They could be used for studying drugs for purposes of diagnosis, treatment, rehabilitation, prognosis, etiology, and prevention of disease. Second, the draft indicated in interventional IITs the test article must be a drug that had received NMPA approval and is used within its approved scope, except if certain criterial are met, primarily: the research is in a level III.A hospital, there is no effective treatment for a life-threatening illness, and there is sufficient evidence of safety and tolerance to support the departure from the approved labeling.[88]

5.12.3 Real-World Evidence

NMPA has been exploring the acceptance of real-world evidence generally, such as observational studies, to support marketing, label expansions (e.g., new indications), and renewal of drugs. In order to implement this system, NMPA has finalized several guidelines on real-world evidence, including a main guideline governing the general requirements for such evidence and one on pediatric studies.[89] The first guideline, which was finalied in 2020, outlines a research process for obtaining real world data which it defines as “real word research means the collecting of clinical data related to the health of research subjects (real world data) or summary data derived therefrom in a real world environment for preset clinical issues and through analysis obtaining

[88] National Health Commission (2020b) Draft on Medical Institutions Conducting Investigator Initiated Clinical Trials.

[89] National Medical Products Administration (2020k) Guidelines on Real World Evidence Supporting Drug Development (Trial Implementation); National Medical Products Administration (2020p) Technical Guidelines Supporting Real World Evidence for Research and Development and Review of Pediatric Drugs.

information about the use and risk and benefits of a drug."[90] Since then, use of drugs in the medical tourism zones have become an important source of this type of data.

5.12.4 Special Approvals in Medical Tourism and Other Pilot Zones Based on Clinical

The DAL contains provisions to expand access to unapproved drugs for patients that need them for life-saving care in several ways, expanded access programs for clinical trials, lesser penalties for importation of unapproved drugs for personal use, and approval for drugs on a named patient basis by individual provinces. Prior DALs contained an emergency access provision, but the approver was NMPA, and such approvals were rarely granted—only in cases of national emergency or when there was a high-profile need.

In 2016, a special zone was put in place in Hainan Province. The Bo'ao Medical Tourism Zone has grown from a rarely used experiment to a regularly used channel for drug makers to obtain early approval for limited quantities of their drugs to be used by specially approved hospital applicants prior to national approval. Since 2018, the Hainan government has had the power, specially granted by the State Council, to approve imported drugs that are approved abroad but not in China for use in its zone.[91] The use data resulting from those approvals is now starting to serve as a partial basis for NMPA to evaluate granting national approval.

In 2020, China established another special medical zone in Guangdong Province. This "Greater Bay Area Zone" sought to give special status to drugs that are approved in Hong Kong or Macao and that meet clinical need.[92] Health care regulators determine whether there is sufficient clinical need for a drug and compile a catalogue of drugs for importation into Guangdong. The Guangdong Government may approve drugs for use in nine cities in the province, provided they meet the criteria above. NMPA does not separately approve the drugs. As of this writing, Guangdong authorities have begun to approve drugs and devices through this channel.[93]

This system is an interesting example of cooperation between Mainland China and Hong Kong and Macao's drug regulatory systems. All three of these places have separate laws on drugs and medical devices. Unlike some other areas, the Mainland China's laws and NMPA regulations do not apply in Hong Kong and Macao. Further, there has been an absence of collaboration between, for example, Hong Kong and Mainland China on drug approvals. Approval in one does not guarantee approval in another. Unlike Mainland China's drug system, Hong Kong approval

[90] National Medical Products Administration (2020k) Guidelines on Real World Evidence Supporting Drug Development (Trial Implementation).

[91] State Council (2018).

[92] Guangdong Government (2022).

[93] State Administration for Market Regulation, et al. (2020e).

relies on approvals of drugs abroad in one or two selected jurisdictions, including the U.S., E.U., Switzerland, Canada and Japan,. Hong Kong also subscribes to the Pharmaceutical Inspection Co-operation Scheme (PIC/S), which harmonizes drug GMP inspection procedures. Mainland China is not a member. Thus, the Greater Bay Area program allows for Hong Kong and Macao's systems to play a stronger role in supporting drug approval on a limited basis in the mainland.

These zones can be a way to introduce drugs to health care professionals in China before they are approved there. But, perhaps more importantly, use of the drug or device there can be a source of data that can be used to support approval by NMPA. Both Hainan and NMPA have supported use of this type of data.[94] CDE has developed a guideline and has communication with the Hainan Institute of Real World Data.[95] CDE has approved drugs (and CMDE devices) relying in part on data generated from use in Hainan.

Two additional pathways deserve mention. First, the DAL permits licensed medical institutions to compound their own small molecule drugs for use on their own patients.[96] Second, NMPA and the NHC issued a work plan for qualifying medical institutions throughout the country to apply for importation of clinical needed drugs on an urgent basis in cases in which the drug is needed for an orphan disease or to prevent or treat a serious life-threatening illness for which there no effective treatment or the drug for import would have clear clinical advantages.[97] This latter system is just beginning, and there do not appear to be published examples of its operation at the time of this writing.

5.13 Expedited Programs

As will be discussed, China does not reward drug innovators with strong and formalized intellectual property benefits, such as data or market exclusivity (other than for generics that successful challenge patents, which can obtain a year of exclusivity, and certain Chinese medicine, which can obtain 20–30 years of exclusivity), but it has given priority to certain applications for drugs that are particularly innovative globally, meet unmet clinical needs, and/or represent significant improvements over existing therapies. That priority typically means that an application receives more attention and guidance from CDE officials. Categories of drugs eligible for priority consideration have existing in drug regulations in China for many years, but in 2020 NMPA restructured those categories and deepened

[94] National Medical Products Administration (2021a) NMPA and Hainan Province Jointly Promoted the Pilot Application of Clinical Real-World Data of Drugs and Devices.

[95] Hainan Institute of Real World Data.

[96] National People's Congress Standing Committee (2019a) Drug Administration Law, Article 76.

[97] National Medical Products Administration, National Health Commission (2022) Import Work Plan for Drugs that have an Urgent Clinical Need.

guidance on the programs to make them more effective at marshaling agency resources to push forward applications.

Again, the categories have always been based around a combination of clinical need and encouraging innovation, although at some points other types of categories such as manufacturing location (i.e., encouraging transfer of the same to China) have existed in policy documents.

Some of the original set of categories for "special approval," dating at least back to 2005, were for innovative drugs (i.e., drugs new to the world), oncology drugs, drugs with geriatric or pediatric indications, and drugs for infectious diseases that were particularly problems for China, such as HIV/AIDs and hepatitis.[98] China has focused on improvements in oncology treatment for a long time, as China has millions of new cases of cancer per year (approximately 4.8 million in 2022)[99] and has been among its leading causes of death in prior years. HIV/AIDS, hepatitis B, and, more recently, SARS and COVID have also had priority categories associated with them. And China has remained focused on its aging population with the category of geriatric drugs.

In 2016, when the registration categories were restructured, NMPA issued new policy documents that articulated various other types of priority categories; it renewed this list in 2017 (China Food and Drug Administration 2017e). These categories were representative of the newfound push to encourage innovation in China, rival other innovative markets, and reduce the lag time between launch in other markets and China.

Drugs with clear clinical value that met any one of the following seven conditions could be expedited:

- Innovative drugs never before approved in the world.
- Innovative drugs that transfer manufacturing to China.
- Registration applications that contain advanced technology preparations, or new treatment techniques and have clear treatment advantages.
- Drugs for which the clinical trial application is filed 3 years prior to patent expiry or for which the manufacturing license application is filed 1 year before patent expiry.
- The applicant holds a clinical trial approval/notification in the U.S. or E.U., drugs for China are produced on the same line as those supplying the U.S. and E.U., the China applications are simultaneous with those jurisdictions, and the manufacturing license have passed U.S. FDA or EMA inspections.
- Applications for Chinese medicines (including ethnic medicines) that are clearly positioned to prevent major diseases.
- New drug registration applications that are under national development or major science and technology plans, and the clinical trials are carried out by the national clinical research center and approved by the central management department.[100]

[98] State Food and Drug Administration (2005b) Article 48.

[99] Xia et al (2022).

[100] China Food and Drug Administration (2017c).

The same document also set forth the following priority disease areas, and provided that the drug had clinical value:

- AIDS
- Tuberculosis
- Viral hepatitis
- Orphan diseases
- Drugs to treat malignant tumors
- Children's medicines
- Drugs to treat disease special or prevalent amongst the geriatric population[101]

5.13.1 Urgently Needed New Drugs Marketed Abroad

Although China focused on new-to-the-world drugs as the foundation of its efforts to promote innovation, it also devised a way to encourage drugs approved abroad to come to China quickly. Shortly after the Innovative Opinion in 2017, NMPA and the NHC devised a procedure to select and then quickly approve certain drugs approved abroad in the United States, European Union, and Japan within the last 10 years.[102] It is not unusual for China to focus on approvals of these three agencies, i.e., FDA, EMA, and PMDA, as particularly reliable and potentially eligible for special treatment. For example, some proposals for awarding regulatory data protection have listed approval in one of these countries as a criterion.

These special procedures applied to drugs that treat rare diseases ("orphan drugs"), drugs that treat serious, life-threatening illnesses for which there is no effective treatment or preventive method in China, and drugs that treat serious, life-threatening illnesses and have clear clinical superiority.

CDE organized a group of experts to review drugs approved in the U.S., E.U., and Japan from 2008 to 2018 and compile lists of drugs meeting these criteria. An expert commission then confirmed the preselected drugs. The lists were specific to certain drugs. Three such lists were released.[103]

Once on a list, the applicant would have access to a meeting with CDE to plan development. In some cases, these drugs could be approved based on foreign approval and data alone, in a condensed approval time of three months. The approvals could be conditional in that the applicant might need to conduct a study once the drug was on the market and submit the results within a specific time to

---

[101] *Ibid.*

[102] National Medical Products Administration (2018g) Notice on Matters Related to Evaluation and Approval of Urgently Clinically Needed Drugs from Overseas.

[103] *See, e.g.*, National Medical Products Administration (2018e) First List of Overseas New Drugs Urgently Needed in Clinical Practice; Center for Drug Evaluation (2019a) Second List of Overseas New Drugs Urgently Needed in Clinical Practice; Center for Drug Evaluation (2020i) Third List of Overseas New Drugs Urgently Needed in Clinical Practice.

maintain the approval. A post-market risk mitigation plan to monitor and address risks once the drug is on the market was also required as part of this pathway.

5.13.2 Breakthrough Applications

Breakthrough approval was a new program under the newly revised Drug Registration Regulation in 2020. NMPA and CDE issued specific procedures for breakthrough applications shortly thereafter.

Breakthrough applications are for innovative or other new drugs (i.e., Registration Categories 1 and 2 above, for drugs that are new-to-the-world). Once a drug is approved abroad it is no longer eligible. In addition to being "new to the world" in some way, the drug must treat a severe and life-threatening illness for which (1) there is no effective treatment in China or (2) the proposed medicine is clinically superior to existing therapies. An applicant submits its application to the Center for Drug Evaluation and, if approved, will receive the opportunity to have a meeting with CDE prior to its pivotal (e.g., Phase III) trial, to submit research materials on a rolling basis to CDE, and to receive its comments before the next submission.[104] In 2021 41 drugs were included in this program and in over 30 were included in 2022, although at the time of this writing statistics are still emerging.[105] Many of the indications of the drugs admitted were COVID-19 or oncology-related, but other therapeutic areas, such as cardiovascular drugs were also represented.

5.13.3 Conditional Approval

Conditional approval has existed in China for some time in different forms, although it was not until recently when these procedures were formally described in detail in the regulations. For 10 years, it has been possible in very select cases to obtain marketing approval, for example, based on foreign-generated data and the promise of conducting a clinical study in China post-approval. After the 2017 Innovation Opinion, policy documents began to mention a conditional approval route, especially for orphan drugs (see above and below). NMPA issued a draft of a specialized rule on conditional approval in 2017, but did not finalize it.[106]

Instead, conditional approval was codified in the 2020 revisions to the Drug Registration Regulation and, shortly thereafter, NMPA released specialized

[104] State Administration for Market Regulation (2020c) Drug Registration Regulation. Articles 59–62; *see also* National Medical Products Administration (2020e) Breakthrough Drug Therapy Examination and Approval Procedures (Trial Version).

[105] Pharnexcloud (2022).

[106] China Food and Drug Administration (2017c) Draft Technical Guideline of Conditional Approval for Marketing of Drugs with Urgent Clinical Need.

procedures. Conditional approval may be granted under three circumstances: (1) the drug treats a life threatening illness and existing data forecasts efficacy, (2) the drug is urgently needed due to a public health issue and the data supports efficacy and (3) the candidate is a vaccine that is urgently needed to deal with a public health emergency or is otherwise deemed urgently needed by the NHC, and the benefits outweigh the risks. As part of a conditional approval, the applicant must negotiate conditions with NMPA, such as post-market safety mitigation measures (e.g., special plans for administration) and/or post-market studies.

Applicants must fulfill conditions within the agreed-upon timeframe with NMPA. If not, then NMPA has the power to act, and has discretion to cancel the drug or vaccine approval.

Conditional approval was employed during the COVID-19 pandemic. NMPA expedited the approval of both drugs, including vaccines, and medical devices intended to prevent, treat, or detect COVID and set conditions for further study. Some of approvals had shortened timeframes that were less than the standard 5-year drug or device approval. Conditional approval can also be combined with the special approval pathway (see below) specifically for public health issues.

5.13.4 Priority Approval

Like other expedited programs, drugs that apply for priority review must have clear clinical value and meet one or more of the following criteria.

- They are in short supply and urgently needed.
- They are innovative or improved new drugs for prevention and treatment of serious infectious or orphan diseases.
- They are new types of pediatric drugs or new dosage forms.
- They are urgently needed or innovative vaccines.
- They have qualified for breakthrough designation or conditional approval.

Admission to the priority review pathway can cut the number of days for approval from 200 to 130. Or, if the drug is an orphan drug that is in urgent clinical demand and that is already approved abroad but not in China (similar to the program that CDE ran in 2018), the approval time can be 70 business days. To apply for priority review, the applicant must communicate with CDE prior to the submission of a marketing application to discuss a priority application.

5.13.5 Special Approval

NMPA may issue "special approvals" for drugs to handle public health emergencies. The Chinese government has various tools available to declare public health emergencies, which it did during the original SARs outbreak and has done in the case of COVID-19. These declarations, issued on national or provincial levels, trigger the availability of special approval procedures for drugs.

The special approval pathway relies on regulations that were issued after the SARs outbreak in 2003.[107] It allows companies with drug products that can respond to health emergencies to submit clinical trial applications that can be approved in a matter of a week or two to NMPA. After the trial is complete (or data is otherwise available for a marketing authorization), NMPA can set conditions on the approval of the drug during the emergency. This is analogous to an emergency use authorization in the United States. During COVID, China approved vaccines made by domestic companies (as of this writing, no foreign company's vaccine has been approved) on a conditional basis.

China has various laws in place to handle public health emergencies, including its Infectious Disease Law and its Emergency Response Law (currently being considered for amendment) and related regulations. These laws and other related policies give the government broad powers to declare public health emergencies at the national and/or provincial levels. To trigger the special approval pathway, there must be (a) state of emergency, (b) activation of emergency response procedures, or (c) a proposal by the drug reserve agency of the government to specially approve drugs made in accordance with an existing national standard.

If these conditions are in place, an applicant can submit a pre-registration application for a determination of feasibility, for which an evaluation will be organized within 24 h. Thereafter, the applicant can submit samples for testing, clinical trial, and marketing applications on an expedited basis. Sometimes testing and clinical trials will not be needed, although they were in the case of COVID vaccines. The laboratory testing at government laboratories is supposed to be carried out in as little as 2 days and the registration application review can be cut from 70–130 days[108] for normal expedited applications to as little as 15–30 days.[109]

In practice, these timelines can be extended, but they show how fast these applications can be approved. In the early days of the COVID pandemic, drugs and trials could be approved in a matter of a few days and, in some cases, investigator-initiated studies supported by the government required no CTA approval at all. China's emergency response system can be quick because it utilizes a bureaucracy that can organize efficiently, once high levels of the system decide on a course of action.

---

[107] State Administration for Market Regulation (2005).

[108] State Administration for Market Regulation (2020c) Drug Registration Regulation. Article 70.

[109] State Administration for Market Regulation (2005).

5.13.6 Orphan Drugs

Drugs that treat rare diseases, or orphan drugs, have been an expedited category for several years in China. However, China has never adopted a conceptual definition of orphan drug nor a set of procedures for evaluating and approving such drugs. Medical experts in China have suggested definitions, including disease with a prevalence rate of less than 1/500,000 in adults and an incidence of less than 1/10,000 in newborns.[110]

Instead of selecting a definition, based on an instruction in the Innovation Opinion to create a positive list of orphan diseases, in 2018 the NHC, MOST, NMPA, the Ministry of the Industry and Informatization, and the State Administration for Traditional Chinese Medicine jointly released a catalogue of orphan diseases based on opinions collected from experts. The catalogue is available on the NHC's website and contains 121 diseases.[111] The National Health Commission added 86 more orphan diseases in 2023.[112]

5.14 Marketing Application

China's marketing application process is similar to that of a clinical trial application. Indeed, until recently the materials required were essentially the same.[113] This may have been in part because China was used to receiving drugs much later than the rest of the world. But since the Innovation Opinion, China has introduced more differentiation between the clinical trial and marketing application. Still, as discussed above, the amount of manufacturing information required is more in certain respects at the clinical trial stage than would be expected.[114] There will also be some differences between applications for drugs that are produced nearly entirely abroad (except perhaps for some repackaging) and imported into China and drugs that are produced domestically within Mainland China, excluding Hong Kong, Macao and Taiwan.

[110] Liu (2023) Chen Wei, deputy to the National People's Congress: It is urgent to clarify the definition of rare diseases and adopt academic implementation programs, Southern Metropolis Daily.

[111] National Health Commission et al. (2018a).

[112] National Health Commission (2023c) Notice on the Issuance of the Second Orphan Drug Catalogue.

[113] State Food and Drug Administration (2007b) Drug Registration Regulation, Appendix 2.

[114] Center for Drug Evaluation (2020g) Notice on the Release of Module M4 on Administrative Documents and Drug Information.

5.14.1 Drug Masterfiles

In addition to the product, its ingredients and packaging must be approved at the time of the marketing application. Prior to reforms in 2016 to 2019, China required separate registration approvals for active and certain excipient ingredients and packaging materials. Under the Innovation Opinion reforms, China put in place a bundled review of drug products, ingredients and packaging. Now, in certain cases (e.g., ingredient/materials for self- or exclusive-use by the MAH), dossier materials necessary for the approvals of these ingredients and packaging can either be submitted with the drug product registration application or they can be submitted to a separate masterfile by the supplier itself (if the supplier is ex-China, the submission is through an appointed agent entity in China). In the latter case, an active masterfile, labeled "A" in the system, could be referenced with a drug marketing application.[115]

In addition, NMPA issued a list of low risk and commonly used ingredients that are exempt from masterfile registration altogether. These are ingredients that have been long-used in food and drugs and are covered by an existing food (e.g., a food standard, see Chap. 8) or drug standard (i.e., the China Pharmacopeia or a foreign pharmacopeia, e.g., US, EU, UK, and Japan). This list was issued in 2019.

The masterfile applies for certain registration categories of small molecule drugs, but does not apply for biologic raw materials, except for excipients and packaging materials.[116]

As noted above, ex-China companies that register active ingredients, excipients, and contact packaging must appoint an agent to register their products and submit a substantial dossier which includes manufacturing information and test results. The agent will again be type of local company that can contain an account with NMPA's platform. The requirements for registration of an ingredient or packaging on the masterfile platform are extensive and have required significant information, including test results and certain process validation information, particularly for sterile ingredients.

5.14.2 Registration Testing

In addition, NMPA requires "registration testing," which includes a specification/ standard review and sample testing. This testing can be done before, during, or after a marketing application.[117] Prior to 2020, NMPA required that there be testing on batches of product for purposes of the clinical trial application, but that requirement

[115] National Medical Products Administration (2017b) Notice on Adjusting the Examination and Approval of Active Pharmaceutical Ingredients, Excipients, and Packaging Materials.

[116] National Medical Products Administration (2019b) Notice on Several Matters Related to the Further Optimization of Drug Bundled Review and Approval and Supervision Work.

[117] National Medical Products Administration (2020j) Drug Registration Testing Work Procedures and Technical Requirements Standard (Trial Implementation).

has been relaxed and registration testing is only required in some cases, at the discretion of the CDE reviewer, usually required in the case of vaccine clinical trial applications.

Sample testing occurs to ensure that a drug meets the specifications in its drug registration standard, which is a master specification document that the applicant develops with its application. The specification document may rely on various existing standards for testing, ingredients, and other items. These standards are located in China's Drug Code or the China Pharmacopeia, which is usually updated every 5 years. If the applicant relies entirely on China standards, then it only needs to conduct sample testing. However, if an applicant uses additional items not covered or deviating from existing China drug standards, then both sample testing and a "standard review" testing are required.[118] The import drug registration standard is an important document that is released with the approval for marketing. Each batch of product must ultimately be released to the market based on the tests set forth in this standard.

5.14.3 Pre-Approval Verifications

As discussed above, a marketing application can also trigger different types of inspections or testing. There are two types of inspections of manufacturing facilities: a registration verification inspection and a GMP inspection prior to marketing. Registration inspections occur prior to marketing because the drug is an innovative small molecule drug, drug with a new modification, or any biologic. It is focused on the making of the drug itself and verification of the appropriate quality system.[119]

The drug GMP inspection is also focused on the facility in which the drug is made and tied to inspections at the local levels of government to license different facilities. As will be discussed below, all drug manufacturers in China must possess drug manufacturing licenses. The provincial and other local medical products administrations issue those licenses for a specific scope of activities and products. For example, some manufacturers may be approved to manufacture drug product and drug substance for small molecule drugs and/or biological products, whereas others may only be approved for drug repackaging and labeling. A number of foreign drug manufacturers still make drug product outside of China, but have purchased repackaging facilities in the country to assist with packaging and labeling for the China market.

A drug manufacturing license is required before a drug can go on the market, but an applicant can first submit the paperwork for the manufacturing license and undergo the GMP inspection during the marketing application phase.

Other types of drug manufacturing inspections, including for-cause inspections, will be discussed below.

[118] State Administration for Market Regulation (2020c) Drug Registration Regulation, Ch. 5.

[119] Center for Food and Drug Inspection (2021) Drug Verification Work Procedures.

5.14.4 Marketing-Authorization Holder ("MAH") and Its Agent ("MAH Agent")

As in Chap. 4, China has unique rules requiring the MAH for a product made overseas to be an overseas entity that holds (or will hold) a marketing approval in another overseas jurisdiction, such as the U.S., E.U., or Japan, in its name; proof of an approval in the name of another entity in the corporate group or a company with which the MAH has a contractual relationship is not sufficient. When it is an imported drug, the DAL as revised in 2019 now requires that the MAH appoint an agent in China to assist it with meeting its drug regulatory obligations, including registration, supplemental registration, renewal, pharmacovigilance, and inspection readiness.

The requirement for some type of agent has existed for many years, but it was not until the revisions to the DAL in 2019 that NMPA began to formally expect the agent to have substantial drug regulatory capacity (i.e., personnel and internal procedures) and bear joint liability with the MAH itself for compliance with drug regulatory obligations. In practice prior to 2019, NMPA and its predecessor agencies previously thought of the agent as a connection to communicate with and even punish the overseas registrant of the drug, the newly revised DAL codified that role and made the agent even more prominent in the regulatory process.

The revised DAL contained provisions that expressly stated that the MAH agent could be jointly liable with the MAH and that MAH agent can be treated like an MAH for purposes of the prohibited acts and penalties under the DAL. In other words, although the prohibited act and penalty provisions of the DAL only spoke in terms of MAHs, manufacturers, distributors, and user entities, the MAH agent could be considered an MAH for purposes of penalties. At the time of this writing, however, the rules setting the specifics of the agent's role, obligations, and appointment mechanism have not been finalized because there is significant debate about the scope of the local agent's responsibility and liability.[120] The proposal for the appointment mechanism is a record-filing of the agent's name and information to the local authorities where the agent is located.[121]

In practice, due to China's restrictions on domestic entities holding the rights to foreign made drugs, the MAH agent has also become a vehicle for the domestic companies to obtain an official role on the marketing authorization of the drug (although not as the main license holder). Being the agent provides them with important experience in managing research and development and compiling

[120] National Medical Products Administration (2020h) Draft Tentative Measures on the Domestic Agent of Overseas Marketing Authorization Holders.

[121] Having a domestic agent to handle local matters for overseas companies is not new. These agents were required to be reported through record filings in the 1990s, but the filings were abandoned. The idea of a domestic agent that would handle life cycle management of the drug in partnership with the overseas MAH emerged with the revision of the DAL in 2019. State Drug Administration (1999b) Notice on the Activation of Special Seal for Record-filings of Domestic Sales Agents for Imported Drugs.

regulatory submissions. In some cases, agent status has prepared companies to receive a technology transfer, i.e., to receive instruction from an overseas originator company to make and manage the drug locally in China in an independent way.

5.14.5 One Drug One Registration One MAH

Another feature of NMPA regulation/policy is its position of one registration per drug without articulating clear standards as to when changes to the features of a drug or its uses could justify splitting it into two registrations. Although certain changes or additions to a new drug require a new application, absent some other difference between two drugs, such as a different formulation or manufacturing process, NMPA prohibits splitting of indications or routes of administration onto separate registrations, even if (for example) safety and effectiveness profiles of the different indications are so different that a separate registration may be is justified. An analogous rule is that each drug may only have one marketing authorization holder; no duplicates attached to other marketing authorization holders are permitted.

5.14.6 MAH Transfers and Technology Transfers

Another changing feature of China's drug landscape is the ability to transfer marketing authorizations between entities. There is also a technology transfer pathway via which manufacturing can be transferred from overseas to China and a new marketing authorization established in the name of a domestic entity.

Previously, companies could not apply to transfer MAs for imported drugs. That type of amendment application did not exist. Instead, NMPA would permit these types of approvals to be transferred when a holder applied to change the name of the company on the application. While changing the name is not the same thing as changing the entity, NMPA and its local PMPAs were willing to accept that type of application as representing such a transfer. The required materials for this application could be simple (e.g., new labels, a certificate of pharmaceutical product, and a letter explaining the entities involved), but NMPA could ask additional questions and conduct more analysis of the receiving entity. As required for the original applicant, if the drug was approved abroad then a certificate of pharmaceutical product evidencing a foreign approval would need to bear the name of the transferee (i.e., the receiving entity).

The situation for domestic drugs was more complex. Prior to the amendment of the DAL in 2019, domestic drug products did not receive licenses. Instead, a company received a new drug certificate and manufacturing license, the latter of which needed to be issued before the drug could be marketed. Ultimately, the drug itself received an approval number that was linked to the facility's manufacturing license. If a domestic drug applicant wanted to contract manufacture, it would need to find another facility with an appropriate drug manufacturing license and bundle the approval number with that facility license. The contract manufacturing arrangement would also need to be noted to the provincial government where the contract manufacturer was located.[122] But there would still need to be two manufacturing licenses. In 2009, NMPA released a rule that would allow a technology transfer to move manufacturing from one company to another (different from contract manufacturing).[123]

A transfer from an imported to domestically produced drug could also rely on the 2009 technology transfer regulations. This procedure would convert the imported drug license into a domestic drug by attaching it to a domestic facility that was qualified to make the drug. The 2009 regulation was general and provided little guidance on what was needed to accomplish the transfer. Local drug manufacturers were also perhaps not ready to engage in a complex process of transferring technology from abroad. Therefore, it was not until more recently that technology transfers of drugs developed abroad became a feature of China collaboration deals.

It was not until 2021—after the revised DAL allowed for MAs attached to products—that NMPA finally released its Administrative Rules on the Post-Market Amendment of Drugs. These rules contained provisions on transferring marketing authorizations from one entity to another via a relatively simple administrative procedure and on transferring manufacturing from one site to another. It also permitted a pathway to submit a supplement or a new drug application to manufacture that relied on an approval of an imported drug. This pathway allowed an applicant to prepare a facility domestically while the imported drug is on the market. Once the domestic facility is qualified it could apply for a marketing authorization for a domestically marketed product. The imported drug license can remain, and the new domestic product will be a generic or biosimilar (because, as discussed, China does not have an innovative drug category for domestic drugs that are not new to the world).[124]

While China seeks to encourage manufacturing to be transferred to China in several ways, including tax incentives (e.g., exemptions from value added tax)[125] and local subsidies, its technology transfer and drug registration pathways do not recognize the drug as innovative once it is transferred to China. In other words, because the drug is not new to the world at the time of the transfer, the domestic version must use a generic pathway. This could have downstream effects in the procurement and reimbursement negotiation processes and appears to be a significant disadvantage of this new system.

The transfer to technology from overseas requires substantial planning, effort, and trust between parties, as well as the freedom to travel. The pandemic made the

[122] China Food and Drug Administration (2014) Provisions on the Supervision of Drug Contract Manufacturing.

[123] State Food and Drug Administration (2009).

[124] National Medical Products Administration (2021e) Post-Market Change Measures (Trial Implementation).

[125] State Tax Administration (2023) Guidelines for Preferential Tax Policy of Mass Entrepreneurship and Innovation.

needed coordination between foreign and China-based manufacturing and quality teams challenging.

Another issue was making changes to pending applications. Until recently, there was not a process for making changes to a clinical trial or marketing authorization application during the review process. In 2022, NMPA issued procedures governing changes to pending applications.[126] These changes may only be administrative in nature, such as the change of an applicant name, an applicant entity or an agent name or entity (for imported drugs). During the review of a clinical trial application, the agency will not accept a change to the applicant entity, although a change to the clinical trial agent is possible. These changes can be made through a supplemental application.

For changes that significantly impact the safety, effectiveness, or quality, the applicant must withdraw and refile its application, starting from the beginning. For changes that relate to technical aspects of the product, the agency may review them in conjunction with the original application but reserves the right to reject the supplement if it is deemed related to a significant change related to safety, effectiveness, or quality. These procedures are one of the conveniences that NMPA established to increase efficiency of the application process.

The process for changing the local agent on a marketing authorization (whether pending or approved) or a pending clinical trial application has evolved to be more protective of the rights of the holder. Under the current rules, the holder is not reliant on the original agent to file the application to essentially fire itself. Instead, the holder issues a letter firing the prior agent and another authorizing the new agent, and the new agent files the papers. This procedure avoids the new agent holding the registration process hostage and demanding concessions from the overseas holder by refusing to engage in the change process.

5.14.7 MAH Amendment and Renewal Processes

Unless under a conditional approval with a shortened time period, each drug is approved for a five-year license term. After at least 4.5 years, the MAH must apply to NMPA to review the drug license, although companies may apply for renewal earlier in order to avoid gaps in the effective period of the license that cause problems with the ability to import the drug. At the time of renewal, NMPA (through CDE, which conducts the renewal) will consider the safety record of the drug, adverse drug reactions, annual reports, and recalls (if any). If there have been changes in standards, for example, governing ingredients or testing, then NMPA may require some additional testing. However, the applicant may not apply for amendments during the renewal process. Once complete, assuming all goes well, the drug will be approved for a new five-year license term. Product that has been made during the prior valid license term can continue to circulate in the market during the time of the

[126] Center for Drug Evaluation (2021f) Changing Work Procedures During Drug Registration Application Review (Trial Implementation).

renewal, provided that there are no serious safety or quality concerns, in which case a recall or re-evaluation may be warranted (see below).

For amendments to a drug registration, the applicant must assess the extent of the change and whether it qualifies as a minor, moderate, or major change. Minor changes can be reported in annual reports, whereas moderate and major changes require some sort of pre-approval. Moderate changes must undergo a notification process in the form of a record-filing (i.e., once the documentation is submitted and complete the change can proceed). Major changes require pre-approval before proceeding.

The Drug Registration Regulation does some initial categorization of changes. Significant changes to the manufacturing process, label changes related to safety and effectiveness, and transfers of the marketing authorization constitute major changes. Less significant changes to the manufacturing process, and changes to the label on the drug packaging and sub-packaging, constitute moderate changes. However, CDE has also guidelines—Post-Market Chemical Drug Pharmacological Change Research and Technical Guiding Principles (for trial implementation), from February 2021—that provide more detailed instructions on determining the changes that fall into each of the three categories and the research and validation of the effectiveness of the change.

The Administrative Measures on Drug Post-Market Changes, discussed above, was the first rule to provide clear guidance on how to transfer marketing authorizations between companies and how to transfer manufacturing sites to China. These measures also provide important procedural guidance and opportunity for dialogue with CDE, such as on applying to NMPA to downgrade a change from major to moderate, implementing changes (with an allowed six-month transition period), and consulting with and resolving debates concerning the classification of a change with CDE (the agency's view will likely prevail). An applicant may seek a Type III meeting with CDE to determine the classification of an amendment. These meetings will be discussed in more detail below.

5.15 Drug Labeling and Naming

5.15.1 Drug Names

As in other places, drug labels and package inserts in China must have a generic name and may in some cases have a trade name. The generic name is formulated by the China Pharmacopeia Commission and filed with the National Medical Products Administration. The generic name is a usually a transliteration of the name from the International Nonproprietary Names for Pharmaceutical Substances (INN), compiled by the World Health Organization. Often the label and package insert will list both the English INN and its Chinese transliteration. This is true in practice for both small molecule drugs and biologics, including generics and biosimilars.[127]

[127] State Food and Drug Administration (2010b) Drug Naming Principles.

Trade names, e.g., trademarked names, may also be listed on a drug label, but may only be listed in the registration documents if the drug has a new chemical structure or active ingredient that is under "administrative protection" (a now defunct form of regulatory protection offered pursuant to China's trade obligations that protected drugs without patent protection from 1986 to 1993 that were approved in China and also met other distribution criteria), a new drug monitoring period (also now defunct), or an existing formulation patent.[128] This is an old regulation and some of these options are no longer in existence. Of these three, imported drugs are most likely eligible for the third category of existing formulation patents.

However, it may be possible to list a trademark on the label and package insert. To do so, the applicant would add it to the sample labeling and package insert submitted with the application but would not list the name on the application form itself. Whatever name is listed should be accompanied by a trademark certificate and, if in English, a translation (usually a transliteration) in Chinese. If the name is part of a logo, it may be possible to not offer a translation.

Listing in English or another foreign language is a significant point because it is often used as a means of promoting the drug: the company will advertise that it is offering the same imported drug of the same quality that is marketed and sold abroad. Some companies may decide to market to a patient's friends and relatives who are in the U.S. or Europe and may be inclined to favor their relative in China using the drug.

5.15.2 Drug Labels

Each drug must have a label that is attached to the smallest sales unit and a package insert, which is a longer booklet that contains more detailed information about the indications, strength and dosing regime, adverse events associated with the drug, and clinical data support it.

China does not clearly distinguish between physician and patient labeling. The package insert is the primary tool for both patients and doctors to understand how to use the drug. NMPA's website states: "The package insert is statutory document that provides important information about drugs, and is also the main basis for doctors, pharmacists, nurses, and patients to choose drugs used for treatment. It is the main medium for popularizing medical knowledge."[129]

Samples of the label and package insert in Chinese must be submitted with the registration application, and if the drug is approved abroad a copy of the foreign drug label/package insert must also be submitted. The foreign and the Chinese labeling are supposed to match, although the rules do not state this very clearly, which can create issues if, *after* approval in China, the company wants the foreign labeling to

[128] State Food and Drug Administration (2007c) Interpretation of Issues Concerning Administrative Provisions on Drug Directions and Labels.

[129] National Medical Products Administration (2017a), Why Read Drug Instructions?

change slightly based on overseas regulatory changes or market conditions. As discussed, foreign-approved drugs served as the predicate for drugs approved in China, whereas domestically made drugs were traditionally considered to be either generics or China-first drugs.

The provisions on drug labeling are quite old compared with other rules. Last revised in 2006, these rules provide the basic content of labels and package inserts and are interpreted in light of other administrative documents and application guidance. The applicant is required to keep the labeling up to date, and NMPA may also require the MAH to revise the label in accordance with emerging adverse drug reactions and safety information and to add warnings.[130]

There may be a negotiation between the applicant and CDE about content *prior* to approval. In this case, there may be differences between the label used overseas and that in China, such as in the wording of the indication and that the content of safety information and the way it is presented, such as the wording of warnings.

5.16 Over-the-Counter Drugs

Over the counter (OTC) drugs are approved in the way described above. There is not a shortened process for them to come to market based on an agreement for them to adhere to a particular formulation.

As discussed below in the sections on distribution, there are two types of OTC drugs: OTC A and OTC B. The distinction is attached to the OTC drug at approval and matters primarily for the method of distribution. OTC A drugs must be distributed at stores with drug distribution licenses that have full-time pharmacists. OTC B drugs are considered lower risk and are therefore, permitted to be sold in normal stores that are approved by the provinces, and although they must have professionalized staff, a registered pharmacist is not required.[131] There are a variety of factors that qualify a drug for OTC B status, including whether it can be used daily, can achieve results in a short period of time, and whether its active ingredients (s) have been marketed widely for 10 years or more in China.[132]

When submitting a drug registration, an applicant can simultaneously apply directly for an OTC designation for the following types of OTC drugs, if they can make the case that consumers can use the drug independently: (1) there is already an OTC drug approved in China with the same active ingredients, dosage form, indications, and strength; (2) the candidate would change the dosage form or strength of the approved OTC in China, but the indications and the route of administration will remain the same as the approved version; (3) the OTC candidate is a new compound preparation of active ingredients that are already approved for

[130] State Food and Drug Administration (2006).

[131] State Drug Administration (1999a).

[132] State Drug Administration (2000).

OTC use in China; or (4) other conditions in which NMPA specifically authorizes for direct OTC applications.[133]

OTC drugs can be registered as generics and can avoid clinical data if, for example, the drug only seeks to change characteristics that are considered irrelevant to effectiveness and quality, such as taste, color, smell, coolness, consistency, and packaging specifications. Bioequivalence data may be required to demonstrate equivalence to the marketed OTC. With any direct application for a non-prescription drug, the applicant must submit to a "suitability" determination for OTC use conducted by the Center for Drug Re-Evaluation. The suitability determination is limited to 30 business days.

NMPA also permits applications to "switch" a prescription (Rx) drug to OTC status. An applicant can accomplish this by submitting an application to CDE. NMPA has established an application for this purpose,[134] including guidelines on when an Rx-OTC switch would be appropriate. These guidelines include a list of indications for which OTC status might be accepted, in various different therapeutic areas, such as neurology, orthopedics, nutritional (some vitamins and supplements can be regulated as drugs), respiratory, gastroenterology, dermatology, gynecology, ophthalmology, otolaryngology (ear, nose and throat), oral care, and infectious diseases. Before the switch is finalized a notice is issued for public comment, and assuming no serious objections, it will be finalized by the issuance of a notice. NMPA also has the authority to switch drugs to OTC on its own. Switches do not happen infrequently. In 2020, there were 40 Rx-OTC switches, of which 31 were OTC-A and 9 were OTC-B.[135]

5.17 Vaccines and Preventive Biologics

China has various types of drugs that are subject to special laws and regulations. Vaccines are one such type. Vaccines in China are regulated as biologics. They are considered drugs but also defined separately as "preventive biologics intended for human immunization to prevent and control the occurrence and prevalence of infectious disease."[136]

Vaccines are governed by a separate set of application requirements and, overall, an emerging and specialized body of vaccine law and regulation. For example, there were separate rules on vaccine development (i.e., vaccine good clinical practice rules) and separate rules on vaccine distribution that were tightened in 2016, after a

[133] State Administration for Market Regulation (2020c) Drug Registration Regulation, Article 36; *see also* State Food and Drug Administration (2004).

[134] State Food and Drug Administration (2010c) Notice on Matters Related to the Conversion of Prescription Drugs to Non-Prescription Drugs.

[135] Sina News (January 21, 2021).

[136] National People's Congress Standing Committee (2019b) Vaccine Administration Law, Article 2.

scandal in which expired, low-quality vaccines were discovered in circulation. Building on that reform, in 2019 China enacted a Vaccine Administration Law (VAL) that elevated specialized provisions on development, manufacturing, distribution, and post-market surveillance to the level of a national law.

The differences between vaccine regulation and that of standard small molecule drugs and therapeutic biologics can be summarized as follows.

Development As with other drugs, vaccines must be studied in clinical trials in China. NMPA is likely to want at least an initial study in China as the basis for approval—perhaps more likely in the case of vaccines—and if the vaccine is approved conditionally, NMPA may insist on a sizable post-market study. Vaccine clinical trials must be conducted in Tier III-A hospitals or Center for Disease Control institutions.[137] Unlike other drugs and biologics, NMPA is more likely to ask for laboratory testing on the vaccine prior to permitting a vaccine clinical trial to proceed. Vaccine trials are subject to separate set of good clinical practice rules articulating subject protections and safety reporting requirements.[138]

Marketing Authorization and Manufacturing Unlike other drugs, where the marketing-authorization holder can contract out manufacturing, vaccine marketing-authorization holders must have their own manufacturing capacity. A vaccine marketing authorization holder is defined under the Vaccine Administration Law as an entity that obtains both a drug registration (i.e., the vaccine marketing authorization) *and* a pharmaceutical product license.[139] They would hold a marketing authorization and possess a facility with a Type A drug manufacturing license (see above/below). Contracting out production is only permitted if approved by NMPA to handle capacity constraints, urgent need, or the production of multivalent vaccines. In 2022, NMPA released a rule on vaccine manufacturing and distribution that heavily restricted the ability to contract manufacture, including prohibiting segmented manufacturing, meaning that both manufacturers must have the ability to produce the whole vaccine (perhaps with the exception of repackaging).[140] There is nothing that limits these restrictions to domestic producers, but in practice, it is not clear whether these restrictions are applicable to imported vaccine manufacturers.

The government scrutinizes the establishment of new vaccine manufacturing enterprises or the the expansion of the scope of manufacturing of existing operations. Due to concerns about GMP compliance related to vaccines, the government has worked to introduce additional controls on manufacturing and to encourage manufacturing only by large-scale professional makers.

---

[137] National Medical Products Administration (2020b) Drug Registration Regulation, Article 22.

[138] China Food and Drug Administration (2013a) Vaccine Good Clinical Practice Principles (Trial Implementation).

[139] National People's Congress Standing Committee (2019b). Vaccine Administration Law.

[140] National Medical Products Administration (2022f) Vaccine Circulation and Manufacturing Provisions.

Additional controls include a specialized appendix to the vaccine GMP, the requirement for testing, and release to the market by specialized testing institutions accredited by the government. These institutions must conduct quality release, which could be via testing (as opposed to paper release). Post-market vaccine makers are required to submit annual quality reports, accounting for, inter alia, out-of-specification manufacturing events.[141] Labeling and quality and safety information must be posted-online.

Vaccine Distribution Distribution and post-market surveillance are perhaps where some of the most significant differences in the vaccine regulatory regime emerge. For purposes of distribution, vaccines are divided into Class I and II vaccines. Under the VAL, however, Class I vaccines are referred to as National Inoculation Program (NIP) vaccines and Class II vaccines are non-NIP vaccines.

China's goal has been broader reach of vaccination programs. There is a national mandatory vaccine program, under which the state pays for vaccines. Mandatory vaccines include those that inoculate individuals against hepatitis B and tuberculosis after birth, as well as an expanded program of other vaccines for children ages birth to 6 years. Those vaccines include hepatitis A, Japanese encephalitis, meningococcal disease, measles/mumps/rubella, pertussis/diphtheria/tetanus, and poliomyelitis.[142]

In 2016, police in Shandong Province discovered that a mother-daughter team were selling hundreds of millions of yuan (RMB) worth of vaccines that had not been stored in proper conditions to other provinces. The incident eventually attracted national attention,[143] and resulted in the revision of the Vaccine Circulation Regulations to tighten requirements for storage and to ensure that long distribution chains were eliminated for all vaccines.[144]

The Vaccine Circulation Regulations now require that vaccine procurement take place through exchanges run at the provincial level of government. The sale must take place between the marketing-authorization holder (i.e., the manufacturer) and the Center for Disease Control, which distributes the vaccines to various vaccination sites, e.g., hospitals. Only one logistics provider is allowed to assist in delivery, provided it meets drug-distribution conditions. But the sale itself is MAH to CDC. In the case of imported vaccines, they must appoint a national distributor to serve as their agent in this process and make the sales on their behalf. The change to the distribution regulations and later in the VAL together significantly increased the restrictions on the distribution of vaccines in an effort to tightly control vaccines under one public market.

Despite these changes, vaccines have remained a sensitive area of regulation and, at times, have set off political issues. In 2018, despite reports from the government

[141] National Medical Products Administration (2020q).

[142] China CDC (2021).

[143] Cao et al. (2018).

[144] State Council General Office (2016).

that showed that vaccine manufacturing was generally compliant with drug good manufacturing practices, reports emerged of a local company that had serious GMP violations. The violations were not connected to serious human harm, but the companies were delisted and nearly shuttered. The problem turned into a political avalanche that led to the resignation of the head and a vice minister at NMPA.[145]

China has been relatively self-reliant in the COVID vaccines approved both compared to other countries and to its approach to vaccines in other areas.[146] Some manufacturers claimed they were told they needed to transfer technology to China before they could obtain approval.[147] In contrast, imported vaccines in other areas have enjoyed success in China, although in some cases they have been subject to other unwritten restrictions, such as approval in the U.S., E.U., or Japan.[148] COVID therapeutics that are produced overseas have been approved for marketing in China.

5.17.1 Vaccine Post-Market Surveillance and Injury Compensation

Vaccines are subject to various requirements for post-market surveillance, both under the general drug rules and the VAL and its implementation regulations. These include adverse drug reaction monitoring, reporting, and compensation, annual quality and drug reporting, recall reporting and implementation, and drug re-evaluation. The VAL also requires that the vaccine MAH release on its website information regarding the product's label, lot release, recall, inspections, and punishments and compulsory vaccine insurance information.

Regarding the safety reporting system, regulation centers on the adverse event following inoculation ("AEFI") reporting and compensation system. Although often translated into English as "adverse events" following inoculation, like the World Health Organization's term, in Chinese these events are literally "abnormal drug reactions following inoculation." They do not focus on a mere temporal connection to vaccine use, but rather on causation—although the approach of courts and administrative bodies to causation is often to find a simple temporal relationship.

An AEFI in China is defined under the VAL (and under prior rules) as:

> Drug adverse reactions to vaccines made to quality standards during the processing of procedure of standardized inoculation or thereafter that cause damage to the inoculated person's tissues, organs or their functions and that are not the fault of any party.[149]

[145] Liu (2018b) Former FDA to Resign Over China's Latest Vaccine Scandal.

[146] Stevenson (February 18, 2022).

[147] Reuters (October 1, 2022).

[148] Liu (April 30, 2018a).

[149] National People's Congress Standing Committee (2019b) Vaccine Administration Law, Article 52.

The VAL and distribution regulations carve-out six separate circumstances that do not qualify as AEFI: (1) common post-vaccination reactions to a vaccine's particular features, (2) harm caused by a quality defect of a vaccine, (3) damage caused by an inoculation site's failure to adhere to vaccination work procedures and guidelines, (4) coincidental disease that occurs because the inoculated person was in the incubation or prodromal period at the time of inoculation (often referred to as "coincidental" reactions), (5) damage caused because the inoculated person has a contraindication that they or their guardian did not truthfully disclose and the contraindicated disease returns or is aggravated after vaccination, and (6) psychogenic reactions caused by psychological factors.[150]

There is little guidance on the meaning of these exceptions, but they are explained further in the National AEFI Monitoring Plan (2010). The first exception is explained as a general, mild reaction, specifically a "transient physiological disfunction," including fever and/or redness and swelling accompanied by other symptoms, such as general discomfort, fatigue, and loss of appetite.[151]

Once a "suspected" AEFI occurs, the medical institution or vaccination site must report it to the local health authority and local drug administration.[152] If that happens, then the MAH still needs to report the event again in annual reporting to NMPA (i.e., in an annual report of adverse events). NMPA's Center for Adverse Drug Reactions issued extensive guidelines for MAHs with definitions and requirements on the content of the reporting of AEFI to regulatory authorities,[153] but there is not a clear requirement to report AEFI to NMPA for purposes of further evaluating the safety profile of the drug. The Good Pharmacovigilance Practice rule, issued in 2021 (discussed below), states that it excludes AEFIs but that it is enacted pursuant to the VAL, which would indicate that it covers vaccine non-AEFI adverse drug reaction reporting. Without further debating the issue here, AEFIs are important both for safety and for purposes of awarding compensation.

Once a suspected AEFI is reported, the local CDC (or potentially a higher level municipal or provincial authority, if it is a death event)[154] will then assemble an expert group to conduct an "investigation and diagnosis" of the case through the local branch of the China medical association. Pursuant to the VAL, if the case is determined to be an AEFI or an AEFI cannot be excluded (i.e., possibly an AEFI), then the inoculated person or their family is owed compensation from the government for NIP vaccines (Class I) and from the manufacturer for non-NIP vaccines (Class II). After an investigation and diagnosis is complete, the parties (the inoculated person, the vaccination site, and the manufacturer) have the right to appeal through what is referred to as an "identification" proceeding by another

[150] National People's Congress Standing Committee (2019b), Vaccine Administration Law, Article 52.

[151] National Health and Family Planning Commission (2010).

[152] National Health and Family Planning Commission (2008).

[153] National Adverse Drug Reaction Monitoring Center (2013).

[154] National Health Commission (2008), Article 11.

expert group assembled by the medical association. A subsequent appeal is allowed from the identification to a higher-level medical association at the provincial-level, and administrative litigation (see Chap. 3) may be allowed to remedy procedural defects.

There are not clear rules for how the expert panels involved in AEFI cases should make determinations about causation. According to some commentary, expert review should be an objective scientific assessment of causation. It should examine whether data shows that those who receive the vaccine are more likely to experience a certain type of event or such event generally or in this specific case is merely coincidental due to perhaps an alternative cause.[155]

However, practice has shown that the assessment is often colored by local pressure to award compensation. Originally, only a finding that the event was an AEFI, i.e., that the vaccine actually caused the injury, would lead to compensation. Some provinces, such as Anhui and Guangdong, enacted local rules that permitted less certain conclusions to also permit compensation. Guangdong has four categories of AEFI that are still used in assessing causation in cases:

- AEFI
- Very Likely an AEFI
- Possibly an AEFI
- Not Very Likely an AEFI
- No AEFI[156]

After court cases exposed frustration with not being able to award compensation to families when causation was possible, but not sufficiently certain, the VAL codified an approach that would allow compensation even if the investigation and diagnosis experts concluded that an AEFI could not be excluded. That sounded like a reasonable compromise, but with no definition of what "cannot exclude" means, some investigation and diagnosis committees have interpreted it as a license to grant compensation if there is any chance that the vaccine caused the event, and denied company appeals even when there is no scientific evidence of a causal association between the illness that caused the injury and the vaccine. Further, the exceptions described above to the AEFI definition are so vague and potentially narrow that even when there is an expected effect of the vaccine, the committees may still award compensation.

Other practices and procedures render the avenues to seek objective scientific determinations challenging. For example, the AEFI Identification rules require that all parties to a proceeding (i.e., inoculated person, manufacturer, and inoculation site) submit documentation related to the incident. The rules also state that if a party does not submit documentation, then it must bear the adverse consequences.[157]

[155] Yao and Yang (2013).

[156] Guangdong Health Commission (2018).

[157] National Health Commission (2008), Article 21.

However, in practice some medical associations have interpreted the requirement that all parties submit documentation as a gating item to start a proceeding. In some cases, if a court orders a medical association to initiate an identification proceeding, the medical association may even try to find grounds to refuse court's request, leaving the company seeking expert review with only an option to lodge further appeals. If the inoculated person intentionally refuses to submit documentation, then the manufacturer could be denied its right to appeal. In other cases, the local CDC, fearful of protests or violence from the injured party or their family (see above discussion of violence against doctors and health care providers), may use its leverage as the sole procuring customer of vaccines to urge the manufacturer to settle for an amount that is even more than would be required under applicable local rules. For example, under the AEFI rules, refusal of an autopsy in a death case can result in the loss of compensation, but in practice the family may protest and force the case or settlement negotiations, which is not an uncommon tactic in China.[158]

Finally, although the AEFI rules prohibit medical associations from accepting cases if a similar case has already been filed in court, they do not require that a court dismiss a case if a proceeding is already pending or concluded before a medical association. Courts can (and do) accept cases from aggrieved families and permit them to fight for compensation contrary to what an expert body has already ruled. The net result of these issues is that AEFI expert proceedings may not be regularly a reliable way to determine causation.

A brief note on the amount of compensation. The provinces have enacted a patchwork of rules and documents to determine the amount of compensation in accordance with determinations made by the medical association. For example, the medical association can determine that the injury ruled to be caused by the vaccine is a level 1, 2, 3, or 4 (there can also be multiple sub-tiers)[159] and local rules will guide compensation payments related to health care and economic compensation based on that conclusion. The compensation is often tied to per capita income in a particular locality in the year prior to the injury. The result may be that compensation is significantly lower in one region as opposed to another, leading to incentives to use duplicative court proceedings, protests, or coercive negotiations to increase the amount that a hospital or drug company may be willing to settle for.

5.18 Other Specially Regulated Drugs

Like vaccines, there are special rules governing other types of drugs. These include substances for which there are special controls, such as narcotic drugs and psychotropic drugs, radiopharmaceuticals, and blood products. These drugs may be subject to additional licensure requirements, specialized good manufacturing practice rules,

[158] Liebman (2016).

[159] *See, e.g.*, Guangdong Government (2011).

manufacturing and distribution restrictions, and outright prohibitions on advertising. This section briefly introduces some of those special features.

Radiopharmaceuticals are defined as any form of radionuclides or radio-marker drugs that are used for clinical diagnosis or radiotherapy.[160] They are regulated in different respects by NMPA, the National Science and Defense Bureau and the Ministry of the Ecology and the Environment. These drugs were originally not clearly part of the overall marketing authorization holder reform introduced nationwide under the 2019 DAL, but subsequently NMPA issued a regulatory document to indicate that they should be reviewed and approved under this framework, i.e., with a marketing authorization holder and potentially third party entrusted manufacturing entities.[161] Another important feature of this regime is that the entities that actually manufacture and distribute these drugs must have a radiation safety license issued by the Ministry of Ecology and the Environment, which requires separate systems, personnel, and facilities capable of handling radioactive material. Various other agencies, including the China National Nuclear Safety Administration, play a role in supervision of these products.

Narcotic drugs and psychotropic drugs contain ingredients that are on a list of ingredients that subject to special regulation. The Catalogue of Varieties of Narcotic Drugs and Psychotropic Drugs is issued jointly by NMPA, the National Health Commission, and the Ministry of Public Security. These are actually two separate lists.[162] The manufacturers and distributors of these drugs must be specially licensed to produce or distribute them. Psychotropic drugs are divided into Class I and Class II, with Class I being those which require heightened scrutiny. A designation of Class I means that the distribution chain is restricted to wholesalers and hospitals with special qualifications, whereas Class II permits broader distribution.[163] In addition to these restrictions, controlled substances are not permitted to be advertised under the Advertisement Law.

Blood derived products are managed differently too. China enacted a law on blood donation in 1997 to regulate the donation and collection system. In 1996, the State Council issued rules (amended again in 2016) on the collection and distribution of raw plasma and blood products that are regulated according to the DAL. The government runs the system for the collection of plasma and apheresis, including credentials for and licensure of healthcare and blood production units engaged in plasma apheresis. Raw or source plasma (i.e., plasma that is collected from a single plasma station for purposes of supplying a blood product manufacturer) is forbidden from exportation from China.

---

[160] China Food and Drug Administration (2017a) Administrative Measures on Radiopharmaceuticals, Article 2.

[161] General Offices of the National Medical Products Administration and the Science and National Defense Bureau (2021).

[162] National Medical Products Administration, National Health Commission, and Ministry of Public Security (2013).

[163] State Council (2016c) Provisions on the Administration of Narcotic and Psychotropic Drugs, Article 25.

Blood products are defined as the various albumin products manufactured from blood plasma. Companies that produce these drugs must obtain the plasma by signing an agreement with a licensed plasma collection station, and that plasma collection station may only provide plasma to one production entity. Producers and distributors of blood products are approved by local authorities in accordance with the DAL and the producers must follow drug GMP.[164] A specialized blood product GMP appendix was issued in 2020 and amendments proposed in 2023.

5.19 Role of Drug Patents and Non-Patent Exclusivities

China does not have data and market exclusivities like those that exist under European Union and United States law. The exceptions are market exclusivity for generics that successfully challenges a relevant patent and a system of protection for certain Chinese medicines. In the former case, a generic applicant that is the first to successfully challenge a patent will receive 12 of market exclusivity, during which no generic drugs of the same categories may be approved. The length of the protection must not exceed the original patent term of the challenged drug.

As to innovators, China committed to implement data exclusivity for innovative drugs with new ingredients pursuant to the Agreement on Trade-Related Aspects of Intellectual Property Rights (TRIPS), and provisions implementing this commitment appeared in the DAL Implementing Rules, the DRR, and in the regulations governing animal health products. These provisions granted 6 years of protection to undisclosed test data for drugs containing a new chemical ingredient (which China has sometimes argued only includes small molecule drugs). During that time, the NMPA will not approve (but could accept) an application for the same type of drug, unless the applicant generates the data itself or obtains consent from the rights holder of the data. The provisions did not distinguish between innovative drugs and follow-on drugs nor introduce the concept of follow-on reliance on data. Data protection also made a rare appearance in the Free Trade Agreement between the People's Republic of China and the Swiss Confederation, which also expressly offered 6 years for new to China medicines, including biologics.[165]

However, these provisions were not implemented in practice. There is no application or form by which an applicant can seek data protection, no system for notifying others of its existence, and no evidence of approval of any application being rejected on these grounds. NMPA does not have an internal system for assessing whether applications merit this protection nor whether they may not be permitted on this ground. Companies do sometimes request regulatory data protection because of the existence of these provisions in the regulations, but NMPA may not respond, and there is no way to ensure that theoretical RDP can block an infringing product.

[164] State Council (2016a) Blood Product Administration Provisions, Article 20.

[165] Ministry of Commerce (2017).

In 2017, NMPA proposed a regulatory data protection system that the government ultimately adopted into the Innovation Opinion. This system would grant regulatory data protection to drugs in five undefined categories: innovative small molecule drugs, innovative therapeutic biologics (presumably excluding vaccines), drugs that treat orphan indications, pediatric drugs, and a peculiar fifth category of generic drugs that have launched a successful patent challenge.

Subsequently, NMPA released a *draft* implementing regulation on the regulatory protection of clinical trial data.[166] That draft was never finalized. That draft, like prior proposals, limited data protection in new and different ways. In general, it required that the data not be disclosed and be used in some way to support a marketing application to NMPA. However, for unknown reasons, it restricted protected data to effectiveness data—leaving the possibility of disclosing safety data. This proposal did not seem to account for how an applicant could realistically separate the two sets of data and it did not seem to account for the fact that conclusions that the applicant reached based on the safety data would be reflected in the labeling and post-marketing commitments.

The draft RDP measures also limited data based on the location of the clinical trial and the timing of the marketing submission in China relative to other countries. This was a measure seemingly purely aimed at reducing the drug lag and encouraging manufacturers to bring their products to China earlier than they had previously done. For example, applicants that conducted a trial in China could get 6 years of data protection for small molecule drugs and 12 years of protection for biologics. If the marketing application was submitted simultaneously in China and other countries, the applicant would get full protection. If not, then the protection could be reduced to 1 to 5 years—penalizing applicants for failing to submit in China in line with other places.

Similarly, if the applicant relied entirely on foreign data, it would only get one-fourth the protection it would otherwise get. If it supplemented a foreign-data submission with "China clinical trial data"—an undefined term that did not make it clear whether the data had to actually come from inside China—then the applicant would get one half the protection it would otherwise have received.

Despite multiple promises to offer data protection, including one in TRIPS in 2002, one during dialogues with the U.S. (called the Joint Commission on Commerce and Trade), one in the Innovation Opinion, and one by the premier, in March 2018, as an effort to promote imported oncology drugs, China has not yet implemented this system. Many of the provisions on data protection have been removed from implementing rules, such as the 2020 Drug Registration Regulation.

In 2022, in a new draft of the DAL Implementing Regulations, there were provisions on data protection and what appeared to be market exclusivity for orphan and pediatric drugs. However, as of the time of this writing, nearly a year later, those

[166] National Medical Products Administration (2018c) Draft Drug Test Data Protection Implementing Rules (Trial Implementation) for Public Comment.

provisions have not been finalized, and even if they are, they are too high level to implement in practice without further rules.[167]

In 2022, CDE appeared to open window for companies to rely on bilateral investment treaties to obtain RDP. While the technical review as set forth in the Drug Registration Regulation is approximately 200 business days (130 days for priority review or 70 days for urgently needed orphan drugs), it is common for CDE to issue a deficiency letter and request additional materials from the applicant and to conduct a verification of the quality system for the drug. The review time does not include time for these processes, such as while an applicant prepares supplemental materials, makes corrections after verifications, and/or checks quality standards and draft package inserts. Applicants can log into CDE's online platform to determine the reason for the review pause and see details regarding resumption of the evaluation.[168] CDE's newly issued rule governs the circumstances in which application review can be halted for various other reasons, ranging from cessation of the process for an expert committee to cessation because there are allegations of patent infringement that will be handled through the patent linkage system (described below in this Chapter). This rule also references halting the application process if a bilateral trade agreement requires regulatory data protection (see below). Therefore, requests for RDP pursuant to the Sino-Swiss Bilateral Investment Treaty may be honored under this rule. By tracking applications, it is possible to infer that CDE has held generic approvals until a six-year RDP period expired.

5.19.1 New Drug Monitoring Period

China had previously implemented a "new drug monitoring period," that might have functioned like data or market exclusivity as it does in Japan, which has a similar construct. The new drug monitoring period was included in the DAL Implementing Regulation[169] and related implementing rules administered by NMPA. The new drug monitoring period allowed NMPA to impose up to a five-year monitoring period on new drugs produced by drug production enterprises. During a monitoring period another drug of the same variety would not be approved for production or import. The phrase "drug production enterprises" has been interpreted to mean that it can only be applied to domestically produced drugs. Further, prior to 2016 and the implementation of a new-to-the-world definition, "new drug" was also interpreted to include new generic drugs. This meant that the monitoring period could allow a first-to-market generic (based on a foreign product) to block the innovator from the market. If there was no drug produced

[167] National Medical Products Administration (2022a) Draft Drug Administration Law Implementing Rules for Public Comment.

[168] Center for Drug Evaluation (2022e) Standard for Suspension and Resumption for the Drug Review Process.

[169] State Council (2019a) Drug Administration Law Implementing Regulation, Article 33.

within 2 years of the approval for production, the monitoring period would lapse and other drugs could proceed with applications for production.

The monitoring period had another significant loophole. Pre-2020 versions of the Drug Registration Regulations provided that if, at the time of the establishment of the monitoring period, others had approved registration applications (clinical trial applications), the clinical trials could proceed and applications for domestic production could later be approved.[170] Therefore, those with pending trials could become part of the monitoring period, and it was not an arrangement exclusive to one company. Applications that were pending but not approved at the time of the monitoring period would be returned.

The monitoring period was removed from the Drug Registration Regulation in 2020 and drafts of the DAL Implementing Rules[171] and draft CDE guidelines[172] also removed it, meaning that, while it is still on the books, it is unlikely to play much of a role going forward.

5.19.2 Early Forms of Patent Linkage

Although there were provisions in prior versions of the DRR that prohibited the acceptance of applications during most of the patent life (at least, not until 2 years prior to expiry),[173] NMPA's predecessors ultimately rejected the idea of linking drug review and patent life. The following opinion, in answer to a question from the Sichuan Province FDA, articulates the reasoning for this, including that drug registration (i.e., the process of determining whether a candidate drug meets requirements for safety, effectiveness, and quality) should remain separate from legal processes for protecting patents.

> Opinion on Intellectual Property Issues Concerning Mannan Peptides
>
> [To the] Sichuan Food and Drug Administration:
>
> Your Bureau's "Request for Patent Protection of Mannan Peptide Series Products of Chengdu Lier Pharmaceutical Co., Ltd." ([Sichuan] FDA [2006] No. 56) was received.
>
> The following comments are made on drug registration and intellectual property issues:

[170] State Food and Drug Administration (2007b) Drug Registration Regulation, Article 70.

[171] National Medical Products Administration (2022a) Draft Drug Administration Law Implementing Rules for Public Comment.

[172] Center for Drug Evaluation (2022a) Draft Drug Registration Examination Guidelines for Public Comment.

[173] State Food and Drug Administration (2005a) Drug Registration Regulation; State Food and Drug Administration (2007b) Drug Registration Regulation.

1. The first paragraph of Article 38 of the Administrative Licensing Law stipulates that if the applicant's application meets the statutory conditions and standards, the administrative organ shall make a written decision on granting the administrative license according to law. The Drug Administration Law, the Drug Administration Law Implementation Regulations and the Drug Registration Regulations stipulate that the conditions for drug registration are safety, effectiveness, and quality control, and do not require patent examination of registered drugs.

2. The implementation of drug registration is to review whether the declared drug meets the requirements of safety, effectiveness, and quality control. If the requirements are met, the [FDA] will issue a certificate of approval for the drug. The holder of the drug approval certificate can produce and distribute the applicant drug and should also abide by other relevant laws and regulations, including the relevant provisions of the Patent Law. Therefore, the issuance of a drug approval certificate by the [FDA]does not constitute a violation of the patent right [related to the] drug.

3. [NMPA] attaches great importance to the intellectual property rights of drugs. In order to protect and respect intellectual property rights, the Drug Registration Regulations stipulates the issues concerning the intellectual property rights of pharmaceuticals, and requires the applicants provide a statement as to the status of their and others patent rights and the ownership circumstances for all of drugs or the prescriptions, processes and uses for which they are applying for registration; if others have patents in China, the applicant should also submit a statement that [its application] does not infringe the patents of others. In addition, during the drug registration process, the [FDA] will also inform the drug registration applicant of the patents involved to remind them of patent issues and remind them of possible patent infringement disputes.

State Food and Drug Administration

June 13, 2006

5.19.3 New Patent Reforms

In 2021, China established various drug-related patent protections and connected them to the drug regulatory process. These reforms were first proposed in a nascent way in the 2017 Innovation Opinion and then affirmed in the U.S.-China Phase One Agreement of 2020. This included a patent linkage system and a patent term extension mechanism. At the time of this writing the patent linkage system is in its infancy. These reforms were included in the revision of the Patent Law in 2021 and rules implementing it are also now effective (National Medical Products Administration, National Intellectual Property Administration 2021). Companies have begun to use it and the first cases in which innovators have challenged generic applicants during the generic review process have begun to emerge. As will be explained below, the patent linkage system (sometimes called the drug early dispute resolution system) originates from Article 76 of the Patent Law and allows innovators, who receive a statement from a generic competitor that the relevant innovator patent is invalid or does fit within the technical plan of the generic product, to launch litigation or administrative lawsuit to resolve their dispute with the generic maker before the generic is approved. Patent term extension (PTE) originates from Article 42, Paragraph 3 of the Patent Law and allows for five extra years of patent life (but cannot extend the total period of 14 years) to compensate for review of the marketing application of new drugs.

The rules for patent term extension (PTE) were in draft form at least from 2021 to the end of 2023. Applicants could file an application within three months of obtaining a marketing authorization but those applications were put on hold, pending finalization of the implementing rules. The final provisions governing PTE are now contained in the newly revised Patent Law Implementation Regulations[174] and the Patent Examination Guidelines (both finalized in December 2023).

It is not entirely clear at the time of this writing whether PTE grants will be discretionary (as language in the Patent Law itself suggests) or whether a company that meets all requirements in the newly issued implementing rules and guidelines shall be granted PTE. The rules go into effect in March 2024. Under the rules, PTE may be available for preparation method patents or method of use patents. Only one patent may be considered per marketing authorization, and PTE is only available for one drug per patent. The date of publication of the patent must be before the date of the marketing authorization and the patent claims much include the technical solution of the drug.

Another potential limiter of PTE is that is supposed to be available for new drugs. The newly revised Patent Examination Guidelines indicate that this is for drugs in the innovative registration category (Category 1) or certain improved new drugs. The Guidelines list certain improvements that qualify, such as new esters of existing active ingredients, new indications for existing small molecule drugs or therapeutic biologics, or new vaccine strains. It remains to be seen how this will work in practice.

5.19.4 Patent Linkage

The patent linkage system was a long time in the making. Proposals for such a system had been percolating since the 1980s, and had been considered and resisted on various grounds, including that the drug regulator did not want to involve itself in patent issues, which were believed to be under the exclusive jurisdiction of the national patent department, previously the State Intellectual Property Office and now the China National Intellectual Property Administration (NIPA). Combining NMPA

[174] State Council (2023b) Patent Law Implementing Regulations.

and NIPA under one roof in the State Administration for Medical Regulation (SAMR) set the stage for cooperation on linkage, although NIPA was removed from SAMR in 2023. The NMPA, NIPA, and the Supreme People's Court issued the two rules[175] and one judicial interpretation that form the core of this system on July 4, 2021.[176]

NMPA and CDE established a platform ("Patent Listing Platform") by which marketing-authorization holders can list certain patents within 30 days after they receive their marketing-authorization approval. MAHs for small molecule drugs can list ingredient compound patents, a composition patent containing an active ingredient and medical use (i.e., indication) patents, while biological product MAHs can list "active ingredient sequence structure" and medical use patents. There is little guidance as to the content of these categories at present, but in practice, MAHs are registering their patents, and this system is becoming a feature of planning for the China drug market.

Listing the patent on the platform means that generic and biosimilar applicants must certify as to the relationship between patents on the platform and their products. These follow-on applicants make the certification at the time of their marketing-authorization application. (Pursuant to China's Patent Law, research and development by follow-on makers for purposes of obtaining an administrative license for a drug or medical device does not constitute an act of infringement, see below.)

There are four possible certifications that a follow-on applicant can make.

- I: There is no patent information for the reference drug on the Patent Listing Platform.
- II: There are relevant patents on the Patent Listing Platform, but those patents have expired or been declared invalid, or a generic applicant has obtained a license from the patentee.
- III: There are relevant patents on the Patent Listing Platform, but the generic applicant promises that it will not market the drug until the patents expire.
- IV: There are relevant patents on the Patent Listing Platform, but they should be declared invalid (4.1) or the generic drugs do not fall within the scope of the relevant patent (4.2).

Within 10 days of acceptance of the generic drug application, CDE will publish notice of the certification. Generic drug applicants must also directly provide the MAH associated with the relevant patents notice of its certification, including a comparison table of the claims if it is a 4.2 certification. The MAH will notify the patentee. Strangely, the regulations do not clearly state when the generic applicant needs to provide that direct notice to the MAH.

CDE's public notice of the accepted generic application (not the notice the generic applicant provides to the MAH) triggers a right for the patentee or MAH to challenge

[175] National Medical Products Administration, National Intellectual Property Administration (2021); National Intellectual Property Administration (2021).

[176] Supreme People's Court (2021).

the follow-on application through Article 76 of the Patent Law. The patentee or MAH has two choices. It can file a lawsuit in the intellectual property court in Beijing to have it declare that the follow-on drug does fall within the scope of the relevant patent, thereby blocking approval of the generic application until expiration of the relevant patent, or it can file a request for an administrative ruling by the CNIPA for the same declaration. In either case, the suit must be filed with 45 days of the public notice of certification by CDE. Within 15 days of acceptance of the case by the court or NIPA the applicant must send the notice of acceptance to CDE. The timeliness of these procedures triggers a stay of generic approval. NMPA/CDE will continue to review the application, but it will not issue an approval until either the case resolves in favor of the generic or a time period of 9 months, whichever is shorter.

The procedure for biological products and traditional Chinese medicine products (not covered in this book) is different. In the case of biologics, the innovator MAH must register patents to obtain rights to early challenge of a biosimilar application. Biosimilar applicants must certify to relevant patents, but the provisions on direct notice do not apply. A biologic patent holder may file an Article 76 suit through the courts or NIPA, but proper filing does not stay the biosimilar approval. Rather, if the biologic patent holder or MAH wins its suit, then the biosimilar cannot be marketed until the relevant patent expires. It is not clear why China, a country with such high expectations for its biologics industry, chose to short-change biologics when it could have created one uniform system. However, as other patent-linkage systems, such as those in the United States, only cover small molecule drugs, China may have decided to wait for experience to instruct how biologics should be handled.

A note on the form of the suit: generics can use a validity suit to challenge an innovator patent prior to marketing. The innovator would apply for a declaration that the generic's product would fall within the scope of protection of its patent, but this is not clearly styled as an infringement suit. Both the Supreme People's Court and NIPA issued procedural rules governing Article 76 suits, but there are still many questions to be answered.

5.19.5 Patents and Procurement

Another potential patent enforcement mechanism is the Opinions on Strengthening the Protection of Intellectual Property Rights in the Centralized Pharmaceutical Procurement System. This document requires that companies involved in the volume-based procurement system certify that their products do not infringe the patents of others. If there is such an issue, the participant may be ordered to resolve their disputes before the National Intellectual Property Administration or in the courts. The National Healthcare Security Administration can remove clearly infringing products from the procurement system.[177] A later opinion issued by the State

[177] CNIPA, NHSA (2022) Opinions on Strengthening the Protection of Intellectual Property Rights in the Field of Centralized Pharmaceutical Procurement.

Council on Further Optimizing the Environment for Foreign Investment and Increasing Foreign Investment emphasized this linkage between procurement and patent protection as an important protection for the intellectual property of foreign invested enterprises."[178]

5.19.6 "Bolar" Exemption

The patent law also contains an exemption from infringement if a generic or biosimilar applicant is making, importing, and/or using a patented product for purposes of obtaining administrative approval. The exemption is broadly stated and interpreted by the courts. For example, it can be read to include not only an exemption from patent protection over the molecule of the originator drug itself but also from patent protection over the active ingredient and manufacturing process.[179] Development can begin many years before patent protection ends. Although the 2007 Drug Registration Regulation contained a prohibition on clinical trial applications until 2 years prior to the expiry of the relevant patent, that prohibition was not regularly enforced and was ultimately removed in the 2020 version of the Drug Registration Regulation.

5.20 Drug Manufacturing

China makes drugs, active ingredients, precursor chemicals, packaging materials and components thereof both for use internally and for export. China has long been home to operations that manufacture both what could be considered generic drugs by international standards and active pharmaceutical ingredients. The generic drugs were traditionally used domestically and active pharmaceutical ingredients and components of those ingredients have been exported. China has also become gradually home to more contract manufacturing of drugs for export, particularly active pharmaceutical ingredients and intermediates, although that has been limited in some respects because of concerns about quality and intellectual property. In some cases, China forbids the manufacture of drugs purely for export, such as in the case of vaccines or blood products.[180]

As discussed, recently there has been a push to manufacture more innovative drugs in China, spurred by policies like Made in China 2025. Therefore, with the reforms ushered in by the Innovation Opinion and its predecessors, other related documents pushed for more high-tech drug manufacturing in China. Some emerging

[178] State Council (2023a), Opinion on Improving the Environment for Foreign Invested Enterprises and Increasing Foreign Investment.

[179] National People's Congress Standing Committee (2020).

[180] State Food and Drug Administration (2005a) Filing Provisions on the Receipt of Entrusted Manufacturing for Drugs from Overseas Companies.

biotech companies in China now have manufacturing operations in based in China and are considering using those to supply the rest of the world, but the pandemic and concerns about China in the overall supply chain generally can slow those plans.[181]

Manufacturing is a big topic, but there are several core regulatory aspects. The first aspect is how location of manufacturing, or the classification of drugs as domestically made or imported is made because, as discussed above, that classification influences a number of other regulatory determinations and factors.

Second is good manufacturing practice rule implementation and issues of quality and safety. Improvement of drug quality was a major factor underlying the revision of the DAL, which created a role for the marketing authorization holder as the entity responsible for drug quality throughout its lifecycle and a role for drug manufacturers that are responsible for implementing the instructions of the MAH and producing in accordance with applicable laws and standards.

The DAL maintains a regime to license production facilities and monitor their quality systems, but that regime has faced difficulties in the past and is improving gradually. China's GMPs apply to manufacturing operations abroad, and the Center for Food and Drug Inspections (CFDI) inspects for compliance. China does not recognize compliance with U.S. or E.U. GMPs as equivalent, although parts of China GMP have been influenced by those rules. As of 2022, China is in the process of joining the Pharmaceutical Inspection Convention and Pharmaceutical Inspection Cooperation Scheme (PIC/S), which could result in further harmonization, but it is not clear when membership will be completed and what that will mean.[182]

The third aspect of manufacturing that we will cover here is supply chain, and that we will only cover lightly. This issue has been highlighted more recently in the pandemic in that it has sharpened an already growing concern that China has a very large share of the supply chain and therefore controls the supply of critical medicines for other countries to such an extent that it could pose national security issues. However, there is some debate about how substantial a share of these supply chains China has.

5.20.1 Manufacturing Local and Pathway Determination

Drug manufacturing location is significant in two ways. First, the location of the facilities that make the drug product and conduct fill-and-finish operations (inside China versus outside) determine whether a drug will be considered imported or domestic. There is no definition under the DAL or the implementing rules that NMPA administers as to how much of a drug substance (i.e., the active ingredient) and product need to made in China for NMPA to no longer consider that drug to be imported, and therefore, domestic.

---

[181] United Nations Office on Drugs and Crime (2020).

[182] National Medical Products Administration (2022b) NMPA Studies and Promotes PIC/S Membership Work.

And, discussed above, the distinction between an imported and a domestically produced drug is very meaningful. Although the DAL no longer has separate provisions for each type as it did from 1984[183] to 2019, NMPA's application forms and databases are still divided in that manner. That distinction generally rests on how much of the manufacturing is done inside China and is applied a bit differently for small molecule drugs and biologics. The latter may need to have raw materials, such as cell lines, made in China to be considered domestic, whereas small molecule drugs may be more likely to have imported active pharmaceutical ingredients. NMPA has substantial discretion to consider a drug domestic versus imported and has never adopted a clear definition in this respect.

China has general customs rules that provide some clues, but those do not entirely guide the inquiry. The customs rules require that the country of origin be determined by the last substantial change in the product. This is determined with respect to tariff nomenclature, meaning that the substantial change is a change in the product that causes a "change of the classification of the items in the Import and Export Tariff Nomenclature of a country (region) for the goods obtained after manufacturing or processing with the materials not originated in this country (region)." This does not include processing of the goods for transportation or storage, loading or unloading, and packaging and other types of treatment need for sale.[184]

Somewhat consistent, generally speaking (and perhaps subject to some exceptions), most manufacturing steps must be conducted outside of China for a drug to be imported and fill-and-finish operations in China could take the drug out of the imported drug category. Repackaging is the clearest exception. It is generally permitted for a drug to be repacked from a larger bulk package into smaller packages, and many foreign drug companies conduct repackaging and labeling activities in China. NMPA's repackaging rule defines repackaging in the following way:

> Sub-packaging of overseas-produced drugs means that the drug has completed the final preparation process overseas, and the size of the large package is changed to a small package in China, or [processes of completing] the outer packaging, placement of instructions, and labeling of a drug that has completed the inner packaging are carried out.[185]

Establishing such an operation requires a qualified facility with a drug-manufacturing license (discussed below) and that repackaging facility must be record-filed with CDE so it can be attached to the relevant imported drug MAH. Everything else, including the cultivation of cell lines for biological products may need to be done in China, absent a very clear exception that is unlikely to be granted.

As discussed above in this chapter and in Chap. 4, the distinction between imported and domestic matters both in terms of regulatory requirements

[183] National People's Congress Standing Committee (1984), Articles 26–30.

[184] State Council (2019b) Regulations on the Origin of Imported and Exported Goods, Articles 3–5.

[185] National Medical Products Administration (2020n) Procedures and Requirements for Record-Filing of Drugs Produced Overseas.

(manufacturing licenses) and the distribution and procurement processes. China's law for government procurement, which is the process by which local hospitals obtain drugs, requires purchasing domestic products when they are available in certain cases.

5.20.2 Drug Manufacturing License and GMP Compliance

One of the chief drug regulatory differences between an imported drug and a domestically produced drug is that a domestically produced drug facility requires a drug manufacturing license, issued by the local provincial governments. At a high level, the process requires two steps: an application to the provincial government for the license itself and the GMP qualifying inspection.[186] Prior to the Innovation Opinion and its related reforms and their codification in the 2019 DAL and its implementing regulations and rules, a GMP qualification used to be subject to a separate administrative process in which the company received a drug GMP certificate.[187] Now GMP qualification is combined with the manufacturing license application process. The application can take place during the clinical trial for the drug, while the inspection by local authorities may take place during the marketing application, along with a site inspection by NMPA as part of the marketing-authorization approval process.

As with other licenses, a holder must renew the drug manufacturing license every 5 years and must submit amendments to update the information on the license, including the scope of products made at the facility. These licenses are non-transferrable as an asset, although they may travel with the entity, including in a merger.[188] Despite the official rule, in some cases, provincial governments have allowed the manufacturing licenses to transfer via an amendment to change the name of the company on the license. This process is usually the result of discussions with local authorities and may be subject to additional conditions.

The DAL used to state that construction of a drug manufacturing facility would be in accordance with industrial plans. Specifically, Article 7 of the 2001 DAL stated that, to construct a drug manufacturing facility, the enterprise would need to comply not only with the requirements on facilities, personnel, and quality systems, but also with the national development plans, in order to avoid redundancy.[189] But, China has removed that provision in later versions of the DAL, and in 2015 China further amended the law to allow companies to apply for business licenses before receiving

[186] State Administration for Market Regulation (2020a) Drug Manufacturing Supervision Measures, Articles 7–9.

[187] State Food and Drug Administration (2011a) Administrative Measures on Drug Good Manufacturing Practice Accreditation.

[188] State Administration for Market Regulation (2020a) Drug Manufacturing Supervision Measures, Article 76.

[189] National People's Congress Standing Committee (2001).

their manufacturing (and distribution licenses), removing a further administrative hurdle.[190] Still, local authorities may be inclined to limit the number of drug manufacturing licenses as China has been pushing to consolidate drug manufacturing under larger professionalized manufacturing operations, and that reason may still be asserted by local authorities when refusing to support additional manufacturing licenses pursuant to policies that are not public.

Whereas previously the drug would be attached to the drug manufacturing license itself and require amendments to change, now the production facility must have the appropriate scope to conduct manufacturing of the drugs that it produces. This means that it must be licensed to produce certain types of small molecule drugs or biologics of certain dosage forms. If the facility were to add a different type of drug, it would need to apply to amend the scope of its manufacturing license and undergo an inspection. The failure to do so can result in penalties.[191]

5.20.3 Marketing-authorization Holders and Manufacturing Licenses

Under the prior system, if a company wanted to contract manufacture, both the entrustor and the entrustee would need manufacturing licenses attached to manufacturing facilities, and the contract manufacturing arrangement would be filed for the record with local authorities. But this did not leave room for smaller-scale research and development companies without their own facilities to hold manufacturing licenses.

The MAH system fundamentally changed the nature of this system, as larger, more professionalized contract manufactures are allowed to develop. Under this system, the licenses will be associated with the products. However, the MAH system did not reform the system as much as one would have expected. Although the MAH would get a product license, to contract out manufacturing it would also need a drug manufacturing license. It would get a Type B Entrustor Drug Manufacturing License, whereas the entrusted manufacturer would get a Type C. For those companies that make their own drugs, i.e., the MAH is the manufacturer, there would be a Type A manufacturing license.[192]

Thus, while the MAH system created a product license, it did not eliminate all the prior bureaucratic hurdles for local drug producers. The following Q&A from Guangdong Province explains the rationale:

[190] Elsinga (2015).

[191] State Administration for Market Regulation (2020b) Drug Manufacturing Supervision Measures, Article 68.

[192] State Administration for Market Regulation (2020b) Drug Manufacturing Supervision Measures, Article 77.

Why Does a Drug Marketing-authorization Holder Need to Obtain a Drug Manufacturing License

[Guangdong Province Medical Products Administration]

The "Drug Administration Law" stipulates that a drug-manufacturing license approved by local provincial drug regulatory authorities is required to engage in drug production activities. The Measures for the Supervision and Administration of Drug Production further clarify that holders of drug marketing authorization (including self-produced or entrusted production) should apply for the "Drug-Manufacturing License," and details work procedures and requirements, and establishes uniform provisions governing the requirements for periodical renewal amendment, cancellation, and revocation [of such licenses].

First, from legal perspective, in accordance with Articles 32 and 41 of the newly revised "Drug Administration Law," the marketing-authorization holder, is the primary body engaged in drug manufacturing, and whether it produces drugs on its own or through entrustment, those are all manufacturing activities. Applying for the drug-manufacturing license conforms to the legislative spirit of the newly revised "Drug Administration Law."

Second, from the perspective of actual supervision work, the law states that the marketing-authorization holder shall be responsible for the safety, effectiveness and quality controllability of the drug from research and development to use and after marketing, and shall be responsible for the continuous improvement of drug production, such as continuous compliance and change management. The marketing authorization applies for the drug-manufacturing license in accordance with the law, and maintains a seamless connection with the current pharmaceutical production supervision policies and requirements, better implements the marketing authorization holder's responsibility, and also clarifies the relevant administrative measures after the holder obtains its license.

Third, from the perspective of advancing the reform of "delegation, control, and service," during the [marketing-authorization holder pilot program,] marketing-authorization holders had a "last mile" problem in various aspects, such as in bidding, sales, and taxation. After the holder applies for the drug production license, it can better release the policy benefits and solve the practical difficulties.[193]

***

[193] Guangdong Medical Products Administration (2020) Why Does a Marketing Authorization Holder Need a Manufacturing License.

This explanation makes sense in context and is not so different from places like Japan, where a local MAH is required to hold another license (a drug marketing license), which is attached to it as a company. China's drug regulatory system is still new (comparatively) and its domestic industry inexperienced in many respects. There may be a fear that allowing R&D-based companies to hold marketing authorizations free and clear without additional scrutiny through licensure procedures at the provincial level of government will encourage entities to hold those licenses without sufficient consideration of the complex regulatory systems needed to ensure safety, effectiveness, and quality. A separate license issued by the province as a gating function to marketing is a more powerful tool than a system of post-market inspections to confirm that the MAH has put in place required systems. As the third point makes clearer, it better tethers the MAH as an entity to the overall administrative system in China, which has previously been dominated by licensed drug manufacturers and distributors.

5.20.4 Manufacturing of APIs/Bulk Chemicals and Drugs and Ingredients for Export

China is one of the largest producers of active pharmaceutical ingredients (API) and precursor chemicals in the world. As of 2019, according to some statistics, China had API production worth approximately 4.4 billion USD, more than twice that of India (another leader in this area), and 77% of its produced APIs were destined for export.[194]

Despite this capacity, there has been some debate about whether API producers are subject to the manufacturing license and GMP regulatory system described above. International criticism focused on this issue in 2008 when counterfeit heparin, adulterated with chondroitin to save money, killed 81 people in the United States and injured at least 785 with severe allergic reactions.[195] The case became an early symbol of growing concern about the quality of both food and drug consumables coming from China. This incident launched advocacy to better control supply chains and traceability (i.e., indications as to whether products contained ingredients from China) and to urge China to better regulate its own API industry.

Although the definition of drug in the prior versions of DAL included "drug ingredients," facilities that made active pharmaceutical ingredients for export did not necessarily obtain drug manufacturing licenses. The newly revised DAL removes drug ingredient from the definition of drug (perhaps because, as discussed above, the ingredients are now either bundled with the application or subject to a Masterfile registration, but do not receive separate drug registrations), but this development means that it is more challenging to argue that ingredients are subject to all of the same rules as drugs. In a more positive development, the newly revised DAL now

[194] Bumpas and Betsch (2009).

[195] Committee on Energy and Commerce (2008).

expressly states that drug API makers are subject to the GMP rules, to the extent applicable.

Whether the quality of drug exports from China is improving is subject to some debate. On one hand, China has a relatively steady level of negative inspection results by U.S. FDA (i.e., official action indicated)—albeit with some decline—over the period from 2015 to 2019, perhaps indicating that there has not been much change.[196] Similar supply chain and safety concerns have arisen with respect to the EU.[197]

There are also domestic factors that complicate quality improvement: China is catching up in terms of personnel and technology with Europe, the United States, and Japan. It has also been working to improve the quality of its GMP inspectors, but issues remain, and many inspectors still lack the experience of their counterparts in other countries. There may also be resistance by local governments, which are the primary regulators of manufacturing facilities, to vigorously enforce GMP standards for fear of causing damage to the local economy if facilities need to close or become too costly to run due to the enforcement of quality standards.[198]

There can be a tendency to try to hide these issues by altering or falsifying quality test data.[199] Some analyses indicate that facilities in China and India are twice as likely to have these issues as those in the United States.[200] In one inspection, an employee at a drug facility in China actually ran away from a U.S. FDA inspector when the inspector asked that the employee hand over a flash drive that had been plugged into laboratory equipment (another employee later returned to offer a drive but it was not clear if it was the same one). In other incidents, FDA inspectors have been barred from the facility, intimidated by local police, or even locked into rooms due to allegations that their credentials are fake.[201]

From my own experience, I can say that things generally look better at the facilities we see (they are shining and new and the employees well-spoken and professional), but when looking beneath the surface there can still be issues. These include inexperience with more sophisticated products, cleaning, facility design, standard operating procedure updates, and recordkeeping.

5.20.5 Counterfeit Drugs

The DAL sets forth the definition of a counterfeit drug. Under the 2019 version, a counterfeit drug is one that has ingredients that are not those specified in national standards, is a non-drug substance representing itself as a drug, a drug that

[196] Brennan (January 20, 2020).

[197] Martuscelli (April 15, 2021).

[198] Lee and Hirschler (2012).

[199] Unger (May 18, 2018).

[200] Eban and Salzman (October 29, 2019).

[201] Eban (August 7, 2019).

misrepresents itself as another, a drug that is deteriorating, or a drug that specifies indications on it that are beyond its [approved scope]. A counterfeit drug can be subject to both administrative and criminal sanctions in China, with the administrative sanctions being seizure of product and income, fines of 15-30 the value of the product, and revocation of the drug license. There are also a set of penalties for serious circumstances that include loss of facility licenses and debarment of the domestic or overseas marketing authorization holder for 10 years. The DAL also contains a category of "inferior" drugs that have lesser penalties attached to them, e.g., lesser fines and no debarment, for example. There are seven circumstances that count as counterfeit drugs, including drugs with no or altered expiration dates or lot numbers, drugs that are expired, and drugs that contain substances that have been added without authorization. Some of the inferior drug categories could overlap with the counterfeit categories, such as drugs with ingredients that do not conform to standards.

Importantly, in contrast to prior DALs, under the 2019 DAL, drugs that are imported or produced without approval are no longer in these categories. The DAL prohibits and attaches penalties to importation or production of unapproved drugs, but those penalties are separate from those for counterfeiting. And, the Law now contains a provision that allows for less or no penalties for importation of small amounts of unapproved drugs that are legally marketed overseas. In the press, this reform was attributed in part to a very popular film, Dying to Survive, in which an individual in Shanghai imports unapproved, low cost life-saving generic cancer drugs from India in order to aid individuals who could not afford them. The film was hugely popular, reflecting popular concern over the high prices for important drugs that individuals may have to pay out of pocket.[202]

The Criminal Code contains crimes of producing, selling, or providing counterfeit or inferior drugs, and in 2022 the Supreme People's Court released a judicial interpretation that further defined aspects of these provisions. For example, synthesizing, refining, extracting, storing, and processing drug raw materials for the purpose of producing, selling, or providing counterfeit or inferior drugs can be covered under the crime of producing counterfeit drugs. This judicial interpretation also sets forth the circumstances that can enhance penalties for counterfeit drugs under the Criminal Code.

In practice, counterfeiting is a problem both in China and from China. The OECD has designated it as one of the jurisdictions in the world from which the most counterfeit drugs originate from, a group which also includes India. However, China's press also contains stories in which 10s of thousands of China's own citizens have been injured by counterfeit drugs annually.[203] Drugs may be sourced from Southeast Asia and brought into China via countries in the Middle East. Drug companies that discover these issues may report to SAMR and the Ministry of

[202] Kuo (2019).

[203] Ziavrou et al. (2022).

Public Security (the police) either in writing or via phone. Sometimes the drugs are available online through China's robust e-commerce channels.

5.20.6 Supply Chain Aspects

There has been broad concern since before the pandemic (and perhaps even before) that places like the U.S. and Europe are too reliant on China for many things, including drug ingredients. This includes both drugs *and* devices, because, as we will discuss in the medical device chapter, China used explicit and in some cases unwritten emergency measures to control the flow of certain personal protective equipment to the outside world. There isn't a single precise concern but a general unease at being too reliant in general on China for drug ingredients or too reliant in areas, such as for certain antibiotics.

Some sources would have you believe that nearly every drug we put in our body comes from China in some measure, but that is not the case. China is among the top 20 countries for drug exporting drugs to the world but according to fairly recent statistics has been further down the list than other countries, such as Germany and Belgium, the United States, and Switzerland.[204]

The problem is one of information, since the U.S. FDA has not necessarily collected information that would reveal all of these sources. The sources that do display information, including a 2019 report from FDA on Drug Shortages, indicate that about 14% of the API facilities in drug applications are in China, whereas 31% are in India and 31% are in Europe.[205] These figures do not include non-application drugs. For example, certain over-the-counter drugs are not the subject of pre-market applications and are simply made in accordance with regulatory monographs. A later report produced during the first 100 days of the Biden administration showed that China accounts for a relatively small portion of finished dosage for drug products flowing into the U.S. China and India may account for 42% (China 16% and India 26%) of the registered facilities that produce APIs for drugs consumed in the U.S., but that data cannot show the volume produced from a given facility.[206] And the overall amount of drugs and APIs received from China in monetary terms is substantial according to trade data ($1.8 billion in APIs from China versus India's $582 million); in certain areas China may have most of the volume.[207] China is undeniably an increasingly important source of drugs. For example, due to a shortage of chemotherapy drugs US FDA began exploring using a reputable China generic manufacturer to provide the drugs.[208]

[204] Organization for Economic Cooperation and Development, European Intellectual Property Office (2020).

[205] United States Food and Drug Administration (2019).

[206] White House (2021).

[207] Ibid.

[208] Choi (2023) Chemotherapy Shortages Push Cancer Centers Toward Crisis.

It also does not show whether other non-API ingredients (i.e., excipients) and/or raw materials and fine chemicals (i.e., the components of APIs, excipients, and packaging materials) come from China. For example, the reports do not how many of the 31% of APIs from Europe rely on chemicals that come from China.

While uncertainties remain, the Biden administration report sets sound recommendations on diversifying the supply chain to ensure that there is sufficient redundancy to avoid critical shortages. In this respect, China is like every other foreign nation: we do not want to put all our eggs in one basket.

However, there is an additional risk to China in that its government has power to put restrictions on exports (or imports) of valued resources in event of public health emergencies, with little room for anyone to negotiate with it. Laws such as the Emergency Measures Law and the Law the Prevention and Treatment of Infectious Disease grant the government broad powers to organize resources (production, natural, and otherwise) to handle public health emergencies. The COVID-19 pandemic has demonstrated the power to control supply chain as an emergency measure, and while other countries could also set such restrictions, China's can be opaque and change rapidly, leading to uncertainties and difficulty in planning.[209]

5.21 Drug Distribution

There are various policies and actors that shape China's drug distribution system. The policies include public health and sanitation, health care quality, reimbursement and procurement (i.e., committing to supply a drug for the health care system), other systems of drug availability and shortage prevention, anti-counterfeiting and anti-drug trafficking, anti bribery and corruption, and public procurement. Related to the latter system, there are goals for state-led development of the domestic drug industry and for ensuring an advantage for large China-based generic and biosimilar makers and emerging innovative research-based companies.

The actors in the drug distribution system are also increasingly varied as China moves from a system in which drug distribution was centered on pharmacies within public hospitals to a system that includes large private pharmacies.[210] The influence of the aforementioned policy drivers has increasingly consolidated importation and wholesale of drugs in the hands of larger drug distributors, some of which may be wholly or partially state owned.

These actors and stakeholders include drug distributors, MAHs, drug wholesalers (including warehouses), and drug retailers, which are divided into various tiers, including stores that sell only over-the-counter drugs. Stores that sell over-the-counter drugs are divided into OTC A retailers (i.e., those which have a pharmacist on hand) and OTC B retailers (i.e., stores that sell OTC drugs that do not require a pharmacist, such as aspirin). Finally, there are the health care authorities, i.e., the

[209] Blankenship (April 6, 2020).

[210] Wang (September 28, 2022).

NHC and the NHSA, and various user entities, including medical institution hospitals and clinics that they regulate. These institutions give drugs as part of medical care or have their own in-house pharmacies that dispense drugs to outpatients.

Local authorities also play a role. Local medical products administrations issue and regulate drug distribution licenses and inspect establishments, and local health commissions regulate hospitals and health care providers, including doctors and nurses. Local health care security authorities (i.e., the local counterparts of NHSA) also run the procurement programs for public hospitals at the local levels of government. In the case of vaccines, local procurement is run by local centers for disease prevention and control. As discussed above, the vaccine MAH, which must also be the manufacturer, must sell directly to the CDC.

The 2019 DAL made several important innovations in the regulation of drug distribution. First, it contained provisions indicating that entities transporting and storing drugs could be considered drug distributors and therefore would be subject to distribution licensing and good supply practice rules, or at least be subject to some part of the distribution regulatory regime. This is an innovation because prior versions of the DAL and central and local rules were not clear on this point, potentially leaving some distribution activities to unlicensed third-party logistics providers or general shipping companies. It's important that these entities have some legal authority requiring that the comply with applicable Good Supply Practice rules, which include important rules about temperature and environment control in storage and transportation. Second, the 2019 DAL called for the establishing of a universal traceability system for drugs, which the government had been calling for many years prior to the 2019 amendment. This has led to standards governing a Drug Traceability Code for all drugs that will be printed on the label or affixed to the packaging. Third, the 2019 DAL incorporated provisions on online distribution, including regulating both the online stores and the web platforms that host them (discussed below).[211]

5.21.1 Drug Importation and Export

Imported drugs must be admitted into China by clearing customs, and that requires, among other things, evidence of permission to conduct a clinical trial, permission to be used as a reference product in a clinical trial (such as for a generic or a biosimilar), or permission to enter the country as a product for commercial marketing, i.e., an approved drug with an imported drug registration. There are certainly other ways in which drugs might be admitted—e.g., for use in one of the medical tourism zones or on another named patient basis—but these are some of the primary ways. Certain drugs, such as controlled substances, require not just evidence of the trial approval or

[211] *See* Modern Drug Logistics Development and Supervision Research Group (2022) W. Zhang and T. Hao, eds. (2022) Current Drug Supervision Research.

the marketing permit but also a separate import filing for the ingredient that is on the catalogues of controlled ingredients.

Generally, drugs that are brought into the country must be handed off to licensed drug distributors, which assist with applying for any necessary testing at the ports. Drugs usually require an imported drug customs clearance and port testing clearance. Depending on the type of drug (new to China or biologic versus small molecule) the port testing may need to occur before the customs clearance.[212] In addition, for certain types of biological products, including vaccines, blood products, intro diagnostics used to screen blood and other biological products designated by NMPA lot by lot release may be required at institutions designated by the National Institute of Food and Drug Control in certain ports.[213] These lot release records are searchable and published in batches on NIFDC's website.[214]

In 2018, as part of the Innovation Opinion reforms, the requirement to test existing small molecule drugs at the border was lifted. Specifically, the notice stated that "[i]mported chemical raw materials and preparations (excluding chemicals sold in China for the first time)" would not require mandatory inspections on a batch-by-batch basis during importation.[215] This one of other measures that China has taken to encourage importation of certain drugs, including removing tariffs on substantial numbers of oncology drugs.[216]

Other laws governing quarantine of human or animal plant materials, such as the Border Health Quarantine Law and its implementing rules may apply depending on the material and its state (finished drug versus intermediate). The Border Health Quarantine Law gives customs the power to inspect and detain persons and articles for the purpose of preventing infectious disease in China. Customs does the inspecting but must notify the local health authorities in the event of an infectious disease risk. After a cleared quarantine inspection, which means there was no risk, the article has been decontaminated, or there is otherwise permission for entry, a quarantine certificate is issued allowing for entry into China.[217] This Law and its related guidance can require inspections of human tissue, biological products, and blood products as they cross the border into China.[218] There is also a body of law governing quarantine of imported plan and animal materials.[219] Importation or exportation of human biospecimens involved in clinical research or other substances may be subject to the Biosecurity Law, see Chap. 7.

[212] National Health Commission, China Customs Administration (2012) Drug Importation Measures.

[213] National Medical Products Administration (2020a) Administrative Measures on Biological Product Batch Release; National Medical Products Administration (2003) Notice on Handling Matters Related to Imported Drug Filings.

[214] National Institute for Food and Drug Control (2024).

[215] National Medical Products Administration (2018a) Notice on Matters Related to Customs Clearance and Inspection of Imported Chemicals.

[216] Mu (2018); Wang (2024).

[217] National People's Congress Standing Committee (2018) Border Health Quarantine Law.

[218] General Customs Administration (2023).

[219] National People's Congress Standing Committee (2009) Entry and Exit Plant and Animal Quarantine Law.

Requirements for the export of drugs depend on the requirements of the destination, and customs may inquire into this compliance and require proof. For example, in 2013 NMPA released a notice on complying with certification requirements for active pharmaceutical ingredients destined for the European Union.[220] These certificates can only be issued to companies have obtained approval for drug product and perhaps an approval for the ingredient itself (no longer issued, see above) after either a data review or an inspection. NMPA also released a general rule that allows issuance of a certificate to prove the legal marketing of drugs that are approved for marketing and production in China.[221] Some drugs or components, such as narcotic drugs, require a special export permit prior to export.[222] NMPA issued a notice urging strict compliance with these rules and the requirements of the importing country during the pandemic in response to concerns about supply leaving China during that time.[223]

5.21.2 Public Health Aspects of Drug Distribution

One key aspect of drug distribution regulation is to ensure that products are moved in accordance with their label and other standards that ensure their safety and quality. For this purpose, NMPA and the PMPAs administer a regime that issues licenses to wholesale and retail entities that engage in drug distribution activities. Part of the licensure process requires that the entity have in place a system to comply with drug good supply practice (GSP), which are the standards governing the personnel, facilities, systems, and records required to engage in drug distribution. There is one set of drug GSP for both prescription and OTC products. There are separate measures on distribution of drugs via the Internet, which is an emerging area for the movement of prescription drugs.

China's laws do not have clear and comprehensive definition of distribution activities, but they generally include drug sale, purchase, storage, and transportation. Importation of drugs also requires a distribution license. Only drug distributors can apply for the necessary import testing or related clearance at customs. In addition to physically transporting drugs, an entity needs a drug distribution license or a marketing authorization to be able to book sales for drugs.

Manufacturers used to be permitted to sell the drugs they produced from their places of business, but that permission was removed under the 2019 DAL. In 2020, NMPA explained that manufacturers could no longer be entrusted as distributors, but provided a grace period for sales contracts between an MAH and a manufacturer

---

[220] China Food and Drug Administration (2013b) Matters Concerning Certification for Exporting Active Pharmaceutical Ingredients to the European Union.

[221] National Medical Products Administration (2018h) Drug Export Sales Proof Administrative Provisions.

[222] China Food and Drug Administration (2016a) Service Guidance for Issuance of Import and Export Permits for Narcotic and Psychotropic Drugs.

[223] National Medical Products Administration (2020c) Ensuring During Export Supervision During the Pandemic.

entered into prior to December 1, 2019 (the effective date of the DAL). Those contracts could proceed to completion, but would not be effective longer than December 31, 2022.[224] Now, a MAH may sell the drugs for which it holds the MA wholesale, provided that it meets the necessary GSP standards, but a MAH may not sell its drugs retail.

The drug distribution license process requires that the applicant demonstrate to local authorities that it has the necessary personnel, facilities, and procedures in place to distribute the drugs within the scope of its license. Thus, there may be specialized requirements for distributing biologics and vaccines versus small molecule drugs. Certain drugs, including controlled substances, such as narcotic or psychotropic drugs, require special distribution licenses. Given the sensitive nature of these drugs, China puts quotas on the number of entities that may hold distribution credentials. Currently, there is only a very small list of ten wholesale distributors of narcotic and Class I psychotropic drugs (see above discussion), for example.[225]

As discussed below, in order to encourage larger, professionalized distributors and curb corruption, China has issued very few drug distribution licenses since 2013 and intensified efforts to shorten the distribution chain.

5.21.3 Distribution to User Entities and Pharmacies

Since 2014, China has gradually moved away from hospitals as the dominant institution of dispensing drugs and allowed an increase in chain pharmacies, including more recently online pharmacies.[226] Revenue from both bricks and mortar and online pharmacies went from 3.6 billion yuan in 2013 to 8.7 billion yuan in 2022. The scale of online pharmacies has been growing around 16.7% per year.[227]

Online pharmacies can sell prescription drugs, subject to a prohibited list. Certain product types, such vaccines, controlled substances (narcotic drugs, psychotropic drugs, and medical use toxic drugs), radio pharmaceuticals, blood products, precursor chemicals may not be sold online. These pharmacies (including online pharmacies) are regulated as retail drug distributors, subject to special rules online sales and the general drug distribution rules above. Both the online drug sales entities and the platforms that host these stores are regulated under special drug internet sales provisions that NMPA recently revised in 2022.[228]

Permission to sell drugs via online platforms is not equivalent to permission to advertise them there, although drug information is listed on these sites and the

[224] National Medical Products Administration (2020m) Notice Regarding Several Current Matters on Drug Distribution Supervision.

[225] State Food and Drug Administration (2007d) Notice on Publishing the List of National Designated Wholesale Enterprises of Narcotic Drugs and Class I Psychotropic Drugs.

[226] Feng et al. (2022).

[227] Minenet (2023) Sales Scale Exceeds 260 Billion! Online Pharmacies Continue to Expand Their Territory.

specific drugs sold there must be reported to local medical products authorities with jurisdiction over the sales entity (either the marketing authorization holder or a distributor). The drug internet sales rules expressly prohibit online giveaways as a form of promotion. Display of drug may be further regulated, e.g., requirements that OTC and prescription drugs be displayed separately and restrictions on providing labeling only to certain users that have gone through a prescription review.

Domestic e-commerce of registered drugs is different from cross-border e-commerce, which is discussed in Chaps. 8 and 9. Cross-border e-commerce is for foreign labeled products, often types of food, cosmetics, some medical devices, and certain OTC drugs. Cross-border e-commerce products come into China in small amounts and are distributed directly to consumers. They are imported into bonded zones that are less regulated or via direct mail. They are considered personal use products and therefore are not subject to many pre-market approval rules in China. Certainly, prescription drugs of the type that are most frequently counterfeited (e.g., lifestyle drugs, antibiotics, and painkillers) are not on the lists permitted through this channel. But, as the Chinese press has reported, the e-commerce and its small online transactions have created opportunities for concealed transactions of counterfeit drugs and other products.[229] That's why China's cross-border e-commerce regulation has become increasingly stringents, including tasking the e-commerce platforms with policing activities of the operators that they permit to sell on their sites.

5.21.4 Anti-corruption in Drug Distribution

Given that this is a regulatory discussion, anti-corruption issues are somewhat out of scope, but concerns about corruption have shaped policy on drug distribution in China to a great extent. Specifically, beginning in 2013, when the government began to more vigorously concentrate on graft in the health care industry due to several high-profile cases, the number of distribution licenses was thought to be too large. Various distributors and links along a distribution chain created potential opportunity for mark-ups and illegal schemes.

As a result, the government slowed the issuance of new distribution licenses. From 2012, the number of licenses (including retail licenses) remained fairly flat, rising only about 30,000 from 2012 to 2017. Recently, it has started to go up more steeply. However, when one looks at the percentage increase over the last few years (2017–2019) wholesale distributors are only about 2% of the total. It seems that most of the new licenses are going to small retail stores or chain stores.[230] Because a distribution license is needed to book sales, companies that want a given affiliate to be involved in the sales process need to either try convincing a local government to

[228] NMPA (2022g) Drug Internet Sales Administrative Measures.

[229] Zhao et al. (2017).

[230] National Medical Products Administration (2020b) Annual Report of Drug Supervision and Administration Statistics.

issue the license or, more likely, purchase an existing company that already holds the license. Distribution licenses travel with the entity, but local governments rarely permit them to be transferred as an asset (with some limited exceptions) between entities, because they are specific to facilities, systems, and personnel, some of which may not be transferred in an asset sale.

Around 10 years ago (in 2013), the government focused on allegations that companies were using third parties to funnel illegal payments to doctors to increase use of their drugs. China and foreign regulators have conducted investigations of the flow of money to doctors for prescriptions and clinical research. China also focused regulatory attention on the medical industry, issuing blueprints for enhancing the ethical conduct of doctors. The NHC released a regulation on "clinical research projects" in hospitals, which required that monies for drug and device research products be directed into institutional accounts after those projects had been reviewed at an institutional level by a special committee.[231]

Perhaps the most succinct summary of the government's concern in this area came in a policy document with Nine Prohibitions on certain corruption issues in the health care industry. The nine prohibitions were directed at health care providers and include:

- No links between health care provider income and drug or medical service profits.
- No rebates for prescribing drugs or for referrals for medical services or drugs.
- No overcharging patients.
- No acceptance of illegal donations.
- No illegal advertisements or promotion of drugs, medical devices, foods, or other products by health care providers or medical institutions
- No gathering of statistics for commercial purposes or personal gain by health care providers.
- No private buying or selling of drugs, medical devices, or other medical equipment by health care providers.
- No acceptance of kickbacks or commissions from health care companies or engagement in entertainment activities provided by those companies.
- No solicitation or acceptance of financial benefits from patients.[232]

Drug companies are arguably under the same obligations through the DAL (which contains its own anti-kickback provisions—prohibiting items of value to doctors or other health care personnel by drug companies to secure prescriptions) and other laws, such as the Criminal Law (criminal bribery) and the Anti-Unfair Competition Law (commercial bribery). Therefore, foreign and, increasingly, Chinese companies developed stronger internal anti-corruption compliance systems. Some of them adopted the code issued by a local association for research-based companies, the Research and Development Based Pharmaceutical Action Committee ("RDPAC"),

[231] National Health Commission (2014).

[232] National Health and Family Planning Commission, et al. (2013).

that foreign research-based companies developed for activities in China. The RDPAC Code is based that of the International Federal of Pharmaceutical Manufacturers and Associations, which is based in Belgium and cooperates with other national associations in Europe, the United States, and Japan.[233]
Other codes of China-based associations, such as China Pharmaceutical Research and Development Association (SinoPHRDA), also have industry codes that address issues of interactions between drug companies and health care providers.

In 2020, various Chinese companies also collaborated through the China Pharmaceutical Industry Association to produce a group standard on Pharmaceutical Industry Compliance Management Practices. A group standard is made by various stakeholders, which commit to adherence to its principles. This standard (PIAC/T 00001-2020) covered the responsibilities of various stakeholders, including MAHs and contract manufacturing organizations and contract sales organizations, in a broad range of areas such as commercial bribery, anti-monopoly, taxation and finance, product promotion, procurement, environmental health and security, adverse drug reaction reporting, and data compliance and cybersecurity. The standard was implemented beginning on February 26, 2021.[234]

Similar to the conduct that was targeted by the Nine Prohibitions, the government began a new anti-corruption campaign amongst doctors for improper benefits received from drug and medical device companies in the summer of 2023. As of this writing, this new campaign is scrutinizing various types of problematic behavior and improper benefits to physicians, from their pocketing the difference between amount received and sales price in sales of medical equipment and drugs to receipt of illegal donations and conference fees. As of the time of this writing, over 170 doctors and local leaders are under investigation, and it is not clear what the long-term consequences will be for industry and for law enforcement in this area going forward.[235]

5.21.5 The Two-Invoice System

Another consequence of the compliance movement in the pharmaceutical industry was the Two Invoice System. The NMPA and the NHC, along with six other agencies, released a document in 2017 to reduce the number of links in a distribution chain to publicly established medical institutions (i.e., public hospitals) in China. Subject to some exceptions, the Two Invoice System only permits two tax invoices (*fapiao*) between the manufacturer and the hospital unit. The stated purposes are to lower drug prices (the reduced number of links would increase transparency in the

[233] RDPAC (2022).

[234] China Small Molecule Drug Producer Industry Association (2021).

[235] McCarthy (2023).

way that prices are set) and reduce fraud and money laundering. This policy has significantly limited the number of entities that can book sales in a distribution chain, making it harder for there to be unnecessary mark-ups but also more difficult to allow revenue sharing between partners and to move inventory around. The Two Invoice System has required that inventory either needs to be moved by something other than a sale or put in the hands of one general distributor.[236]

The system is subject to some important exceptions. First, it does not cover transfers between companies in the same corporate group, including "transfers" from a distributor to its wholly-owned subsidiary. And it does not cover transfers from the MAH to an exclusive general distributor. Importantly, some of the provinces enacted their own rules in which they have exempted transfers between the MAH and the MAH agent (an important link). There is also an exception to permit transfers to additional distributors that can reach areas that some of the general distributors cannot, such as remote townships or villages.

The Two Invoice System has been an influential policy in the planning of deals, because companies need to plan carefully which companies can "book sales" for purposes of tax and determination of royalties.

5.21.6 China's Health Care System, Pricing and Reimbursement, and Procurement

The focus of this chapter is drug regulatory issues, but a basic understanding of downstream pricing and reimbursement concerns is an important piece in the design of an overall strategy. This section breaks down the mechanisms that the government uses to control drug prices. It then moves on to give a brief picture of the reimbursement system for the state-run insurance plans, which is a separate system from China's system for drug price regulation.

Before 2015, China set or guided the price for many patented drugs. Manufacturers would have to apply for ex-factory (wholesale) and maximum retail prices, and a complex set of central and local regulations governed the process by which the government obtained material on the manufacturing and distribution costs that assisted it in determining these prices. In 2015, the National Development and Reform Commission, which then regulated this area, released the Opinion on Advancing Drug Price Reform.[237] This Opinion eliminated the government maximum retail prices for most drugs, with the exception of narcotic drugs and Class I psychotropic drugs. (For these two types of drugs, ex-factory and maximum retail prices would continue to be set by the government.)

Since the 2015 Opinion, although the government no longer sets the prices for many drugs, China has pursued other methods to keep the price lower. These include

[236] National Health Commission (2016a) Implementation Opinion on the Implementation of a "Two-Invoice" System in Public Medical Institutions.

[237] National Development and Reform Commission et al. (2015).

using negotiations to add the drug to reimbursement rolls for the state health plans or the procurement system, such as through Volume-Based Procurement methods that award a certain portion of sales in a locality to a company with the lowest bid. The volume-based procurement system applies to drugs that have a generic substitute that has been demonstrated to be equivalent and often-times results in a price reduction of approximately 50–60%[238] for nearly 400 varieties of drugs. The program is currently implemented nationally, but provinces will soon be able to run their own programs.[239] The volume based procurement ("VBC") system also influences the reimbursement price because the VBC prices are referenced when setting reimbursement rates in different provinces.[240] What is not covered by reimbursement may be passed on to individuals as a copay.[241]

Another system that influences distribution and prices is the reimbursement system. China has urban and rural state-run insurance plans (the private insurance market plays a very small role). Drugs that are on the National Drug Reimbursement List (NRDL) are reimbursable in China. This list includes two sub-lists, A and B: the first for essential drugs (fully reimbursable; mostly generic) and the second for a combination of innovative and generic drugs that may be partially reimbursable.[242] Although in the past it could take years for an update to the NRDL, it is now updated on a regular basis,[243] pursuant to a process that the National Healthcare Security Administration established in 2020.[244] Companies negotiate to add their drug to the list and, in doing so, may need to significantly reduce the price. In 2020, the average price cut for the 119 drugs that were added was approximately 50%.

In addition, NMPA launched or in some cases threatened to launch antitrust investigations into drug companies to determine whether they were abusing their market position.[245] This was part of a larger effort by China to use antitrust mechanisms to control the costs of drugs and medical devices.[246]

---

[238] Xinhua (2023) The Procurement of 41 Kinds of Drugs Successfully Covers Multiple Treatment Categories - Interpretation of the Highlights of the Ninth Batch of National Centralized Drug Procurement.

[239] National Healthcare Security Administration (2021) Notice on the Continuation of Work After the Expiration of the Centralized Drug Procurement Agreement.

[240] Chongqing Municipal Healthcare Security Bureau (2022).

[241] National Healthcare Security Administration (2019) Expansion of the Regional Scope of the Pilot Program for Centralized Drug Procurement and Use Organized by the State.

[242] National Healthcare Security Administration, et al. (2020b) National Basic Medical Insurance, Work Injury Insurance and Maternity Insurance Drug Catalog.

[243] National Healthcare Security Administration, et al. (2022) National Basic Medical Insurance, Work Injury Insurance and Maternity Insurance Drug Catalogue.

[244] National Healthcare Security Administration (2020a).

[245] Lane (May 9, 2016).

[246] World Health Organization (2017).

5.21.7 Prescription of Drugs By Healthcare Providers

Another piece of distribution the prescription of drugs. In general, doctors in China are required to prescribe drugs in accordance with rules and hospital policies. The NHC released rules on drug prescription in 2007 that remain in force today.[247] Under these rules, doctors must prescribe drugs by their common name (i.e., not by a brand name) and must prescribe them in accordance with their registrations or treatment guidelines issued by the NHC. These treatment guidelines may include other indications beyond those in the approved registration.

There have been debates about whether doctors are permitted to more freely prescribe drugs off-label. Chinese law has not provided for flexibility similar to that in the United States, where doctors have freedom to prescribe off label, even if there is not an officially issued government treatment guideline. There have been discussions for some time about how much freedom to give doctors in China, and whether off-label prescription would require an internal hospital approval or evidence from an official compendium. Importantly, China in 2021 amended its Physicians Law to provide as follows:

> Article 29: Physicians shall adhere to the principles of safe, effective, economical, and reasonable use of medication, and follow the guidelines for clinical application of drugs, clinical diagnosis and treatment guidelines, and drug package inserts and other principles of the rational use of drugs.
>
> In special circumstances, such as if there is no effective or better treatment, after the doctor obtains the patient's clear informed consent, the doctor can use the drug usage that is not defined in the drug insert but has medical evidence to implement treatment. Medical institutions shall establish a management system to review the suitability of doctors' prescriptions and medication orders, and strictly regulate doctors' medication behavior.[248]

With this provision comes more clarity and flexibility than existed previously. However, it is important to recognize that this limited freedom to prescribe off label is different from the freedom to advertise or promote off label, both of which are not permitted in China (see discussion below and in Chap. 10).

There are not strictly mandatory blanket substitution requirements for generics or biosimilars, but there are policies that encourage use of equivalent follow-on medicines by healthcare providers when possible and seek to facilitate that use.[249] Under the 2007 prescription regulations, doctors generally must prescribe medicines by their non-proprietary name unless there is a patent on a new-to-the-world active ingredient, or it is a medicine with two or more active ingredients marketed under a trade name approved according to NMPA's rules. If a physician were to prescribe using the trade name of an innovative drug, a pharmacist may not substitute a generic

247 Ministry of Health (2007).

248 National People's Congress Standing Committee (2021).

249 National Health Commission (2018c) Interpretation of Opinion on Reforming and Improving the Policy on Supply Guarantee and Use of Generic Drugs.

or biosimilar in its place. Some of this may depend upon whether the drug under that trade name is available. Although the National Reimbursement Drug List (which includes drugs that are reimbursable to some extent on China's state insurance plans), now includes more innovative drugs, it remains dominated by generics, and centralized procurement policies still favor lower cost drugs.

5.22 Advertising and Promotion

5.22.1 Drug Advertising

As will also be discussed in Chap. 10, of all the different types of advertisements for regulated products, drug advertisements have the most restrictions. This is particularly true for prescription drugs. To advertise prescription drugs, the content must be approved by a local medical products administration and the advertisement must be placed in a medical journal that has been approved to contain such content by NMPA and the NHC. Non-prescription drugs may be advertised directly to consumers, but only after a local MPA reviews and approves the content.[250] The advertisement issued must be exactly the same as that approved, and copies of the approved content are posted online in databases. Further, the DAL contains provisions that suggest that all promotion is restricted to the advertisement context.[251]

China's Advertisement Law (see Chap. 10) and NMPA regulations on drug, device, and special food advertisements contain various content restrictions, including that the content must be consistent with the advertisement. Although China increasingly permits off-label prescription, there is no context in which manufacturers are permitted to advertise their drug for indications that are not approved as part of the registration (which includes the label and package insert). Other content restrictions include prohibitions on endorsements by doctors or hospitals, scare-tactics implying that failure to use the drug could aggravate patients' conditions, and comparisons with the safety and effectiveness of other pharmaceuticals.

Drug advertisement approval requires a valid drug marketing approval, meaning no advertisements are permitted for unapproved drugs or drugs for which approval is pending. Both the local authorities and patients can seek enforcement (e.g., through complaints) and compensation for harm caused by the advertisement rules. Once an advertising violation is established and posted, more cases from patients may ensue and, depending on the province, further collateral consequences, such as exclusions from local procurement processes.

---

[250] State Administration for Market Regulation (2020d) Tentative Administrative Measures on Advertising Review for Drugs, Medical Devices, Health Foods, and Foods Formulated for Special Medical Purposes.

[251] National People's Congress Standing Committee (2019a) Drug Administration Law, Article 90.

As these rules are fairly restrictive, there is often a question of how to comply with them and still promote drugs in reasonable way. What about all the detailing aids, pens, pads, posters, and booths utilized in other parts of the world? Are they not permitted because they cannot be published in an approved journal? What about websites that discuss the mechanism of action of the drug? Is it not possible to compare drugs together in any context even if there is solid data underlying the comparison?

The answer is, no, there is no safe harbor for these types of promotion, but they do occur. They are justified as common industry practice, scientific exchange (i.e., non-promotion), or non-advertisement promotion (i.e., promotion not subject to the Advertisement Law). But there is no law or regulation specifically providing such an exemption for these different types of communication and activity. While NMPA has enforced against unapproved or illegal drug advertisements, it has not attempted to stamp out all of the forms of "promotion" that are not approved as drug advertisements.

5.22.2 Drug Information

In addition to promotional content, there is also a regime governing provision of drug information online. Individual websites that provide such information must receive a special approval number. This approval number does not cover any advertisements on the website. It is issued to a website, and not each individual webpage.[252]

5.22.3 Medical Representatives

In 2020, NMPA created a registration system for "medical representatives." These are individuals employed or authorized by MAHs to speak with health care professionals about drugs and engage in "academic promotion." Information about individual medical representatives must also be record-filed with an online platform run the by China Pharmaceutical Association. They are allowed to pass along information about the drug (on label) to health care professionals through face-to-face meetings, conferences, or remote tele- or video conferences, and they can also collect feedback about the drugs. They cannot engage in sales of the drug to the health professionals, nor can they inquire about their prescription behavior.[253] Looking at the responsibilities for medical representatives on paper (e.g., explaining academic concepts and features of the drug to doctors), might lead to the conclusion

[252] State Drug Administration (2004) Administrative Measures on Drug Internet Information Services.

[253] National Medical Products Administration (2020l) Medical Representative Record-Filing Measures (Trial Implementation).

that they are professionals with medical backgrounds, similar to medical science liaisons in other jurisdictions, but in practice they are similar to sales representatives that are trained on a given product or products. Some companies may have medical science liaisons on staff in China, which handle more technical communications.

5.22.4 Patient Support Programs

Since 2003, companies have invested in various types of patient assistance and patient support programs. These programs have varied in form and content, and because China has not issued specific guidelines that bridge the gaps in drug, health care, and charitable giving law, there is quite a bit of confusion about how to structure them in a safe and relatively non-promotional way.

5.22.4.1 Patient Assistance Program

A patient assistance program is generally considered a program via which drug companies holding marketing authorizations to drugs can give in-kind drugs to patients in need either for no cost or at a reduced rate. There are many such programs in China, growing from just one in 2003 to over 81 in 2018 and to over 160 in 2021.[254]

To establish a program, the drug company must enter into a donation contract with a designated charitable organization. These are the government-controlled organizations through which donated goods and services are generally dispensed. Such arrangements are subject to requirements under China's Public Interest Donation Law and its Charity Law. There are also specialized rules governing the circumstances under which medical institutions can *receive* donations and the procedures they must follow to approve such arrangements, as well as content requirements for donation agreements. Some arrangements have relied upon industry associations to handle these programs, but there may not be support for such organizations to engage in charitable activities.

Consistent with this structure, in 2016 NMPA established specialized rules governing the importation of drugs designated for donation. They set forth conditions for donation and for recipients, as well as established a specialized record-filing procedure for donated drugs.[255]

Once the program has an approved structure, the drug company must establish a standard for dispensation of the drug, which should be need-based and generally applicable. Conditions that require prior or future purchases of drug in exchange for the donation can be disfavored, although there is not evidence of enforcement against these cases. Further, the drugs should not be dispensed in a way that compromises their quality, including compliance with drug good supply practice.

[254] Jiang et al. (2019); Deloitte (2021).

[255] National Health and Family Planning Commission (2016).

5.22.4.2 Patient Support Programs

While the first of these programs primarily relate to in-kind donations, there are some broader programs that provide disease awareness education and pre- and post-prescription counseling for patients. Although the drug regulatory and health care agencies have not offered any specific guidance on this issue, one option is to use a charitable organization structure. However, the involvement of these organizations means companies will not have close control over these programs. Disease awareness information may be provided but can be questioned by authorities if it intrudes into diagnosis, which is reserved for licensed medical institutions, or if it strays into promotion, which is subject to strict rules (see above).

5.23 Post-Market Surveillance

5.23.1 Pharmacovigilance Systems and Individual and Periodic Event Reporting

As indicated in the DAL, the MAH is responsible not only for development and registration of the drug, but also for post-market surveillance activities. Under prior versions of the DAL, post-market activities were simpler, primarily including maintenance of the license and reporting of adverse drug reactions and recalls.

The revised 2019 DAL envisions a more comprehensive life-cycle approach to risk management. MAHs must still amend their licenses and renew them every 5 years, report adverse drug reactions, and report and implement recalls. They must now submit more comprehensive annual reports on each drug, including quality and safety information and "develop a post-market risk management plan and proactively carry out post-market drug studies" to strengthen drug safety and effectiveness.[256]

Furthermore, pursuant to new, extensive, Good Pharmacovigilance Practice (GPV) rules issued in 2020, the MAH must create an internal infrastructure, including procedures, personnel, and systems to implement post-market responsibilities and audit compliance. This includes appointing an executive -level pharmacovigilance responsible person (with a medical, pharmacy, or epidemiology-related background) for the MAH itself and establishing a safety committee and a pharmacovigilance department.

At the time of this writing, the GVP is fairly new, and still accruing practical implementation history, but it calls for more than has been required in the past. China and overseas-based pharmaceutical companies have been incorporating China post-market surveillance requirements into their operating structures for several years, but this rule calls for a higher level of strategy and sophistication in implementation. Furthermore, NMPA and CFDI are embarking on a plan to conduct pharmacovigilance inspections for compliance with the rule in China (and have

[256] National People's Congress Standing Committee (2019a) Drug Administration Law, Article 77; Center for Drug Evaluation (2021c) Public Comment Draft of the Principles on Drafting Risk Analysis and Management Plans.

already begun to inspect). NMPA and CFDI issued guidelines with a checklist of items for those inspections in 2022.[257]

5.23.2 Adverse Drug Reaction Reporting

Even before the GVP, at the core of pharmacovigilance is reporting on safety incidences with the drug on the market in the form of adverse drug reactions. During clinical trials and post-market studies (whether interventional phase IV studies or non-interventional observational studies), adverse drug reactions are reported between parties (i.e., the site and sponsor) according to protocol, the GCP, separate rules released by CDE, and now under the GPV rules (described above). For marketed drugs, regulations released in 2011 (still effective and overlapping with the GPV) require the reporting of non-serious and serious adverse drug reactions and "group" or "mass" adverse drug events on an individual and annual basis.[258] There are also reporting requirements applicable to drug distributors and to user entities.

The reporting requirement is for reactions, and not events. China defines an adverse drug reaction as a harmful reaction to a drug that meets quality standards, which does not relate to the purpose of the drug. In other words, there is some potential causal connection suggested between the drug and the reaction. Although China does not define what this connection must be (e.g., tentative or confirmed) in existing rules, in reporting guidelines it has subscribed to the notion that MAHs should report "suspected" ADRs, meaning there only needs to be a possibility that the drug caused the event. While the definition states that ADRs relate to drugs that meet quality standards, new rules require reporting even if the reaction may be related to a quality problem or exceeding indications or dosages, among other such reasons.[259]

Until recently, China's policy was to monitor new imported drugs and domestically produced drugs that were under a "monitoring period" more closely. As discussed above, the monitoring period of 3 to 5 years was imposed on domestically produced drugs of a new type. But this concept has been removed from most regulations. For new imported drugs and domestic drugs under a monitoring period, all adverse drug reactions regardless of their severity had to be reported. Non-serious events were to be reported within 30 days, whereas serious death events needed to be reported immediately (i.e., within 24 h) and serious injury events within 7 days.

Previously, drug MAHs could report such events through written forms or electronically. The new GPV rule changes this arrangement. MAHs must register

[257] National Medical Products Administration (2022c) Principles on Drug Pharmacovigilance Inspections.

[258] State Food and Drug Administration (2011b) Adverse Drug Reaction Reporting and Monitoring Measures.

[259] National Medical Products Administration (2021c) China Drug Good Pharmacovigilance Practice Rules, Articles 47 and 132.

with the National Adverse Drug Reaction Monitoring System, meaning these interactions must be electronic. It requires all serious adverse drug reactions must be reported within 15 days of discovery (i.e., the time at which the MAH has enough information to determine that the incident meets minimum reporting requirements), and non-serious adverse drug reactions within 30 days. The requirement to report non-serious ADRs is no longer limited to drugs under a monitoring period or those new varieties of imported drugs during their first license term. Follow-up information about each event should also be reported within the same timeframes of receipt. Serious adverse drug reactions occurring overseas for the same drug as registered in China must be reported according to this timeline, and if an overseas health authority requires removal of the drug from the market, the MAH must inform China medical products authorities within 24 h of receiving that information.[260]

The new GPV rule also requires reporting of clusters of ADRs with the same features immediately upon discovery, precisely defining them as multiple cases with similar symptoms suspected to be related to the same (or an adjacent) batch of the same drug, occurring with the same timeframe and with similar features.[261]

The timing of periodic reports remains focused on the newness of the drug. Specifically, new drugs (i.e., innovative and improved drugs) must submit a periodic safety update report every year for the first 5 years, whereas new drugs in the second license term, Category 5.1 drugs, or follow-on drugs only need to submit those reports every 5 years.

5.23.3 Quality Issue Reporting

Drug manufacturers are obligated to report "quality related major safety incidents." These incidents must be immediately handled by sealing up the drug and its active ingredients and excipients and immediately reporting to the provincial-level authorities.[262] The term "quality related major safety incidents" is not defined in the manufacturing rules, but certain provinces have issued guidance on defining and grading different types of drug safety incidents. For example, Shandong Province has issued a grading system, which has "especially severe drug safety events," covering circumstances in which the drug injured 50 or more people in the same place or serious unexpected adverse drug reactions in which 10 or more people are injured, among other circumstances. The next grade is severe drug safety accidents, which are events in which thirty or more (but less than 50) people have similar clinical manifestations, serious unexpected adverse drug reactions in a concentrated geographic area and time, particularly serious adverse reactions (those possibly causing permanent or life threatening damage to the human body or organs) in up to five people, or deaths of 2-5 people with reactions in others from the same batch of

[260] Ibid.

[261] Ibid., Article 132.

[262] NMPA (2021f) Drug Manufacturing Measures, Article 65.

drug, among others. The grading system also includes two lower grades: larger events and ordinary events.[263]

5.23.4 New Risk Mitigation Systems and Post-Market Studies Under the Drug Administration Law

The DAL requires continued surveillance of the drug once on the market. MAHs are required to develop a risk-mitigation plan. This requirement found its way into earlier rules, such the rule governing the expedited approval of drugs serving a clinically urgent need (see above) but (although there is a draft) no comprehensive guidance on how to formulate such a plan has been finalized.

Similarly, the newly revised DAL carries forward a long-time requirement for "re-evaluation" of a drug's safety and effectiveness profile based on very serious safety signals. Prior DALs called for manufacturers to re-evaluate their drug voluntarily or on government order if evidence emerged that would undercut the risk-benefit profile, based on which the drug was approved. This requirement is now in DAL Article 83, which provides that MAHs must conduct continuing post-market evaluation of the safety, effectiveness, and "quality controllability" of their drugs. Local MPAs may order such an evaluation with cause. If the evaluation shifts the safety and effectiveness profile, then the registration may be canceled.

The DAL places new emphasis on the importance of conducting post-market studies as part of life-cycle management of the drug. China's drug regulations call out both interventional Phase IV studies and post-market non-interventional (observational) research as fulfilling this purpose. In some cases, conducting these studies may be a condition of approval of the drug (see above), but in others they can simply be expected as a part of a push for continued safety monitoring. The drug rules phrase these obligations generally and do not account for drugs with well-established safety profiles.

5.23.5 Annual Reporting

The DAL requires both drug and vaccine MAHs to file annual reports on the drugs for which they hold marketing authorizations. The reports include basic information about the MAH itself and information covering three aspects of each drug: sales and manufacturing, post-marketing research, and risk management. Sales and manufacturing refer to different strengths of the drug inside China.

Post-market research refers to summaries of post-market commitments (i.e., progress toward conditional approval studies) and generic quality and efficacy equivalence evaluations. It also includes information about major and moderate

[263] Shandong People's Government (2021).

changes to drug registration information that required separate approval or record-filing, or minor changes required only in an annual report.

The DAL and recently finalized rules required that this report be submitted prior to March 31, 2022. That report covers data for the prior calendar year (December through January), although the basic information about the MAH itself should be the current information as of the filing of the annual report.[264] The annual report includes a public part with "information on the MAH, the overall situation of the products, an overview of quality management, the construction and operation of the pharmacovigilance system, the circumstances of overseas entrusted processing, and inspections by overseas drug regulatory agencies."[265] There is also a product part with "basic product information, production and sales status, post-launch research and changes to the registration, and risk management status."[266]

Risk assessment information includes information about recalls and handling of out-of-specification batches, as well as pre-market risk control measures for commercial-scale batches. Individual and periodic adverse drug reaction information is reported separately.

The failure to file such reports can lead to considerable fines, into the tens of thousands or even more than one hundred thousand US dollars. However, such fines should only be imposed after a cure period of 20 business days has run, with no rectification by the MAH. As discussed separately, there is also an annual quality report due for vaccine MAHs that covers a similar, but in certain respects broader, scope of information. Whereas standard drug annual reports would be due to provincial level MPA regulators, vaccine reports are submitted both to CFDI and provincial level regulators.[267]

5.23.6 Drug Dossiers

In addition to emphasizing the role of the MAH in continually collecting and evaluating safety and effectiveness information over the life cycle of the drug, the DAL and relevant policies also include a role for the government in collecting and evaluating both drug and MAH information to enhance regulation of the industry. For example, Article 105 of the DAL states that NMPA will establish a drug safety credit file for MAHs, manufacturers, distributors, clinical trial institutions, and non-clinical testing institutions. The purpose of this file is to collect information about the licenses that these entities hold and the results of supervision and testing/inspection (i.e., the results of government enforcement) and related penalties. The Center for Food and Drug Inspections (see Chap. 2) collects this information for

[264] National Medical Products Administration (2022d) Provisions on Drug Annual Reports.

[265] Ibid.

[266] Ibid.

[267] Center for Food and Drug Inspection (2020).

clinical and non-clinical testing institutions, and the provinces collect it for manufacturers and distributors.

The healthcare bureaucracy also gathers data for analysis about drugs. For example, in 2020 the National Health Commission issued a guideline on the "comprehensive clinical evaluation" of drugs, seeking to use real world clinical data from medical institutions in China to evaluate technical and policy issues associated with safety, effectiveness, economics, innovation, suitability, and accessibility of drugs.[268] The general goal is to develop a knowledge base for policies on drug use (NHC has develops treatment guidelines on clinical application of certain types of drugs, for example) and ensuring appropriate supply of drugs with important clinical value.

5.23.7 Recalls

Until recently, the recall rule (previously amended in 2007) provided that drugs must be recalled if the manufacturer (now the marketing-authorization holder) determined the drug to pose a safety hazard. A safety hazard is defined to be "an unreasonable danger that may endanger the safety of human health and life due to research and development, manufacturing, or other reasons. . . ." The rules then go on to state that hidden safety dangers can include non-compliance with national mandatory standards, drug GMP, and other provisions and nonconformity of the manufacturing process to the registration.[269]

However, since that time, SAMR and NMPA have gradually expanded the concept of a defect to include not simply unreasonable risk of human harm but also quality issues, including in the medical device and food spaces where, technically, any non-compliance with a mandatory national standard can constitute a "defect" requiring a recall. This expansion can be seen in the device recall rules discussed in Chap. 6.

A new drug recall rule was issued in 2022. Although the prior rules clearly included non-compliances with quality standards and GMP as triggers, the new 2022 rule provides that a recall is triggered for a product marketed in China by a "quality problem or other safety hazard," which can be caused by non-compliance with requirements governing research and development, production, storage, transportation, and labeling, or the trigger can be another "unreasonable hazard" that "may" render the medicine injurious to human health.[270] There is not a definition as to what constitutes an unreasonable hazard in the new recall rule.

As with all of the spaces, there can be voluntary/affirmative recalls, in which a marketing-authorization holder determines that a drug issue meets the standard for

---

[268] National Health Commission (2020c) Guideline on the Comprehensive Clinical Evaluation of Drugs.

[269] State Food and Drug Administration (2007a) Administrative Measures on Drug Recall.

[270] National Medical Products Administration (2022e) Provisions on Drug Recalls.

recall and affirmatively reports to the government that it will recall the product at the patient level, through a public announcement on its website and, for certain classes, a government website. There are no distributor-level recalls. In other words, once a marketing-authorization holder discovers a quality or safety problem that could cause a recall, it is required to conduct an investigation and recall at the individual level.

In preparing to report a recall, a marketing-authorization holder must conduct an investigation and classify it. The recall classes are: Class I (drug has caused or may cause serious harm to human health), Class II (has caused or could cause temporary or reversible harm to health), or Class III (unlikely to cause harm, but should be recalled for other reasons). At this point the marketing-authorization holder should prepare an investigation and evaluation report of the safety issue, publicly post an announcement, notify distributors and user entities (i.e. hospitals) through a letter to cease sales." At the same time," the MAH should submit a recall notice for filing and review, which includes various details about the decision and plan for the recall, to the provincial regulatory authorities either where the MAH is or where its agent is located, if the MAH is overseas. Companies may seek government buy-in for their recall plan, even though no approval is required.

The government also has the power to order a marketing-authorization holder to conduct a recall if the government determines there is an issue on its own. Failure to follow that instruction can lead to penalties. In practice, it is much more likely that the marketing-authorization holder will affirmatively conduct the recall without an order either because it determines it is necessary on its own or because the government has requested that it do so without an order.

During the recall, the marketing-authorization holder must submit periodic reports based on the progress. The intervals depend on the class: every day for Class I, every three for Class II, and every seven for Class III. Once complete, the marketing-authorization holder will submit a final report on its progress, based on which the local medical products authorities may ask questions and have follow-up items.

Even if the product in China is not directly affected, a recall conducted overseas must be reported under circumstances, such as if the products are similar or produced on the same production line.

5.24 Perspectives on Drug Inspections and Regulatory Enforcement

5.24.1 Inspections

NMPA and the local MPAs enforce their regulations through a combination of review and inspection. Through its review function of initial applications, supplements, and renewals, NMPA can detect issues with investigational and marketed drugs. Reviews of ADRs and annual reports provide regulators with an additional window into issues. In addition, as discussed elsewhere, NMPA and the local MPAs run a reporting system (*jubao*) through which individuals and entities can report issues with drugs, resulting in an investigation, including a re-evaluation, of the drug.

Inspections (including product testing) are another important tool for NMPA, and one that the agency is increasingly utilizing. Inspections can take many forms.

As discussed above, NMPA (through CFDI) can conduct clinical trial data verification inspections.

NMPA (through CFDI) or local medical products authorities inspect licensed manufacturing and distribution facilities on a routine and for-cause basis. Routine inspections can include initial good manufacturing (or supply) practice compliance qualifying inspections, pre-market verification inspections, inspections to expand the scope of a manufacturing or distribution license, or other periodic inspections based on the risk-level of the facility, including its inspection history and the inherent risk of the product made there (e.g., higher risk biological products). Routine inspections can also include border or in-market testing of drugs at user entities or distribution facilities.

For-cause inspections can be announced or conducted as surprise ("flying") inspections.[271] These can occur because of a reported safety issue, recall, or recent enforcement.

NMPA also has a system of regular or one-off self-inspections. The procedure for these inspections is to send stakeholders a notice and require that they investigate certain criteria or questions. The 2015 self-inspection for GCP compliance is discussed above, and as discussed in the device chapter, there was a 2016 self-inspection of device distributors. The new drug manufacturing rules that NMPA released in 2020 require that manufacturers conduct annual self-inspections of GMP compliance and submit them to provincial regulators.[272] These self-inspections are an important tool for NMPA because failure to submit a report or a deficient report can result in sanctions on the license (i.e., suspension of the license) or further investigation. This system allows NMPA to local issues with expending substantial resources.

Overseas inspections of the various types discussed above have become an increasingly regular feature of the enforcement landscape. NMPA has been conducting these on drugs since 2011,[273] and intensified those efforts in the years leading up to the pandemic, with a procedural rule on drug and device overseas inspections emerging in 2018.[274] These overseas inspections are a significant undertaking.

[271] National Medical Products Administration (2021b).

[272] National Medical Products Administration (2020b) Drug Manufacturing Supervision Measures, Article 38.

[273] State Food and Drug Administration (2011c) In 2011 China Conducted its First Overseas Drug Manufacturer Site Inspection.

[274] National Medical Products Administration (2018b) Administrative Provisions on Drug and Medical Device Overseas Inspections.

As the pandemic limited travel, NMPA and CFDI have implemented a virtual inspection system, whereby CFDI inspectors and others working with them from the provinces proceed to the local offices of the agent of the overseas MAH and conduct a full inspection of an overseas facility via camera. These inspections have been very challenging, as they require substantial coordination and translation and communication resources. These inspections have resulted in penalties and temporary import bans.

Regardless of the inspection, NMPA has made efforts, increasingly since the Innovation Opinion, to introduce procedural rigor. This includes requiring an inspection plan, advance document submissions, training programs to professionalize inspection teams, and use of clear credentials to ensure that inspections are legitimate and not abused to exact unsupported fines. In 2021, NMPA issued general Administrative Measures on Drug Inspections (Trial Implementation). These are general rules intended to standardize practices across various different types of inspections, including those of different entities, manufacturers, distributors and user entities, and different types, licensure inspections, routine inspections, for-cause inspections, and other inspections.[275] Even so, (as discussed in Chap. 3), inspectors and local officials still have a great deal of discretion to investigate informally without viable channels for companies to challenge unreasonable document submissions.

During inspections, inspectors have broad (seemingly unlimited) authority to copy and take documents, photos and videos, and product samples. While inspected entities have some room to politely challenge requests if they include clearly irrelevant commercially sensitive information, they risk being deemed uncooperative or obstructive if they flatly refuse to provide information, and the mechanisms to launch such a challenge are not always fair and impartial.

Once an inspection is concluded, inspectors will give an initial readout of their findings, perhaps during a final meeting, followed by a report. Findings will be divided into critical and non-critical (or core and non-core) items, with failure on core items potentially leading to suspensions of import and distribution and requirements for the inspected entity to conduct an analysis of whether a recall is required. The inspected entity will have the opportunity to respond to inspection reports and make initial and promised future changes. Failure to effectively remediate in a timely way can lead to further enforcement, including import bans. Another inspection may be required to determine if remediation is effective.

5.24.2 Penalties and Compensation

Penalties for drug regulatory violations are set forth in the "legal liability" section of the DAL and Vaccine Administration Law. The legal liability section describes

[275] National Medical Products Administration (2021b) Administrative Measures on Drug Inspections (Trial Implementation).

prohibited acts (e.g., GCP non-compliance) and potential non-compliance for the entities involved. Some prohibited acts can be violated by any person or entity, and some can only be violated by the named types of entities, such as MAHs, manufacturers, distributors, or user entities. Importantly, MAH is separated out from a manufacturer, whereas before the law mainly subjected drug manufacturers, distributors, and user entities to penalties. The prohibited acts and penalties are also now more tightly tied to the substantive provisions in the DAL than they were before, thanks to a push to make "legal liability" sections in laws more precise. Also, a greater range of entities with important roles in the lifecycle of the drug can be penalized, such as contract research organizations and clinical and non-clinical research institutions.

Another important piece is that the liability of the agent of an overseas MAH is directly tied to the acts for which an MAH itself may be held liable. In other words, an MAH agent can be held liable as if it were an MAH, consistent with the general principle enunciated elsewhere in the DAL that an MAH and its agent should be jointly liable for regulatory obligations.

The types of penalties vary between a warning and an order to correct at one end of the spectrum to significant fines tied to multiples of the value of the product at issue, seizure of product and related implements, and debarment of entities and responsible individuals at the other. If a penalty is issued, practice has shown that a fine is the most likely one. Debarment is an extreme sanction, and while the DAL references criminal liability, it does not itself contain criminal provisions. Those are under the Criminal Code, which has provisions for fake and substandard drugs.

As discussed in Chap. 3, in any penalty proceeding an important question is whether the violation will be deemed "serious." For example, the Drug Manufacturing Regulations (2020) list various circumstances in which a company will be considered to have a serious quality system violation.[276] A serious violation is problematic because it can increase the fines and chance of more severe penalties against the company and against individuals. It can disqualify the company from procurement processes, because provinces may debar companies from public procurement if they have "serious" legal violations on their record. It can reduce the company's overall credit score in China (every company has a social credit system number), which can affect its ability to apply for grants, loans, and special programs. And it is overall a reputational hit, as the penalty decision will likely be published.

As in other areas, NMPA has also pushed to hold individual personnel more accountable for FDD violations. The DAL enshrines this concept of "liability at the individual level." Legal representatives, managers, and directly responsible persons can be held responsible for serious violations. They can be subject to fines that are multiples of their income, individual debarment from participating in drug regulatory activities, and administrative detention in certain cases.

Another important principle of liability is that of penalizing the "first responsible person." Like the Tort Law and now the Civil Code, distributors can be held liable

[276] State Administration for Market Regulation (2020a) Drug Manufacturing Regulations, Article 69.

for damage caused by drugs, being left to defend a case in the first instance and later seek indemnification from the party that truly at fault, such as the MAH or the manufacturer.[277]

Finally, the Criminal Law is available for various violations, including selling counterfeit or substandard drugs. In 2022, the Supreme People's Court released an interpretation that explained the use of the provisions in the Criminal Code that were revised in 2020, defining the circumstances under which a manufacturer could be deemed to have violated the Criminal Law (see above). This interpretation included the circumstances in which offenders would be eligible for enhanced penalties for the production or sale of counterfeit or substandard drugs, including that the drug is primarily used for pregnant women, children or critically ill patients or the drug is being used to respond to a natural disaster or public health incident. In those cases, among others, the circumstances would be deemed serious, allowing for enhanced penalties.

The interpretation also applies more general provisions of the Criminal Law to drug-related activity. This includes selling drugs that have been purchased by the individual using state medical insurance funds and false drug advertisements that cause serious consequences, such as human harm.[278] The Criminal Law can be used as a deterrent by using its severe penalty provisions to make examples of unacceptable drug regulatory behavior.

References

BigMoleculeWatch (2023) Approved Biosimilars in China. https://www.bigmoleculewatch.cn/%E4%B8%AD%E5%9B%BD%E8%8E%B7%E6%89%B9%E7%9A%84%E7%94%9F%E7%89%A9%E7%B1%BB%E4%BC%BC%E8%8D%AF-approved-biosimilars-in-china/

Blankenship K (April 6, 2020) FDA's Hahn: no sign China has affected U.S. drug supply during coronavirus pandemic. FiercePharma. https://www.fiercepharma.com/manufacturing/fda-s-hahn-no-sign-china-will-turn-off-tap-for-u-s-drugs-during-coronavirus-pandemic

Brennan Z (January 20, 2020) OAI classifications for drug, device facilities decline sharply over last 5 years, FDA data show. Regulatory Focus. https://www.raps.org/news-and-articles/news-articles/2020/1/oai-classifications-for-drug-device-facilities-de

Bumpas J, Betsch E (2009) Exploratory study on active pharmaceutical ingredient manufacturing for essential drugs. World Bank, p 12. https://documents1.worldbank.org/curated/en/848191468149087035/pdf/530750WP0APIEx10Box345594B01PUBLIC1.pdf

Cao L et al (2018) Evaluation of the impact of Shandong illegal vaccine sales incident on immunizations in China. Hum Vaccin Immunother 14(7):1672–1678. https://doi.org/10.1080/21645515.2018.1473697. https://www.ncbi.nlm.nih.gov/pmc/articles/PMC6067851/

Center for Drug Evaluation (2018) Standards and Procedures for Rapid Reporting of Safety Data During Drug Clinical Trials. https://www.cde.org.cn/main/news/viewInfoCommon/f86be6d655db5c711fe660bef22c3bf1

[277] National People's Congress Standing Committee (2019a) Drug Administration Law, Article 144.

[278] Supreme People's Court, Supreme People's Procuratorate (2022b) Judicial Interpretation on Issues on Applicable Law in Criminal Drug Safety Cases.

Center for Drug Evaluation (2019a) Second List of Overseas New Drugs Urgently Needed in Clinical Practice. https://www.cde.org.cn/main/news/viewInfoCommon/82f3bf94dc2c38d1a24d851f0e44914b

Center for Drug Evaluation (2019b) Small Molecule Generic Reference Product Selection and Confirmation Procedures. https://www.cde.org.cn/main/policy/view/058ddb35127bbd69e306855e4442f664

Center for Drug Evaluation (2020a) Administrative Measures for Communication and Exchange Regarding Drug Research and Development and Technical Review. https://www.cde.org.cn/main/news/viewInfoCommon/b823ed10d547b1427a6906c6739fdf89

Center for Drug Evaluation (2020b) Guidelines for clinical trials of Adalimumab Injection Biosimilars. https://www.cde.org.cn/zdyz/domesticinfopage?zdyzIdCODE=827e569a9b7c011e598e902d84a231d8

Center for Drug Evaluation (2020c) Guidelines for clinical trials of Rituximab Injection Biosimilars. https://www.cde.org.cn/zdyz/domesticinfopage?zdyzIdCODE=1c8cbfee534239f519cf1b976041321c

Center for Drug Evaluation (2020d) Guidelines for clinical trials of Trastuzumab Biosimilars for Injection. https://www.cde.org.cn/zdyz/domesticinfopage?zdyzIdCODE=aa6f11df97765a072f0d2de8b8e1ad16

Center for Drug Evaluation (2020e) Guidelines for the Design of Clinical Trials of Liraglutide Injection Biosimilars. https://www.cde.org.cn/zdyz/domesticinfopage?zdyzIdCODE=d25fa0b76abfd60ff936f30cbbe03591

Center for Drug Evaluation (2020f) Guiding Principles on the Non-Inferiority Design of Drug Clinical Trials

Center for Drug Evaluation (2020g) Notice on the Release of Module M4 on Administrative Documents and Drug Information. https://www.cde.org.cn/main/news/viewInfoCommon/99593ac731ab51b87ad7892a84e17924

Center for Drug Evaluation (2020h) Regulations for Safety Update Reports During Research and Development (Interim Version). https://www.cde.org.cn/main/news/viewInfoCommon/afced30f3c45431f04b47a7f3faee971

Center for Drug Evaluation (2020i) Third List of Overseas New Drugs Urgently Needed in Clinical Practice. https://www.cde.org.cn/main/news/viewInfoCommon/08818b168ccc85db9a42a0f6623b5688

Center for Drug Evaluation (2021a) Guidelines for clinical trial design of Pertuzumab Injection Biosimilars. https://www.cde.org.cn/zdyz/domesticinfopage?zdyzIdCODE=26ce2fdeccefb9e0bda17d5d0ccfff03

Center for Drug Evaluation (2021b) Guidelines for clinical trials of Tocilizumab Injection Biosimilars. https://www.cde.org.cn/zdyz/domesticinfopage?zdyzIdCODE=6665a0be4f667721bb425c731580c93c

Center for Drug Evaluation (2021c) Public Comment Draft of the Principles on Drafting Risk Analysis and Management Plans. https://www.cde.org.cn/main/news/viewInfoCommon/81793076ad0771061c5637b03159d474

Center for Drug Evaluation (2021d) Technical Guidelines for Similar Evaluation and Indication Extrapolation of Biosimilars. https://www.cde.org.cn/zdyz/domesticinfopage?zdyzIdCODE=5e15021b72cb2d884afded50850ed1f0

Center for Drug Evaluation (2021e) Procedures for the Commencing a Drug Registration Verification and Inspection. https://www.cde.org.cn/main/news/viewInfoCommon/c1dd9f7df30d686a2adab91f7f34587e

Center for Drug Evaluation (2021f) Changing Work Procedures During Drug Registration Application Review (Trial Implementation) https://www.cde.org.cn/main/policy/regulatview/b4944d722d9611acda9c41921f384cbf

Center for Drug Evaluation (2022a) Draft Drug Registration Examination Guidelines for Public Comment. https://www.cde.org.cn/main/news/viewInfoCommon/c284d30b8839cea6f4af5787511ec8da

Center for Drug Evaluation (2022b) Guidelines for Clinical Trial Design of Cetuximab Injection Biosimilars (Trial). https://www.cde.org.cn/zdyz/domesticinfopage?zdyzIdCODE=9f165d098ffe1c8da6dff7f847064f18

Center for Drug Evaluation (2022c) Guidelines for Clinical Trial Design of Once-daily Basal Insulin Biosimilars. https://www.cde.org.cn/zdyz/domesticinfopage?zdyzIdCODE=aecffed3805ceed486e530474774b017

Center for Drug Evaluation (2022d) Guiding Principles on Organizing Patients to Participate in Research and Development. https://www.cde.org.cn/main/news/viewInfoCommon/41c7a683e4d0dcca28bccadc47096d2a

Center for Drug Evaluation (2022e) Standard for Suspension and Resumption for the Drug Review Process, https://www.cde.org.cn/main/policy/regulatview/bdcdad4461f89b5c2268375794a210ee

Center for Drug Evaluation (2023a) Guidelines on the Summary Analysis and Reporting of Safety Information During Drug Clinical Trials. https://www.cde.org.cn/main/news/viewInfoCommon/837db9784c3a549973c34d9ca16624f6

Center for Drug Evaluation (2023b) Notice regarding the public solicitation of ICH "E19: Selective collection of safety data in specific pre-market or post-market clinical trials" implementation suggestions and Chinese version. https://www.cde.org.cn/main/news/viewInfoCommon/f141e2b5618619b89407ee49907c93cf

Center for Drug Evaluation. Office Services. https://www.cde.org.cn/main/guide/contentpage/545cf855a50574699b46b26bcb165f32

Center for Food and Drug Inspection (2020) Annual Vaccine Marketing Authorization Holder Quality Report Compilation Requirements (Trial Implementation). https://www.cfdi.org.cn/resource/attachment/ueFile/2020/11/V2VkIE5vdiAxMSAxNjo0Mzo0MyAyMDIwMjg0MzE=.doc

Center for Food and Drug Inspection(2021) Drug Verification Work Procedures. https://www.cfdi.org.cn/resource/news/14199.html

Center for Food and Drug Inspection. Clinical Trial Institution Filing Database. https://beian.cfdi.org.cn/CTMDS/apps/pub/drugPublic.jsp

Center for Food and Drug Inspections (2021) Procedures for Drug Registration Verification and Inspection (Trial Implementation) and Five Documents. https://www.cfdi.org.cn/resource/news/14199.html

ChemLinked (July 25, 2022) China Announces Final Result of the 7th Volume-based Procurement (VBP) for Drugs. https://www.reach24h.com/en/news/industry-news/pharma-industry-news/china-announces-final-result-of-the-7th-volume-based-procurement-vbp-for-drugs.html

China CDC (2021) Childhood Immunization Schedule for National Immunization Program Vaccines — China. https://weekly.chinacdc.cn/fileCCDCW/journal/article/ccdcw/2021/52/PDF/210299.pdf

China Food and Drug Administration (2013a) Vaccine Good Clinical Practice Principles (Trial Implementation) http://www.gov.cn/gongbao/content/2014/content_2580994.htm

China Food and Drug Administration(2013b) Matters Concerning Certification for Exporting Active Pharmaceutical Ingredients to the European Union, https://www.nmpa.gov.cn/xxgk/fgwj/gzwj/gzwjyp/20130507120001529.html?type=pc&m=

China Food and Drug Administration (2014) Provisions on the Supervision of Drug Contract Manufacturing. https://www.nmpa.gov.cn/zhuanti/ypqxgg/ggzhcfg/20140814110901454.html

China Food and Drug Administration (2015a) Drug Clinical Trial Data Site Inspection Main Points. https://www.nmpa.gov.cn/xxgk/ggtg/qtggtg/20151110203701981.html.

China Food and Drug Administration (2015b) Interim Guidelines on International Multi-Center Clinical Trials. https://www.nmpa.gov.cn/yaopin/ypggtg/ypqtgg/20150130120001641.html?type=pc&m=

China Food and Drug Administration (2015c) International Multi-Center Drug Clinical Trial Guideline. https://www.nmpa.gov.cn/yaopin/ypggtg/ypqtgg/20150130120001641.html

China Food and Drug Administration (2015d) Technical Guidelines for the Research and Development and Evaluation of Biosimilars (Trial Implementation). https://www.cde.org.cn/zdyz/domesticinfopage?zdyzIdCODE=f044cdf4b7d7286aa12ffb85fc81a74c

China Food and Drug Administration (2016a) Service Guidance for Issuance of Import and Export Permits for Narcotic and Psychotropic Drugs, https://www.gov.cn/bumenfuwu/2016-11/30/content_5140257.htm

China Food and Drug Administration (2016b) Explanation of the Small Molecule Drug Registration Category Reform Plan. https://www.nmpa.gov.cn/xxgk/zhcjd/zhcjdyp/2016031615090153 5.html

China Food and Drug Administration (2016c) Notice on the Work Plan to Reform the Registration Categories for Chemical Drugs. https://www.nmpa.gov.cn/directory/web/nmpa/xxgk/ggtg/qtggtg/20160309151801706.html

China Food and Drug Administration (2017a) Administrative Measures on Radiopharmaceuticals

China Food and Drug Administration (2017b) The Decision Concerning Matters on the Adjustment of Imported Drug Registration. http://www.gov.cn/xinwen/2017-10/10/content_5230906.htm

China Food and Drug Administration (2017c) Draft Technical Guideline of Conditional Approval for Marketing of Drugs with Urgent Clinical Need. https://www.ccfdie.org/zryyxxw/yaop/jgdt/webinfo/2017/12/1515083849132335.htm

China Food and Drug Administration (2017d) Good Laboratory Practices. https://www.nmpa.gov.cn/directory/web/nmpa/xxgk/fgwj/bmgzh/20170802160401550.html

China Food and Drug Administration (2017e) Opinion on Encouraging Drug Innovation by Implementing Priority Evaluation and Approval. https://www.nmpa.gov.cn/directory/web/nmpa/xxgk/fgwj/gzwj/gzwjyp/20171228172301452.html

China Small Molecule Drug Producer Industry Association (2021) China Pharmaceutical Industry Compliance Management Practices. http://www.glo.com.cn/UpLoadFile/Files/2021/3/31/16485185341770b59-1.pdf

Choi J (2023) Chemotherapy shortages push cancer centers toward crisis. The Hill. https://thehill.com/policy/healthcare/4043094-chemotherapy-shortages-push-cancer-centers-toward-crisis/

Chongqing Municipal Healthcare Security Bureau (2022) Notice on the Implementation of the Follow-up Procurement of the Second and Fourth Batches of Drugs Upon Expiration of the National Centralized Procurement Agreement. https://ylbzj.cq.gov.cn/zwgk_535/gstz/202209/t20220919_11124020.html

Committee on Energy and Commerce (2008) The Heparin Disaster: Chinese Counterfeits and American Failures. United States House of Representatives. 110th Congress, Second Session. Serial No. 110-109. https://www.govinfo.gov/content/pkg/CHRG-110hhrg53183/html/CHRG-110hhrg53183.htm

Daxue Consulting (July 19, 2022) China's pharmaceutical industry will be the world's largest in less than 10 years. https://daxueconsulting.com/pharmaceutical-industry-china/

Deloitte (2021) Yearbook of Patient Assistance in 2021. https://www2.deloitte.com/content/dam/Deloitte/cn/Documents/life-sciences-health-care/deloitte-cn-lshc-yearbook-of-patient-assistance-2021-en-211029.pdf

Eban K (August 7, 2019) 8 Ways overseas drug manufacturers dupe the FDA. Wired. https://www.wired.com/story/8-ways-overseas-drug-manufacturers-dupe-fda/

Eban K, Salzman S (October 29, 2019) In generic drug plants in China and India, data falsification is still a problem. Stat. https://www.statnews.com/2019/10/29/data-falsification-still-problematic-china-india-generic-drug-plants/

Elsinga S (2015) China lowers barriers for foreign pharmaceuticals, medical devices. China Briefing. https://www.china-briefing.com/news/china-lowers-barriers-to-entry-for-foreign-pharmaceuticals-medical-devices/

Feng Z, Ye Y et al (2022) Changes in pharmaceutical retail market and regional inequality of community pharmacists accessibility in mainland China: a retrospective cross-sectional study. BMJ Open 12(11) https://www.ncbi.nlm.nih.gov/pmc/articles/PMC9680178/

General Customs Administration (2023) Customs Government Affairs Service Guide (000129013000) "Entry and Entry Special Items Health and Quarantine Approval" https://www.biochains.com/wp-content/uploads/2019/12/%E2%80%9C%E5%87%BA%E5%85%A5%E5%A2%83%E7%89%B9%E6%AE%8A%E7%89%A9%E5%93%81%E5%8D%AB%E7%94%9F%E6%A3%80%E7%96%AB%E5%AE%A1%E6%89%B9%E2%80%9D%E6%94%BF%E5%8A%A1%E6%9C%8D%E5%8A%A1%E4%BA%8B%E9%A1%B9%E5%8A%9E%E4%BA%8B%E6%8C%87%E5%8D%97.pdf

General Offices of National Medical Products Administration and the Science and National Defense Bureau (2021) Notice on Completing the Work on Evaluation and Approval of Radiopharmaceutical Production and Distribution Enterprises and Supervision. https://www.nmpa.gov.cn/directory/web/nmpa/xxgk/fgwj/gzwj/gzwjyp/20210825170157175.html

Guangdong Government (2011) Guangdong Inoculation Abnormal Reaction Compensation Measures (Trial Implementation). http://www.gd.gov.cn/zwgk/wjk/zcfgk/content/post_2711934.html

Guangdong Government (2022) Tentative Provisions on the Urgent Clinical Needs Drugs and Devices Imported from Hong Kong and Macao into the Guangdong Greater Bay Area Zone. http://mpa.gd.gov.cn/xwdt/tzgg/content/post_3498383.html

Guangdong Health Commission (2018) 2018 Surveillance of Suspected AEFI in Guangdong. Name Explanations. http://wsjkw.gd.gov.cn/zwgk_zwwgk_jggk/content/post_2572936.html

Guangdong Medical Products Administration (2020) Why Does a Marketing Authorization Holder Need a Manufacturing License. http://mpa.gd.gov.cn/dawenku/yaopin/content/post_2966513.html

Hainan Institute of Real World Data. https://hnrws.cn/en

Hawkins L (April 8, 2022) Top five countries running the most clinical trials. PharmaIQ. https://www.pharma-iq.com/pre-clinical-discovery-and-development/articles/top-five-countries-running-the-most-clinical-trials

International Trade Administration (2023) Healthcare. https://www.trade.gov/country-commercial-guides/china-healthcare

Jiang R, Sun Y et al (2019) Analysis of implementation and supervision system of patient assistance programs in China. Chinese Pharm J 54(19):1626–1630. http://journal11.magtechjournal.com/Jwk_zgyxzz/CN/abstract/abstract32462.shtml

Jiang Y, Huang Z, Chen L, Cui Y, Zhu H, Shi T, Tan Z (2022) How well informed of participants in clinical trials: a case study of China. Scientific Research. https://www.scirp.org/journal/paperinformation.aspx?paperid=116341

Kuo L (2019) Popular cancer drug film prompts China to speed up price cutes. The Guardian. https://www.theguardian.com/world/2018/jul/19/cancer-drug-film-prompts-china-speed-price-cuts

Lane E (May 9, 2016) China gearing up for antitrust pricing drive against drugmakers. FiercePharma. https://www.fiercepharma.com/pharma-asia/china-gearing-up-for-anti-trust-pricing-drive-against-drugmakers

Lee M, Hirschler B (2012) China's wild east drug store. Reuters. https://www.reuters.com/article/uk-china-pharmaceuticals/special-report-chinas-wild-east-drug-store-idUKBRE87R0OE20120828

Lew L (2019) China cuts prices for 70 more drugs after talks with pharmaceutical firms. South China Morning Post, https://www.hsph.harvard.edu/china-health-partnership/2019/12/11/china-cuts-prices-for-70-more-drugs-after-talks-with-pharmaceutical-firms/

Liebman B (2016) Law in the shadow of violence: can law help to improve doctor patient trust. Colum J Asian Law 30:113. https://scholarship.law.columbia.edu/faculty_scholarship/2201/

Liu A (April 30, 2018a) 9 days for Gardasil 9: China hands out landmark nod with lightning speed**.** FiercePharma. https://www.fiercepharma.com/pharma-asia/9-days-for-gardasil-9-china-hands-out-landmark-conditional-nod-lightning-speed

Liu A (2018b) Former FDA to resign over China's latest vaccine scandal. FiercePharma. https://www.fiercepharma.com/vaccines/former-china-fda-chief-to-resign-over-country-s-latest-vaccine-scandal

Liu Y (2023) Chen Wei, deputy to the National People's Congress: it is urgent to clarify the definition of rare diseases and adopt academic implementation programs. Southern Metropolis Daily. https://m.mp.oeeee.com/a/BAAFRD000020230303769743.html

Liu Y, Schmitt S (February 7, 2017) Storm surrounds clinical trial data in China. PDA Letter. https://www.pda.org/pda-letter-portal/home/full-article/storm-surrounds-clinical-trial-data-in-china

Loftus P (February 9, 2022) FDA raises concerns about China-developed drugs. The Wall Street Journal. https://www.wsj.com/articles/fda-raises-concerns-about-china-developed-drugs-11644408180

Martuscelli C (April 15, 2021) Made-in-China pills come with unwanted side effects for the EU. Politico. https://www.politico.eu/article/europe-china-medicines-drug-imports-health/

Mathias T, Maddipatla M (November 14, 2019) China's BeiGene gets FDA approval for drug to treat rare form of lymphoma. Reuters. https://www.reuters.com/article/us-beigene-fda/chinas-beigene-gets-fda-approval-for-drug-to-treat-rare-form-of-lymphoma-idUSKBN1XO2UI

McCarthy S (2023) China is launching an 'unprecedented' crackdown on corruption in its health industry as economic woes pile up. CNN, August 25, 2023, https://www.cnn.com/2023/08/24/china/china-healthcare-corruption-crackdown-intl-hnk/index.html

MedSci (2022) 50 innovative drugs approved by China's NMPA in 2022. https://m.medsci.cn/article/show_article.do?id=33e9e55836c1

Minenet (2023) Sales Scale Exceeds 260 Billion! Online Pharmacies Continue to Expand Their Territory. https://news.pharmnet.com.cn/news/2023/08/02/582867.html#:~:text=%E5%90%8C%E6%AF%94%E5%A2%9E%E9%95%BF16.7%25%20%E7%BD%91%E4%B8%8A%E8%8D%AF%E5%BA%97%E2%80%9C%E5%9C%B0%E7%9B%98%E2%80%9D%E6%8C%81%E7%BB%AD%E6%89%A9%E5%BC%A0&text=2022%E5%B9%B4%E4%B8%AD%E5%9B%BD%E7%BD%91%E4%B8%8A%E8%8D%AF%E5%BA%97,%25%EF%BC%8C%E6%81%A2%E5%A4%8D%E5%BF%AB%E9%80%9F%E5%A2%9E%E9%95%BF%E6%80%81%E5%8A%BF%E3%80%82

Ministry of Commerce (2017) Free Trade Agreement between the People's Republic of China and the Swiss Confederation. http://fta.mofcom.gov.cn/topic/enswiss.shtml

Ministry of Health (2007) Prescription Administrative Measures. http://www.gov.cn/flfg/2007-03/13/content_549406.htm

Ministry of Science and Technology (2020) Circular of the State Council on Implementing the "Notice on Regulating the Clinical Research of Medical Institutions on the Covid-19 Pneumonia Drug Treatment," the Scientific Working Group of the State Council on the Prevention and Control Mechanism of Covid-19 Pneumonia Epidemic. http://most.gov.cn/tztg/202004/t20200403_152882.htm

Ministry of Science and Technology (2023) Science and Technology Ethics Review (Draft for Comment). https://www.most.gov.cn/tztg/202304/W020230404640742617855.docx

Modern Drug Logistics Development and Supervision Research Group (2022) Zhang W, Hao T (eds) (2022) Current Drug Supervision Research

Mu X (ed) (2018) China removes tariffs on all cancer drugs. Xinhua, http://www.xinhuanet.com/english/2018-05/01/c_137149493.htm; Wang G (2024)

National Adverse Drug Reaction Monitoring Center (2013) Vaccine Manufacturer Adverse Event Reporting Guideline. https://www.cdr-adr.org.cn/drug_1/zcfg_1/zcfg_zdyz/201307/t20130726_46751.html

National Health and Family Planning Commission (2008) AEFI Identification Measures. http://www.nhc.gov.cn/fzs/s3576/201808/e4f7a5704a994a14bd28cd82f94d27e2.shtml

National Health and Family Planning Commission (2010) National Suspected AEFI Monitoring Plan. http://www.nhc.gov.cn/zwgk/wtwj/201402/5dd5633d93174a7c8e93d8af7579a613.shtml

National Health and Family Planning Commission, et al (2013) Notice on Nine Prohibitions for the Construction of the Medical and Health System. http://law.pharmnet.com.cn/laws/detail_2553.html
National Development and Reform Commission, et al (2015) Opinion on Advancing Drug Pricing Reform. https://www.ndrc.gov.cn/xxgk/zcfb/tz/201505/t20150505_963815.html
National Development and Reform Commission (2016a) Implementation Opinion on the Implementation of a "Two-Invoice" System in Public Medical Institutions. http://www.nhc.gov.cn/tigs/s2906/201701/b64ca4c3d5c64a4c860316437d6eb787.shtml
National Development and Reform Commission (2016b) Measures on the Ethical Review of Human Biological Research. http://www.nhc.gov.cn/fzs/s3576/201610/84b33b81d8e747eaaf048f68b174f829.shtml
National Development and Reform Commission, China Customs Administration (2012) Drug Importation Measures. https://www.gov.cn/gongbao/content/2012/content_2283042.htm
National Health and Family Planning Commission, China Food and Drug Administration, Ministry of Civil Affairs, General Customs Administration (2016) Administrative Measures on the Importation of Drugs for Donation. http://www.nhc.gov.cn/yzygj/s3593/201606/ca402a2ba84848208d56175180a3db2a.shtml
National Health Commission (2014) Administrative Measures on the Commencement of Clinical Research Projects in Medical Institutions. http://www.nhc.gov.cn/yzygj/s3593g/201410/9bd03858c3aa41ed8aed17467645fb68.shtml
National Health Commission et al (2018a) Notice on the Publication of the First List of Orphan Diseases. http://www.nhc.gov.cn/yzygj/s7659/201806/393a9a37f39c4b458d6e830f40a4bb99.shtml
National Health Commission (2018b) Administrative Measures on the Clinical Application Medical Technologies. http://www.nhc.gov.cn/fzs/s3577/201809/e61d0999c95d4eb7b8a6658bf6af149c.shtml
National Health Commission (2018c) Interpretation of Opinion on Reforming and Improving the Policy on Supply Guarantee and Use of Generic Drugs. http://www.nhc.gov.cn/tigs/s3574/201804/be0f0fcac42a4e2ab3a6c094f345dfb0.shtml
National Health Commission (2020a) 2019 China Health Industry Development Statistical Report. http://www.nhc.gov.cn/guihuaxxs/s10748/202006/ebfe31f24cc145b198dd730603ec4442.shtml
National Health Commission (2020b) Draft on Medical Institutions Conducting Investigator Initiated Clinical Trials. http://www.nhc.gov.cn/cms-search/xxgk/getManuscriptXxgk.htm?id=630fa2bf316d48a4856f8727450c429b
National Health Commission (2020c) Guideline on the Comprehensive Clinical Evaluation of Drugs, http://www.nhc.gov.cn/yaozs/s7656/202011/d11ddc32fae84121a0dfca36b015a31d/files/863665c55fd74f29a656990d1a4ea5ce.pdf
National Health Commission (2022) 2021 China Health Industry Development Statistical Report. http://www.nhc.gov.cn/guihuaxxs/s3586s/202207/51b55216c2154332a660157abf28b09d.shtml
National Health Commission (2023a) Interpretation of Measures on Ethical Review in Human Life Sciences and Medical Research. http://www.gov.cn/zhengce/2023-02/28/content_5743660.htm
National Health Commission (2023b) Measures on Ethical Review in Human Life Sciences and Medical Research. http://www.gov.cn/zhengce/zhengceku/2023-02/28/content_5743658.htm
National Health Commission (2023c) Notice on the Issuance of the Second Orphan Drug Catalogue, http://www.nhc.gov.cn/yzygj/s7659/202309/19941f5eb0994615b34273bc27bf360d.shtml
National Health Commission, National Medical Products Administration (2019) Management Provision on Drug Clinical Trial Institutions. http://www.gov.cn/gongbao/content/2020/content_5496785.htm

National Healthcare Security Administration (2019) Expansion of the Regional Scope of the Pilot Program for Centralized Drug Procurement and Use Organized by the State. http://www.nhsa.gov.cn/art/2019/9/30/art_53_1818.html

National Healthcare Security Administration (2020a) Interim Measures for the Administration of Use of Drugs Covered by the Basic Medical Insurance. www.nhsa.gov.cn/art/2020/7/31/art_37_3387.html

National Healthcare Security Administration et al (2020b) National Basic Medical Insurance, Work Injury Insurance and Maternity Insurance Drug Catalog. http://www.nhsa.gov.cn/art/2020/12/28/art_38_4219.html

National Healthcare Security Administration (2021) Notice on the Continuation of Work After the Expiration of the Centralized Drug Procurement Agreement, http://www.nhsa.gov.cn/art/2021/11/4/art_37_7315.html

National Healthcare Security Administration et al (2022) National Basic Medical Insurance, Work Injury Insurance and Maternity Insurance Drug Catalogue. http://www.nhsa.gov.cn/art/2023/1/18/art_104_10078.html

National Institute for Food and Drug Control (2024) Biological Batch Release Product Search. https://bio.nifdc.org.cn/pqf/search.do

National Intellectual Property Administration (2021) Drug Patent Disputes Early Resolution Mechanism Administrative Adjudication Measures. https://www.cnipa.gov.cn/art/2021/9/18/art_66_170253.html

National Intellectual Property Administration, National Healthcare Security Administration (2022) Opinions on Strengthening the Protection of Intellectual Property Rights in the Field of Centralized Pharmaceutical Procurement https://www.gov.cn/zhengce/zhengceku/2023-01/02/content_5734611.htm

National Medical Products Administration (2003) Notice on Handling Matters Related to Imported Drug Filings. https://www.nmpa.gov.cn/xxgk/fgwj/gzwj/gzwjyp/20031230010101791.html

National Medical Products Administration (2017a) Why Read Drug Instructions? https://www.nmpa.gov.cn/directory/web/nmpa/xxgk/kpzhsh/kpzhshyp/20171024095801907.html

National Medical Products Administration (2017b) Notice on Adjusting the Examination and Approval of Active Pharmaceutical Ingredients, Excipients, and Packaging Materials. https://www.nmpa.gov.cn/xxgk/ggtg/ypggtg/ypqtggtg/20171130163301730.html?type=pc&m=

National Medical Products Administration (2018a) Notice on Matters Related to Customs Clearance and Inspection of Imported Chemicals. https://www.nmpa.gov.cn/xxgk/ggtg/ypggtg/ypqtggtg/20180426144301235.html?type=pc&m=

National Medical Products Administration (2018b) Administrative Provisions on Drug and Medical Device Overseas Inspections. https://www.nmpa.gov.cn/directory/web/nmpa/xxgk/ggtg/qtggtg/20181228214201582.html

National Medical Products Administration (2018c) Draft Drug Test Data Protection Implementing Rules (Trial Implementation) for Public Comment. https://www.nmpa.gov.cn/xxgk/zhqyj/zhqyjyp/20180426171801468.html

National Medical Products Administration (2018d) First Domestic PD-1 Antibody Drug Toripalimab Injection Approved for Marketing. (https://www.nmpa.gov.cn/directory/web/nmpa/yaowen/ypjgyw/20181217161101989.html

National Medical Products Administration (2018e) First List of Overseas New Drugs Urgently Needed in Clinical Practice. https://www.cde.org.cn/main/news/viewInfoCommon/21de8acd6c395746b041b2ad93eb5c43

National Medical Products Administration (2018f) Matters Concerning the One-time Import of Reference Drugs for Biological Products Used in Clinical Trials. https://www.cde.org.cn/main/policy/view/da582086b6cad2fd3dcc3c4f967472ca

National Medical Products Administration (2018g) Notice on Matters Related to Evaluation and Approval of Urgently Clinically Needed Drugs from Overseas. https://www.nmpa.gov.cn/directory/web/nmpa/zhuanti/ypqxgg/ggzhcfg/20181030171201646.html

National Medical Products Administration (2018h) Drug Export Sales Proof Administrative Provisions. https://www.gov.cn/zhengce/zhengceku/2018-12/31/content_5434051.htm
National Medical Products Administration (2019a) The First, Biosimilar Drug Rituximab Approved for Marketing. https://www.nmpa.gov.cn/zhuanti/ypqxgg/gggzjzh/20190225171901769.html
National Medical Products Administration (2019b) Notice on Several Matters Related to the Further Optimization of Drug Bundled Review and Approval and Supervision Work. http://www.gov.cn/gongbao/content/2019/content_5442292.htm
National Medical Products Administration (2020a) Administrative Measures on Biological Product Batch Release. https://www.gov.cn/gongbao/content/2021/content_5588831.htm
National Medical Products Administration (2020b) Annual Drug Supervision and Administration Statistics Report. https://www.nmpa.gov.cn/directory/web/nmpa/images/1624869232805095741.pdf
National Medical Products Administration (2020c) Ensuring During Export Supervision During the Pandemic. https://www.nmpa.gov.cn/xxgk/fgwj/gzwj/gzwjyp/20200403175601498.html?type=pc&m=
National Medical Products Administration(2020d) Biological Product Registration Categories and Application Material Requirements. https://www.nmpa.gov.cn/directory/web/nmpa/yaopin/ypggtg/ypqtgg/20200630175301552.html
National Medical Products Administration (2020e) Breakthrough Drug Therapy Examination and Approval Procedures (Trial Version). https://www.nmpa.gov.cn/directory/web/nmpa/zhuanti/ypqxgg/ggzhcfg/20200708151701834.html
National Medical Products Administration (2020f) Chemical Drug Registration Categories and Application Material Requirements. https://www.nmpa.gov.cn/directory/web/nmpa/zhuanti/ypzhcglbf/ypzhcglbfzhcwj/20200630180301525.html
National Medical Products Administration (2020g) Clinical Trial Data Safety Monitoring Committee Guideline (Trial Implementation). https://www.nmpa.gov.cn/directory/web/nmpa/images/1602831452803080313.pdf
National Medical Products Administration (2020h) Draft Tentative Measures on the Domestic Agent of Overseas Marketing Authorization Holders. https://www.nmpa.gov.cn/xxgk/zhqyj/zhqyjyp/20200803190645124.html
National Medical Products Administration (2020i) Drug Records and Data Management Requirements (Trial Implementation). https://www.nmpa.gov.cn/yaopin/ypggtg/ypqtgg/20200701110301645.html
National Medical Products Administration (2020j) Drug Registration Testing Work Procedures and Technical Requirements Standard (Trial Implementation). https://www.ccfdie.org/cn/yjxx/yphzp/webinfo/2020/07/1592159028955609.htm
National Medical Products Administration (2020k) Guidelines on Real World Evidence Supporting Drug Development (Trial Implementation). https://www.nmpa.gov.cn/yaopin/ypggtg/ypqtgg/20200107151901190.html
National Medical Products Administration (2020l) Medical Representative Record-Filing Measures (Trial Implementation). https://www.nmpa.gov.cn/xxgk/ggtg/qtggtg/20200930163955170.html
National Medical Products Administration (2020m) Notice Regarding Several Current Matters on Drug Supervision. http://www.gov.cn/zhengce/zhengceku/2020-04/03/content_5498608.htm
National Medical Products Administration (2020n) Procedures and Requirements for Record-Filing of Drugs Produced Overseas. https://www.ccfdie.org/zryyxxw/zxdt/webinfo/2020/08/1592162762177272.htm
National Medical Products Administration (2020o) Republishing the Fee Standards for Drug Registration. https://www.nmpa.gov.cn/xxgk/ggtg/qtggtg/20200630211101986.html
National Medical Products Administration (2020p) Technical Guidelines Supporting Real World Evidence for Research and Development and Review of Pediatric Drugs. https://www.nmpa.gov.cn/xxgk/ggtg/qtggtg/20200901104448101.html

National Medical Products Administration (2020q) Announcement on Issuing the Requirements for Writing Quality Annual Reports of Vaccine Marketing Authorization Holders (Trial). https://www.cfdi.org.cn/resource/news/12783.html
National Medical Products Administration (2021a) NMPA and Hainan Province Jointly Promoted the Pilot Application of Clinical Real-World Data of Drugs and Devices. http://english.nmpa.gov.cn/2021-12/28/c_736395.htm
National Medical Products Administration (2021b) Administrative Measures on Drug Inspections (Trial Implementation). https://www.nmpa.gov.cn/xxgk/fgwj/gzwj/gzwjyp/20210528171603115.html?type=pc&m=
National Medical Products Administration (2021c) China Drug Good Pharmacovigilance Practice Rules, Articles 47 and 132
National Medical Products Administration et al (2021d) The 14th Five Year Plan for National Drug Safety and Promoting High-Quality Development. https://english.nmpa.gov.cn/2021-12/30/c_736377.htm
National Medical Products Administration (2021e) Post-Market Change Measures (Trial Implementation). http://www.gov.cn/gongbao/content/2021/content_5602034.htm
National Medical Products Administration (2021f) The Food and Drug Administration issued the "Drug Inspection Management" Notice of Measures (Trial Implementation). http://www.gov.cn/gongbao/content/2021/content_5627704.htm
National Medical Products Administration (2022a) Draft Drug Administration Law Implementing Rules for Public Comment. https://www.nmpa.gov.cn/xxgk/zhqyj/zhqyjyp/20220509222233134.html
National Medical Products Administration (2022b) NMPA Studies and Promotes PIC/S Membership Work. https://www.nmpa.gov.cn/yaowen/ypjgyw/20220629093952150.html
National Medical Products Administration (2022c) Principles on Drug Pharmacovigilance Inspections. https://www.cdr-adr.org.cn/drug_1/zcfg_1/zcfg_zdyz/202204/t20220415_49622.html
National Medical Products Administration (2022d) Provisions on Drug Annual Reports. https://www.nmpa.gov.cn/xxgk/fgwj/xzhgfxwj/20220412172455115.html
National Medical Products Administration (2022e) Provisions on Drug Recalls. http://www.gov.cn/zhengce/zhengceku/2022-10/28/content_5722317.htm
National Medical Products Administration (2022f) Vaccine Circulation and Manufacturing Provisions. https://www.nmpa.gov.cn/xxgk/fgwj/xzhgfxwj/20220708185734126.html
National Medical Products Administration (2022g) Drug Internet Sales Administrative Measures. https://www.gov.cn/gongbao/content/2022/content_5717002.htm
National Medical Products Administration (2023) Clinical Trial Use Drug Good Manufacturing Practice. https://www.nmpa.gov.cn/directory/web/nmpa/images/uL28qO6wdm0ssrU0enTw9KpzuJ+rL61srBv7ncwO255re2o6jV98fz0uK8+7jlo6kuZG9j.doc
National Medical Products Administration, National Health Commission (2022) Import Work Plan for Drugs that have an Urgent Clinical Need. http://www.gov.cn/zhengce/zhengceku/2022-06/30/content_5698580.htm
National Medical Products Administration, National Health Commission, and Ministry of Public Security (2013) Catalogues of Varieties of Narcotic Drugs and Psychotropic Drugs. https://www.nmpa.gov.cn/xxgk/fgwj/gzwj/gzwjyp/20131111120001419.html
National Medical Products Administration, National Intellectual Property Administration (2021) Drug Patent Disputes Early Resolution Mechanism Implementation Measures (Trial Implementation). https://www.nmpa.gov.cn/xxgk/fgwj/xzhgfxwj/20210703223942131.html
National People's Congress Standing Committee (1984) Drug Administration Law. https://zh.wikisource.org/wiki/%E4%B8%AD%E5%8D%8E%E4%BA%BA%E6%B0%91%E5%85%B1%E5%92%8C%E5%9B%BD%E8%8D%AF%E5%93%81%E7%AE%A1%E7%90%86%E6%B3%95_%281984%E5%B9%B4%29
National People's Congress Standing Committee (2001) Drug Administration Law. http://www.gov.cn/gongbao/content/2001/content_60707.htm

National People's Congress Standing Committee (2009) Entry and Exit Plant and Animal Quarantine Law. https://cdagri.chengdu.gov.cn/nyxx/c109534/2017-04/27/content_cee0c1dcced3419db7b895bea8f6e33d.shtml
National People's Congress Standing Committee (2018) Border Health Quarantine Law. http://www.nhc.gov.cn/fzs/s3576/201808/e8aaf43cf448488391ec62a3c8fdf374.shtml
National People's Congress Standing Committee (2019a) Drug Administration Law
National People's Congress Standing Committee (2019b) Vaccine Administration Law. http://www.gov.cn/xinwen/2019-06/30/content_5404540.htm
National People's Congress Standing Committee (2020) Patent Law
National People's Congress Standing Committee (2021) China Physicians Law. http://www.npc.gov.cn/npc/c30834/202108/d954d9fa0af7458aa862182dc50a0d63.shtml
Organization for Economic Cooperation and Development, European Intellectual Property Office (2020) Trade Counterfeit Pharmaceutical Products, p. 24, https://www.oecd-ilibrary.org/docserver/a7c7e054-en.pdf?expires=1707274441&id=id&accname=guest&checksum=33A55D6A567B5ACD628459D89DD9AC5F
Pharnexcloud (2022) Summary of Drugs Included in the Breakthrough Therapy List. https://www.pharnexcloud.com/zixun/sd_5133
Poli A (August 3, 2018) China isn't yet ready to conduct clinical trials for the pharma industry. Stat. https://www.statnews.com/2018/08/03/china-clinical-trials-infrastructure-transparency/#:~:text=Problems%20with%20protecting%20clinical%20trial,to%20conduct%20a%20clinical%20trial
Pu X (October 24, 2018) The pilot period of the drug marketing authorization holder system is planned to be extended. http://www.npc.gov.cn/zgrdw/npc/cwhhy/13jcwh/2018-10/24/content_2063503.htm
RDPAC (2022) Code of Practice. http://en.rdpac.org/index.php?r=site%2Fnews&id=162
Reuters (October 1, 2022) Moderna Refuses Request to Reveal Vaccine Technology in China. Financial Times Reports. https://www.reuters.com/business/health care-pharmaceuticals/moderna-refused-china-request-reveal-vaccine-technology-ft-2022-10-02/
Sarpong D, Tjong-A-Hung I, Botchie D (2018) Breaking new grounds: deployment of electronic data capture (EDC) for clinical trials in China. https://onlinelibrary.wiley.com/doi/10.1002/jsc.2199
Shandong People's Government (2021) Drug Safety Event Emergency Response Plan, Drug Safety Event Grading Standard. http://www.shandong.gov.cn/module/download/downfile.jsp?classid=0&filename=e421714e1938495eaf52740912b3b5e4.pdf
Sina News (January 21, 2021) In 2020 a Total of 40 Prescription Drugs Will be Converted to OTC. Although medication is convenient, it should be more cautious. https://news.sina.com.cn/c/2021-01-11/doc-iiznctkf1636289.shtml
State Administration for Market Regulation (2005) Special Approval Measures for Drugs. http://gkml.samr.gov.cn/nsjg/bgt/202106/t20210630_331796.html
State Administration for Market Regulation (2020a) Drug Manufacturing Supervision Measures
State Administration for Market Regulation (2020b) Drug Manufacturing Supervision Measures. https://gkml.samr.gov.cn/nsjg/fgs/202003/t20200330_313672.html
State Administration for Market Regulation (2020c) Drug Registration Regulation. https://gkml.samr.gov.cn/nsjg/fgs/202003/t20200330_313670.html
State Administration for Market Regulation (2020d) Tentative Administrative Measures on Advertising Review for Drugs, Medical Devices, Health Foods, and Foods Formulated for Special Medical Purposes. https://gkml.samr.gov.cn/nsjg/fgs/201912/t20191227_309564.html
State Administration for Market Regulation et al (2020e) Notice of the State Administration for Market Regulation and other departments on printing and distributing the "Work Plan for the Innovation and Development of Drug and Medical Device Supervision in the Guangdong-Hong Kong-Macao Greater Bay Area." https://www.nmpa.gov.cn/directory/web/nmpa/xxgk/fgwj/gzwj/gzwjzh/20201020145834142.html

State Council (2015) Opinion on the Reform of the Drug and Medical Device Evaluation and Approval System. http://www.gov.cn/zhengce/content/2015-08/18/content_10101.htm
State Council (2016a) Blood Product Administration Provisions
State Council (2016b) Opinion on Conducting the Quality and Therapeutic Equivalency Evaluation for Generic Drugs. http://www.gov.cn/zhengce/content/2016-03/05/content_5049364.htm
State Council (2016c) Provisions on the Administration of Narcotic and Psychotropic Drugs
State Council (2018) Guiding Opinion on the Complete Liberalizing Reform of Hainan. http://www.gov.cn/zhengce/2018-04/14/content_5282456.htm
State Council (2019a) Drug Administration Law Implementing Regulation. http://www.gov.cn/zhengce/2020-12/26/content_5573533.htm
State Council (2019b) Regulations on the Origin of Imported and Exported Goods. http://www.gov.cn/gongbao/content/2019/content_5468935.htm
State Council (2023a) Opinion on Improving the Environment for Foreign Invested Enterprises and Increasing Foreign Investment, https://www.gov.cn/zhengce/content/202308/content_6898048.htm
State Council (2023b) Patent Law Implementing Regulations, https://www.gov.cn/zhengce/content/202312/content_6921633.htm
State Council General Office (2008) Catalogue of Key Technology and Independent Intellectual Property Rights in China's Information Industry. http://www.gov.cn/ztzl/kjfzgh/content_883829.htm
State Council General Office (2016) The State Council Standing Committee Passes "Decision of the State Council on Amending the Regulations on the Administration of Vaccine Circulation and Vaccination (Draft)." (http://wjw.hubei.gov.cn/zfxxgk/fdzdgknr/qtzdgknr/hygq/202008/t20200807_2753287.shtml
State Council General Office, Central Committee General Office (2017) Opinion on the Deepening of Reform of the Examination and Approval System and Encouraging Drug and Medical Device Innovation. http://www.gov.cn/zhengce/2017-10/08/content_5230105.htm
State Drug Administration (1995) Notice on Adjusting Drug Approval and Inspection Fee Standards. https://www.nmpa.gov.cn/directory/web/nmpa/xxgk/fgwj/qita/19950406010101596.html
State Drug Administration (1999a) Administrative Measures on the Categorization of Prescription and Non-Prescription Drugs (Trial Implementation). https://gkml.samr.gov.cn/nsjg/bgt/202107/t20210703_332221.html
State Drug Administration (1999b) Notice on the Activation of Special Seal for Record-filings of Domestic Sales Agents for Imported Drugs. https://www.nmpa.gov.cn/xxgk/fgwj/gzwj/gzwjyp/19991019010101869.html?type=pc&m=
State Drug Administration (2000) Non-Prescription Drug Classification Measures (Interim). https://gkml.samr.gov.cn/nsjg/bgt/202107/t20210703_332221.html
State Drug Administration (2004) Administrative Measures on Drug Internet Information Services. https://gkml.samr.gov.cn/nsjg/bgt/202106/t20210630_331799.html
State Drug Administration (2007) Drug Registration Regulation (Trial Implementation). https://www.nmpa.gov.cn/directory/web/nmpa/xxgk/fgwj/bmgzh/20021031010101123.html
State Food and Drug Administration (2003) Drug Good Clinical Practice Rules. https://www.nmpa.gov.cn/xxgk/fgwj/bmgzh/20030806010101443.html
State Food and Drug Administration (2004) Provisions on OTC Supplemental Examination and Approval. https://www.nmpa.gov.cn/directory/web/nmpa/xxgk/fgwj/gzwj/gzwjyp/20040316010101330.html
State Food and Drug Administration (2005a) Drug Registration Regulation. https://www.nmpa.gov.cn/xxgk/fgwj/bmgzh/20050228010101137.html?type=pc&m=
State Food and Drug Administration (2005b) Filing Provisions on the Receipt of Entrusted Manufacturing for Drugs from Overseas Companies. http://www.gd.gov.cn/zwgk/wjk/zcfgk/content/post_2722551.html

State Food and Drug Administration (2006) Provisions on Drug Labels and Package Inserts. https://gkml.samr.gov.cn/nsjg/bgt/202106/t20210630_331795.html
State Food and Drug Administration (2007a) Administrative Measures on Drug Recall. http://www.gov.cn/flfg/2007-12/13/content_832615.htm
State Food and Drug Administration (2007b) Drug Registration Regulation. https://www.nmpa.gov.cn/xxgk/fgwj/bmgzh/20070710010101795.html?type=pc&m=
State Food and Drug Administration (2007c) Interpretation of Issues Concerning Administrative Provisions on Drug Directions and Labels. https://www.tid.gov.hk/english/aboutus/tradecircular/cic/asia/2007/files/ci200762a.pdf
State Food and Drug Administration (2007d) Notice on Publishing the List of National Designated Wholesale Enterprises of Narcotic Drugs and Class I Psychotropic Drugs. https://www.nmpa.gov.cn/directory/web/nmpa/xxgk/fgwj/gzwj/gzwjyp/20070108010101792.html
State Food and Drug Administration (2009) Administrative Provisions on the Registration of Drug Technology Transfer. http://www.gov.cn/gongbao/content/2010/content_1585447.htm
State Food and Drug Administration (2010a) Drug Clinical Trial Ethical Review Work Guidelines. http://llwyh.nmgfy.com/newShow.aspx?IID=393&NID=362
State Food and Drug Administration (2010b) Drug Naming Principles. https://www.ccfdie.org/zryyxxw/cfdazsjg/gjydwyh/webinfo/2017/01/1485614811879567.htm
State Food and Drug Administration (2010c) Notice on Matters Related to the Conversion of Prescription Drugs to Non-Prescription Drugs. https://www.nmpa.gov.cn/xxgk/fgwj/gzwj/gzwjyp/20100630120001753.html
State Food and Drug Administration (2011a) Administrative Measures on Drug Good Manufacturing Practice Accreditation. https://www.nmpa.gov.cn/directory/web/nmpa/xxgk/fgwj/gzwj/gzwjyp/20110802120001536.html
State Food and Drug Administration (2011b) Adverse Drug Reaction Reporting and Monitoring Measures. http://www.gov.cn/flfg/2011-05/24/content_1870110.htm
State Food and Drug Administration (2011c) In 2011 China Conducted its First Overseas Drug Manufacturer Site Inspection. https://www.ccfdie.org/zryyxxw/zxdt/webinfo/2017/01/1485585741900069.htm
State Tax Administration (2023) Guidelines for Preferential Tax Policy of Mass Entrepreneurship and Innovation. http://www.chinatax.gov.cn/n810341/n810755/c2576894/content.html
Stevenson A (February 18, 2022) These Vaccines Have Been Embraced by the World. Why Not in China? The New York Times. https://www.nytimes.com/2022/02/18/business/china-coronavirus-vaccines.html
Sun H, Gao L (eds) (2022) Four "fast lanes" including the Breakthrough Therapy Program facilitate drug R&D and marketing. People's Daily Online. http://health.people.com.cn/n1/2022/0705/c14739-32466152.html
Supreme People's Court (2021) Judicial Interpretation on Several Questions on the Law Applicable to Civil Patent Disputes Related to Drug Registration Applications. https://www.court.gov.cn/zixun-xiangqing-311771.html
Supreme People's Court, Supreme People's Procuratorate (2017) Several Questions on the Law Applicable to Handling Drug and Medical Device Application Falsification Material Criminal Cases. https://www.cthousegop.com/mccarty/dmv-deadlines-extended-due-to-covid-19/
Supreme People's Court, Supreme People's Procuratorate (2022a) Interpretation on Several Questions on the Law Applicable to the Handling of Criminal Drug Safety Injury Cases. https://www.court.gov.cn/zixun-xiangqing-348451.html
Supreme People's Court, Supreme People's Procuratorate (2022b) Judicial Interpretation on Issues on Applicable Law in Criminal Drug Safety Cases. https://www.court.gov.cn/fabu-xiangqing-348451.html
Tollefson J (January 18, 2018) China declared world's largest producer of scientific articles. Nature. https://www.nature.com/articles/d41586-018-00927-4

Unger B (May 18, 2018) An analysis of 2017 FDA warning letters on data integrity. Pharmaceutical Online. https://www.pharmaceuticalonline.com/doc/an-analysis-of-fda-warning-letters-on-data-integrity-0003

United Nations Office on Drugs and Crime (2020) COVID-19 and the Drug Supply Chain: From Production and Trafficking to Use. https://www.unodc.org/documents/data-and-analysis/covid/Covid-19-and-drug-supply-chain-Mai2020.pdf

United States Food and Drug Administration (2019) Drug Shortages: Root Causes and Potential Solutions. https://www.fda.gov/media/131130/download

Wang G (September 28, 2022) China Pharma Retail Market: Pharmacy Chains and E-commerce are Growing. BaiPharm. https://baipharm.chemlinked.com/news/china-pharma-retail-market-pharmacy-chains-and-e-commerce-are-growing

Wang G (2024) China Waives Import Tariffs for Active Ingredients of Anti-cancer and Rare Disease Drugs in 2024. ChemLinked. https://baipharm.chemlinked.com/news/china-to-waive-import-tariffs-for-active-ingredients-of-anti-cancer-and-rare-disease-drugs-in-2024#:~:text=According%20to%20the%202024%20Tariff,starting%20from%20January%201%2C%202024

Wang Y, Zhou F, Zhang D et al (2020) Evaluation of the efficacy and safety of intravenous Remdesivir in adult patients with severe COVID-19: study protocol for a phase 3 randomized, double-blind, placebo-controlled, multicentre trial. Trials 21:422. https://doi.org/10.1186/s13063-020-04352-9.

White House (2021) Building Resilient Supply Chains, Revitalizing American Manufacturing, and Fostering Broad-Based Growth: 100-Day Reviews under Executive Order 14017. https://www.whitehouse.gov/wp-content/uploads/2021/06/100-day-supply-chain-review-report.pdf

Wikipedia. List of Medical Injuries in China. https://zh.wikipedia.org/wiki/%E4%B8%AD%E5%9B%BD%E4%BC%A4%E5%8C%BB%E4%BA%8B%E4%BB%B6%E5%88%97%E8%A1%A8

World Health Organization (2017) China policies to promote local production of pharmaceutical products and protect public health. https://apps.who.int/iris/rest/bitstreams/1316049/retrieve

Xi K (March 2, 2020) The Rise of China Biosimilars. Pharma Boardroom. https://pharmaboardroom.com/articles/the-rise-of-chinese-biosimilars/

Xia C, Xuesi D et al (2022) Cancer statistics in the United States and China, 2022 profiles, trends and determinates. Chin Med J (Engl) 135(5):584–590. https://www.ncbi.nlm.nih.gov/pmc/articles/PMC8920425/

Xinhua (2023) The Procurement of 41 Kinds of Drugs Successfully Covers Multiple Treatment Categories - Interpretation of the Highlights of the Ninth Batch of National Centralized Drug Procurement. https://www.gov.cn/yaowen/liebiao/202311/content_6914047.htm

Xinhua News Agency (November 11, 2014) Decision of the Standing Committee of the National People's Congress on Authorizing the State Council to Carry Out the Pilot Program of the Drug Marketing Authorization Holder System in Some Localities and Related Issues. http://www.gov.cn/xinwen/2015-11/04/content_2960422.htm

Yao K, Yang Y (2013) Suspected abnormal reactions to vaccination and assessment of its causal relationship with vaccination. Chinese J Contemp Pediatr 2014, 16(1):1–4. http://www.zgddek.com/CN/abstract/html/2014-1-1.htm

Yeoh S (2021) Harnessing China's global opportunity. Pharma Exec. https://www.pharmexec.com/view/drug-developments-in-china-a-global-opportunity

Zhao G et al (2017) Counterfeit drugs are Rampant online and the amount involved can easily exceed millions. Xinhua, http://www.xinhuanet.com//politics/2017-03/24/c_1120687257.htm

Ziavrou K et al (2022) Trends in counterfeit drugs and pharmaceuticals before and during the pandemic. Forensic Sci Int 338:111382. https://www.ncbi.nlm.nih.gov/pmc/articles/PMC9277998/

Medical Device Regulation 6

6.1 Introduction

Like the drug industry, the story of China's medical device industry has been one of substantial growth over the last 10 years, both in size domestically and influence globally. The size of the market, in monetary terms, has more than doubled between 2015 and 2019, from 308 billion to 629 billion RMB, and by some measurements China makes up 20% of the global market.[1] Imported devices compromised two-thirds of China's market in 2019, for a value of 19.2 billion USD, after having grown 11% from the year prior.[2]

As in other areas, the expansion in many respects has gone from a number of lower risk, lower value products to higher tech spaces, such as robotics, implantable devices, and combination drug-device products. Statistics for 2022 show 16,400 Class I (low risk) device manufacturers, 13,400 Class II (moderate risk), and 2140 Class III (higher risk).[3] Like with drugs, the government has adopted policies to streamline the process to approve medical devices for market and encourage innovation. Document No. 44 and the Innovation Opinion, discussed in depth in Chap. 5, also applied to medical devices.

It is also, like other areas, a story of domestically produced devices versus imported ones. While in many respects imported devices have been well-received in China, the government has adopted policies to encourage the greater use of locally produced devices. This includes restrictive procurement policies that favor bulk orders of lower-priced goods and other policies that expressly forbid or restrict public hospitals from using imported devices in certain areas.

As with other chapters, this chapter will be a discussion of regulatory trends, not industrial or economic. China's medical device regulations have progressed to be a

[1] Deloitte (2021).

[2] International Trade Administration (2022).

[3] National Medical Products Administration (2022g).

J. C. Balzano, *China Food and Drug Law*, Law for Professionals,
https://doi.org/10.1007/978-3-031-61901-4_6

combination of general rules and a wide array of product-specific guidelines in an expanded number of areas, including digital health and robotics. The National Medical Products Administration (NMPA), as the device regulator, has also imposed increasingly rigorous quality standards for clinical development, manufacturing, and distribution. China's medical-device regulations have followed somewhat behind the growth of the drug regime, but they have developed significantly since a major legislative revision in 2014, as China considers this a key area in which to advance innovation.

6.2 Early Device Regulation

China established agency-level medical device rules in the early 1990s that contained a definition of medical device, which has gradually been expanded over time.[4] Before that time, scattered regulations covered medical equipment, without comprehensive organization. Certain products that now qualify as devices did not have product or production licenses. In 2000, China's State Council enacted a comprehensive framework regulation called the Medical Device Supervision and Administration Regulation (MDSAR).[5] However, as of the time of this writing, there has not been a law made by the National People's Congress that covers medical devices.

The MDSAR adopted a framework similar, in certain respects, to those governing devices in other countries. It divided devices into three risk-based classes, all of which required some form of pre-market approval for the production. As with drugs, there were two broad categories of imported and domestically produced devices. The approval for domestically produced devices was tied to a manufacturing site; later, this was changed to a marketing authorization tied to the product.

The MDSAR required that Class III medical devices be clinically verified, but other classes of devices came to market through submission of existing literature and testing reports. In 2004, NMPA issued regulations that outlined high-level rules covering medical device clinical trials and verification studies, including pre-requisites for trial commencement, design and personnel requirements, and basic human-subject protections, such as informed consent, ethics committee approval, and safety reporting.[6]

Under the original MDSAR system, applicants were required to register products according to a product standard and in writing that product standard, comply with broader mandatory standards. China has built up a substantial body of government-promulgated standards at the national (standard code, GB) and agency levels (standard code for medical standards, YY), which dictate many features of device

[4] State Administration of Medicine (1996).

[5] State Council of the People's Republic of China (2000).

[6] State Food and Drug Administration (2004).

design, testing, manufacturing, distribution, and use.[7] A greater focus on standard compliance emerged post-2014, as China concentrated on enforcement of its own standards on overseas applicants of imported medical devices, rather than allowing them rely on compliance with standards in the country of manufacture.

Various other features of device regulation were established in China between 2000 and 2014. China developed a classification catalogue (2002)[8] and classification rules to determine what constitutes a device and what class it falls into (I, II, or III), adverse event reporting (2008)[9] and recall rules (2011)[10] to monitor for and remediate post-market safety issues, and good manufacturing practice (GMP) rules to set basic parameters for quality systems (2011).[11]

NMPA has continually worked to determine the categories of products subject to medical device regulation and to develop specific rules for these regulated categories. For example, NMPA has established rules and guidelines for the development of software as a medical device (SaMD), including artificial intelligence applications, and a specialized body of regulation governing in vitro diagnostic reagents.[12] As the number of medical instruments and equipment subject to regulation has increased, NMPA issued a revised and expanded medical device Classification Catalogue in 2017, which reorganized the categories of medical devices, such as implantable, medical use software, and in vitro diagnostics, explained their features and intended uses, and gave examples of each type. This Catalogue continues to expand and change as practice develops.[13]

Overall, the original MDSAR served as a basis for various initial pieces of device regulation. The government began to revise the MSDAR around 2008 as the regulatory commitment to the healthcare industry grew. The culmination of 14 years of legislation and practice was the 2014 MSDAR, which was completely rewritten in many respects. Thereafter, NMPA quickly restructured the implementing rules for registration, manufacturing, distribution, and post-market surveillance, among other areas.

6.2.1 The 2014 MDSAR

In 2014, after the establishment of the China Food and Drug Administration, the amended MDSAR was released. All devices required clinical evaluation materials and clinical trials were required for class II and class III devices, unless an exemption applied. The amended MDSAR established a clinical-trial exemption catalogue

[7] National Medical Products Administration (2019f).

[8] State Drug Administration (2002).

[9] State Food and Drug Administration (2008).

[10] Ministry of Health (2011).

[11] State Food and Drug Administration (2011).

[12] See, e.g., State Food and Drug Administration (2007b).

[13] National Medical Products Administration (2022a).

(a list of devices for which no clinical trials would be required), including all Class I devices. The 2014 MDSAR implemented changes to product permissions, record-filings for Class I devices (see Chap. 3), and registrations for Class II and III Devices. It also focused more on risk-based regulation of products. Production facilities would still need separate licenses, but those could be associated with device products after registration. It also established a more comprehensive system for post-market surveillance, including good practice rule (i.e., manufacturing, clinical, supply; referred to generally as "GxP") enforcement, a system of different types of inspections, initial inspections, random routine and for-cause inspections, self-inspections for local producers, and ex-China inspections. The 2014 MDSAR down-graded the distribution permission for Class II device distributors from a license to a record-filing.

The 2014 MDSAR also increased penalties, a trend that continued into the next revision of the MDSAR, which was smaller in scale and occurred in 2017. The next full revision of the MDSAR did not occur until 2021.

6.2.2 The 2017 Revision and Implementation of the Innovation Opinion

The 2017 MDSAR contained some important changes. For example, it added some provisions covering large-scale medical devices (e.g., proton and ion radiation therapy systems), which is a separate category of devices that require special permits to install in medical institutions (i.e., hospitals). The National Health Commission (NHC) plays a role in determining what constitutes a large-scale medical device. Although the NHC had promulgated rules governing these devices earlier, in 1995 and 2004,[14] the 2017 MDSAR amendment brought these installations more clearly under the structure of medical devices and incorporated the requirement for an installation permit issued by the NHC into the MDSAR.[15] The NHC issued new rules[16] and a catalogue of covered large-scale devices in 2018.[17]

The 2017 MDSAR amendment also implemented a record-filing requirement for institutions that conduct medical-device clinical trials.[18] As discussed below, in the research and development section, all medical-device trials must be conducted at qualified medical institutions with appropriately accredited departments for the type of device at issue. The 2017 MDSAR, in accordance with emerging policies on streamlining the development process (later officially codified with the Innovation Opinion, as detailed in Chap. 5), changed the requirement for institutional "approval" to one for institutional "filing." This meant that institutions could file

[14] Ministry of Health (2004).

[15] State Council of the People's Republic of China (2017), Article 34.

[16] State Council of the People's Republic of China (2018).

[17] National Health Commission (2018).

[18] State Council of the People's Republic of China (2017), Article 19.

complete documentation and begin hosting trials but would be subject to inspections as those trials occur over time.

At the same time, new drafts of the MDSAR began to emerge, culminating in the passage of an amendment late in 2020. The amendment, the most significant since 2014, was released publicly in 2021.

6.2.3 2021 MDSAR

The 2021 amendment included the implementation of the marketing-authorization holder system (also referred to as the registrant and filer system). For class II and III devices, the marketing-authorization holder is called the registrant; for class I devices, it is called the record-filer or filer (collectively, R/F). An R/F, whether for an imported or domestic device, now has a set of obligations to perform while contracting out manufacturing and distribution to others. As in other spaces, the R/F is required to have its own capacity to supervise manufacturing and quality as well as to perform post-market obligations. However, NMPA still does not broadly recognize the use of different affiliated entities within a corporate group to fulfill regulatory obligations. While some portions of this policy existed in 2014 and in a pilot program prior to the 2021 MDSAR, this amendment implemented the change nationwide, including for imported devices.

The MAH system for medical devices is similar to that for drugs. The place of production (inside or outside of China) determines the place of the marketing-authorization holder. Thus, devices produced outside of China must have ex-China R/Fs and hold a product permission in their own name to apply for registration or, in the case of a new-to-the-world device, they must have a product permission in their name at least by the time of renewal. Overseas R/Fs must appoint domestic agents to help them fulfill their obligations as an R/F, including the establishment of a quality system that is compliant with China's GMP and standards.[19] Prior to the amendment of the 2021 MDSAR, NMPA had released drafts of rules that more fully articulated the role of the agent and its joint responsibilities.[20] Those drafts raised concerns about the extent of the liability of local agents, which to that time could be very small-scale entities with basic staffing, yet jointly liable for obligations that some would argue only an R/F could fulfill.

The 2021 MSDAR also included a new approach to submission of clinical evaluation materials. Instead of a clinical trial being the default and providing different exemptions from that requirement, the 2021 MDSAR exempts some devices from a clinical evaluation altogether. Those devices are now permitted to rely the predicate of a well-established device on the market or to prove safety and effectiveness via non-clinical means. In addition, the 2021 MDSAR notes that even

[19] State Council (2021), Article 20.

[20] National Medical Products Administration (2018b).

if a clinical evaluation is required, that requirement can be fulfilled through literature, existing or market data, or a clinical trial.[21]

The 2021 amendment also reduced a rigid and burdensome requirement for registration type testing (discussed below) of Class II and III devices at state-run laboratories prior to clinical trials. A shortage of competent laboratories able to test a broad range of devices, including novel devices that did not fit traditional paradigms for technical testing, made this system difficult for applicants. Applicants are now permitted to conduct registration type testing in-house or contract out certain parts to laboratories in China, provided that these meet China's standards, including potentially showing accreditation by the China National Accreditation and Conformity Assessment Administration (CNAS). If the applicant entrusts a laboratory to conduct the testing, that laboratory must have China Inspection Body and Laboratory Mandatory (CMA) approval for the specific type of medical device at issue. CMA approval comes from the China National Certification and Accreditation Administration. CMA laboratories are only in China, and while foreign laboratories may be eligible for CNAS accreditation, it's not clear that any foreign laboratory has obtained it.

6.3 Medical Device Definition and Classification

6.3.1 Medical Device Definition

As previously mentioned, a uniform definition of medical device emerged in the 1990s, but was vague in certain respects and arguably narrower than the current one in terms of intended uses that would qualify an item as a medical device: "instruments, equipment, devices, appliances, implants, materials and related items used to diagnose, treat, and prevent human diseases, regulate human physiological functions, or replace human organs."[22] This definition also did not expressly mention software as included.

A similar definition exists today. It defines a device by its form (instruments or equipment, software, in vitro diagnostic), mechanism of action (achieving its effects through physical and not pharmacological means), and its intended use, such as the prevention, diagnosis, or treatment or alleviation of disease or injury. The specific definition is as follows:

> instruments, equipment, utensils, in vitro diagnostic reagents and calibrators, materials, and other similar or related items that are used directly or indirectly on the human body, including required computer software; their effects are mainly obtained through physical methods, not by pharmacological, immunological or metabolic means, or although involved in these ways but only plays an auxiliary role; and their purpose is to:

[21] State Council (2021), Article 24.

[22] State Drug Administration (1991).

(1) Diagnosis, prevention, monitoring, treatment, or alleviation of diseases;
(2) Diagnosis, monitoring, treatment, mitigation, or functional compensation of injuries;
(3) Examination, replacement, adjustment, or support of physiological structures or physiological processes;
(4) life support or maintenance;
(5) Pregnancy control;
(6) Providing information for medical or diagnostic purposes by examining samples from the human body.[23]

The primary source of whether a certain type of instrument with one or more intended uses constitutes a device is the Device Classification Catalogue, a list of instruments and their intended use, description of designs, and examples.

6.3.2 Device Classification

Like in the United States, Europe, and Japan, devices are divided into risk-based classes. Different levels of regulatory controls apply depending on the class. Like in the United States, medical devices are divided into three classes I, II, and III. Class III is for devices that present the highest risk, and Class I is the lowest. The requirements for developing and registering a Class II and Class III device are similar. Class I devices have different requirements, although the applicant must complete pre-approval formalities and may not simply the market the device, as they might in the U.S.

6.3.2.1 Self-Determination of Classification

NMPA's regulations provide that a manufacturer can either default to Class III or can make its own determination as to classification. To make a self-determination of classification, the applicant may examine a few sources. First, as noted, NMPA has issued a detailed Classification Catalogue in 2017, which it continues to update based on recommendations from experts in the field. The Catalogue is the most detailed and extensive list of equipment regulated as devices and therefore likely the most helpful resource.

The prior device classification catalogue, from 2002, was high level and included a limited number of categories. However, in 2015 NMPA formed a commission of experts to engage in revisions to the classification catalogue. NMPA released a substantial revision for public comment in late September of 2016, and subsequently convened a meeting of experts in November of that year to discuss the draft. The final revised catalogue was released in September of 2017 and became effective in August 2018.[24]

The revised device Classification Catalogue came with several innovations. First, it incorporated all Class I devices in the 2014 Class I Device Catalogue and from

[23] State Council (2021), Article 103.

[24] National Medical Products Administration (2017a).

previous notices on the classifications of specific devices that NMPA had issued between 2002 and the 2017 revision. While those notices would no longer have effect (unless included in the Catalogue), from an analytical point of view they are of value in understanding the 2017 Catalogue.

Second, separate from the Catalogue, NMPA issued a set of general rules on classification, which was last revised in 2015. Those rules provide that classification is determined by the device's intended uses, characteristics, and degree of risk, including the nature, time, and extent of contact with the human body. For example, whether the device uses electricity, is implantable, and/or is sterile makes a difference in classification. The general rules also include some special sections, including one on software, which is regularly classified as Class II unless it is intended to make diagnostic determinations.

Table 6.1 comes from the classification rules and provides a shorthand for determining the possible class of a device based on the aforementioned characteristics.

In addition to the aforementioned sources, NMPA and the Center for Medical Device Evaluation (CMDE) issue guidance documents on specific types of devices that contain classification information. For example, there are guidance documents, in the form of review guidelines, on developing medical device software, one on developing glucose test strips and glucometers, and one on developing absorbable hemostats. CMDE has issued a number of these documents over the last 10 years as device regulation in China has become increasingly specific to different types of products. A recent NMPA catalogue of such guidelines shows 399 issued since 2009.[25]

These guidelines include a variety of topics, such as registration unit determination (i.e., the determination as to whether products of a similar technical structure, performance, and intended use can be part of the same registration), naming, research materials, applicable standards, manufacturing, clinical evaluation, labeling, and testing information, among other types.[26]

These guidance documents may also discuss classification directly and the expected features of a device in a way that further enables classification determinations. In other words, although the classification section of a guideline may simply call out the code from the Classification Catalogue, the product-specific guidelines describe the device and expectations for evaluation in greater detail. For example, a guideline would include information to help to define the device, such as its intended use, principles of naming, and principles of operation, and expected structure and composition of the device, including the different types of models and components/parts of the instrument. A product-specific guideline may also provide information on the expected registration unit, i.e., what parts are permitted under one common registration.

[25] National Medical Products Administration (2021o).

[26] See, e.g., National Medical Products Administration (2018a).

Table 6.1 Device classification determination table (China Food and Drug Administration 2015c)

Devices with direct bodily contact											
Non electric	Use conditions →		Temporary use			Short term use			Long term use		
	Use form ↓		Skin/ cavity (mouth)	Wounds/ organs	Blood circulation/ central nervous	Skin oral cavity	Wounds/ organs	Central nervous and circulatory system	Skin/ oral cavity	Wounds/ organs	Circulatory central nervous
	1	Liquid delivery equipment	II	II	III	II	II	III	II	III	III
	2	Changes in blood body fluids	–	–	III	–	–	III	–	–	III
	3	Medical dressings	I	II	II	I	II	II	–	III	III
	4	Invasive instruments	I	II	III	II	II	III	–	–	–
	5	Multiple-use surgical instruments	I	I	II	–	–	–	–	–	–
	6	Implantable devices	–	–	–	–	–	–	III	III	III
	7	Contraception and family planning equipment (Not including re-use of surgical instruments)	II	II	III	II	III	III	III	III	III
	8	Other non-electric devices	I	II	III	II	II	III	II	III	III
Electric devices	Use condition →		Minor injuries			Moderate injuries			Severe injuries		
	Use form ↓										
	1	Energy therapy equipment	II			II			III		
	2		II			II			III		

(continued)

Table 6.1 (continued)

Devices with direct bodily contact					
		Diagnostic monitoring equipment			
	3	Liquid delivery equipment	II	II	III
	4	Radiation equipment	II	II	III
	5	Implantable devices	III	III	III
	6	Other electrical devices	II	II	III
Non-contact devices					
Non electric	Use condition →		Basically no influence	Minor influence	Significant influence
	Use form ↓				
	1	Nursing equipment	I	II	–
	2	Cleaning and disinfection	–	II	III
	3	Other non electric devices	I	II	III
Electric	Use condition →		Basically no influence	Minor influence	Significant influence
	Use form ↓				
	1	Clinical laboratory equipment	I	II	III
	2	Standalone software	–	II	III
	3	Medical equipment disinfection and sterilization equipment	–	II	III
	4	Other active devices	I	II	III

Guidelines also describe other principles of design and development. For example, they may provide a list of medical (YY) and national (GB) standards that are relevant to development and design, expectations for product research and clinical evaluation, labeling and contraindications, and the expected segments of the manufacturing process, among other elements.[27]

Finally, NMPA's National Institute for Food and Drug Control (NIFDC see Chap. 2) releases individual decisions on the classification of certain devices, as well as separate notices amending the Classification Catalogue. The decisions may speak directly to the classification issue at hand or include distinctions (or dicta) that may be relevant. These decisions may be available as notices on the agency's website or as entries on the website for seeking pre-registration classification guidance (discussed below).

6.3.2.2 Interactive Classification

An applicant may seek a pre-determination of classification from NMPA—more specifically, the Medical Device Standards Management Center under NIFDC; see Chap. 2. NIFDC at the central level administers a procedure for imported medical devices, and the provincial-level counterparts do the same for domestically produced products. This procedure allows applications for a determination of classification, including whether the instrument or software is regulated as a medical device.[28]

As part of this process, the applicant submits a photograph and description of the device, including the technical requirements (i.e., performance indicators) and instructions for use, which would also describe its components and intended uses. NMPA typically considers these determinations binding during the registration process. High-level details of these determination decisions are also publicized, although publicly available determinations do not include detailed information. Prior decisions can provide useful guidance for other companies making determinations.

6.3.3 Medical Device Standards

Like other spaces, there are government issued standards that apply to different aspects of medical device design, development, testing, manufacturing, and distribution. Some of these standards are in the form of good practices, such as good clinical, manufacturing, distribution, and clinical use practices (GxP). We will discuss the content and enforcement of those standards in other sections on manufacturing, distribution, and inspections.

As noted above, devices also must meet national, industrial (i.e., agency-level), local, and group (i.e., industry association) standards. Unlike drugs, there is no medical device code of standards, so these standards are generally standalone

[27] See, e.g., Shanghai Medical Products Administration (2020).

[28] China Food and Drug Administration (2017e).

documents. Some are national mandatory standards (GB), some are voluntary national standards (GB/T), some are agency level medical standards (YY or YY/T), and some are standards from other areas. ("T" stands for *tuijian*, or recommended.) The YY standards are in a category of "industrial" standards, a name that can be confusing because it implies that they may be adopted by government-sponsored industry associations (other than foreign associations, there are no private industry associations in China), but "industrial standards" means they are adopted by the administrative agencies under the State Council.

There are many device standards, some of which are prescriptive, closely regulating the design or method of accomplishing a goal. While many standards are based on international models, including those of the International Standards Organization (ISO), they are not carbon copies of those standards and may differ in some ways, or not be based on the most current ISO standard. In 2021, NMPA released a catalogue of medical device standards, which contained 1791 entries.[29] There are several reasons, discussed later, why these standards present significant challenges for the device industry and enforcement.

In 2017, China revised its national Standardization Law, which is a general procedural law for the categorization, enactment, and enforcement of standards in all areas. As part of that revision, all agency-level standards were to be voluntary, and only GB standards would be mandatory. However, the MDSAR still states that when there is no national standard industry (i.e., YY) standards must be followed. Therefore, in most contexts the YY standards are still considered mandatory.[30]

Furthermore, some standards labeled "recommended" are still enforced as mandatory either because they are cited in a rule or guideline by NMPA (GMP inspection checklists are a good example); because they are cited by the applicant in its product technical requirements (the product standard submitted with registration applications) or other registration documentation, thereby committing to follow the standard; or, in some cases, because they were the only available government document for officials to assess a stakeholder's conduct.

However, NMPA has issued documents committing to reduce the number of mandatory standards to ensure that there is flexibility for applicants in the development and design of their devices. For example, the Opinion on Further Promoting the High-Quality Development of Medical Device Standardization Work, issued by NMPA and the National Standardization Commission of China in 2021, stated that the agencies would focus on "optimizing recommended standards" that play a "leading role in industry" and "encourage the formulation and revision of recommended industrial standards in emerging technology fields. . . ."[31] NMPA has also committed to the internationalization of standards in the device area, as it has in other areas. This includes deeper involvement in the International Standardization Organization and the International Medical Device Regulators

[29] National Medical Products Administration (2021m).

[30] State Council (2021), Article 7.

[31] National Medical Products Administration (2021f).

Forum (IMDRF).[32] As discussed in Chap. 4, NMPA views this as a two-way exchange, not only to better understand international practice and standards, but also to insert "Chinese wisdom" into the efforts of those international bodies.[33]

NMPA has also worked to improve the process for making medical device standards. NMPA has a procedural regulation governing the medical device standard-drafting and enactment processes, Administrative Measures on Medical Device Standards, last amended in 2017.[34] In 2021, it seemingly generally relaxed the mandatory nature of standards by providing in the rule on the device registration process that, if no existing standards apply, an applicant could provide data to support use of other standards, e.g., in product design or production.

Despite these efforts, there remain challenges with standard enactment and enforcement in China. In practice, the process of enacting standards may favor companies that produce products in China. Only those companies may participate in or have access to the relevant government or industry association committees or are favored in other ways, such as having additional weight given to their views on the content of drafts. These local producers may use their position on standard-making committees to adopt designs that favor their products.

When international standards are adopted, they are often adjusted or adapted to China practice, including during enforcement. Inspectors may not be willing to listen to other practices or adapt their ideas to other interpretations, even if those interpretations may be reasonable.

The standards that China adopts are often very prescriptive and detailed, and at times, there is not room for design innovation. Despite recent innovations in the rules, unless there is no applicable standard, applicants may not be permitted to provide data to support new designs that vary from standards.

China's system of medical device standards can be opaque and confusing, and some of the standards are outdated. It can be challenging for applicants to sift through this system prior to registration, and they may rely on consultants to write the product technical requirements and submit the application. Using inapplicable standards or misreading them can have downstream consequences. As will be discussed below, the standards used in a device registration application are those against which inspectors will later assess the device when conducting sampling of product in the distribution chain or evaluating the compliance of the manufacturing operations.

[32] Ibid.

[33] Ibid.

[34] State Administration for Market Regulation (2017).

6.4 Medical Device Registration and Record-Filing Process

Medical devices come to market after they complete either a registration process or a record-filing process (the latter for Class I only). The following sections discuss the stages and materials that make up those processes, including non-clinical/pre-clinical evaluation materials, clinical evaluation or exemption therefrom, and information about the structure composition and mechanisms of action of the device, its components, and related manufacturing processes and quality-control systems. The amount of information and granularity (e.g., individual case report forms from clinical trials) required have only increased over time.

6.4.1 Medical Device Pre-Clinical and Clinical Evaluations

A medical device registration or record-filing application must include supporting research materials, including literature, testing, and design information, as well as a testing report, in which product performance is tested against a product standard. This testing must occur before any clinical trial in China is conducted.

Whereas the 2004 clinical trial regulations divided device trials into (1) clinical trials and (2) confirmatory verification studies, the newer MDSARs and their implementing rules speak in terms of a clinical evaluation. One type of clinical evaluation is a clinical trial, but the device could be clinically evaluated in other ways too, such as through in vitro analysis of samples or using existing data.

Clinical trials are currently defined as trials conducted in institutions certified to conduct medical device clinical trials, for purposes of confirming the safety and effectiveness of a device that is subject to a registration application.[35] Thus, under the device rules, trials are specifically designed for registration purposes and other studies do not meet that definition, although, as will be discussed below, NMPA now supports the use of different types of real-world evidence and, as discussed in Chap. 5, NMPA may consider investigator-initiated studies as supportive of an application, too.

The 2014 MDSAR included multiple ways to be exempted from a clinical trial requirement, such as relying on existing devices on the market that are similar and have a strong safety record, using non-clinical evaluation materials, and using existing clinical data about the device in question.[36] However, companies still needed to present a clinical evaluation report if their application is based on these other sources. As of 2019, there were 1003 devices and 416 in vitro diagnostics exempted from clinical trials,[37] and more were added in 2021.

The 2021 MDSAR expands upon the 2014 MDSAR's exemption for clinical trials and now includes two circumstances in which a clinical evaluation is not

[35] State Administration for Market Regulation (2022a).

[36] State Council of the People's Republic of China (2014), Article 17.

[37] National Medical Products Administration (2019b, 2021u).

required: (1) the device has a clear mechanism of operation, a finalized design, a mature production process, and a long ("many years") history of clinical use without serious adverse events or changes in routine use; (2), non-clinical evaluation materials can prove that the device is safe and effective.[38] As of 2021, NMPA had released a list of 1010 types of devices that are exempt from clinical evaluation.[39]

If a clinical evaluation is required, the 2021 MDSAR provides that there are different types of materials that meet the clinical evaluation requirement. These include existing literature research materials or clinical data from sufficiently similar devices and controlled clinical trials.[40] In reality, NMPA has exempted several devices in all three classes from clinical trials and, in contrast to drugs, less medical device registration applications likely require new clinical trials in China.

NMPA has released multiple guidelines on the design of clinical evaluations and level of data required, based on available clinical evidence.[41] A clinical evaluation must be designed to answer the question of whether the device can achieve its expected performance, while minimizing foreseeable risks in comparison to expected benefits. It is defined as "the use of scientifically sound methods to analyze and evaluate clinical data to verify the safety, performance, and/or effectiveness of medical devices within their scope of use."[42] Clinical evaluation is now a concept spanning the lifecycle of the product, and data should be updated by the registrant or filer throughout the life of the device.

If existing data is sufficient, then no new studies should be required and existing data can be the clinical evaluation. For this to apply, an applicant would need to establish whether its device is the same type of medical device to the one that is the subject of the data. In guidelines, medical devices are divided into equivalent or comparable devices:

> When the scope of application, technical and/or biological characteristics of the comparator device are widely similar to the product that is the subject of the application ["candidate product"], they can be regarded as the same type of medical device. Medical devices of the same type [may be divided into] comparable devices and equivalent devices. In some cases, the scope of application of the [candidate product] is the same as that of [the comparator product], and the degree of similarity between the technical characteristics and/or biological characteristics of these medical devices of the same variety makes the safety, clinical performance and/or effectiveness of the two not significantly different. The two devices may be considered equivalent. Equivalent devices are the ideal situation for medical devices of the same type. When the comparator device is not equivalent to the [candidate] product, but the two have broad similarities in scope of application, technical characteristics and/or biological characteristics, the comparison device can be regarded as a comparable device.[43]

[38] State Council (2021), Article 24.

[39] National Medical Products Administration (2021c).

[40] State Council (2021), Article 25.

[41] National Medical Products Administration (2021t).

[42] National Medical Products Administration (2021w).

[43] Ibid.

New clinical data in the form of a new clinical trial may be required if existing data is insufficient or other circumstances considered higher risk.[44] NMPA issued guidelines on determining when a clinical trial for a medical device is necessary. There are generally three circumstances in which this would likely be the case (although the guidelines allow applicants to make a case otherwise). The first is when the device is deemed to be a high-risk Class III medical device. There is a catalogue of these devices (discussed above in this chapter). The second circumstance is when a candidate is a new type of device. A new type of device is one that has clear differences with devices on the market in China in terms of its scope of use, technical characteristics and/or biological features. The third and final circumstance in which a clinical trial may be necessary is when based on non-clinical data and existing clinical data collected it is not possible to confirm the safety and effectiveness of the medical device.[45]

There is no well-established channel for companies to vet their plans for meeting the clinical evaluation requirement prior to submission of the application. So, it may not become apparent until the application for registration is submitted to CMDE that a clinical trial is required.

The registrant/filer is generally required to be the sponsor of the clinical trial; if the registrant/filer is based outside of China, it would appoint a local agent to assist in the running of the trial. In practice, like drug companies, device companies use a range of support, including clinical research organizations and central laboratories for sample analysis to assist with running the trial at the hospitals.

Medical device clinical trials typically do not require pre-approval from the NMPA before they can begin, except for a subset of Class III devices that are considered to be higher risk. NMPA periodically releases a list of the Class III devices requiring clinical trial approval,[46] and in the 2021 MDSAR the approval was changed to a kind of notification procedure (sometimes referred to as an implicit approval), similar to that in the drug space.[47] To commence a trial, the sponsor must complete testing of the device according to its formulated product technical requirements and submit that report, the protocol, the investigator's brochure, the informed consent form, and recruitment materials, among other items to the ethics committee(s) at the site(s) that will conduct the trial. In many cases in which biospecimens, such as blood or tissue samples, are taken during the study, human genetic resource approval will often also be required (see Chap. 7).

Even though NMPA does not require approval of most trials, some applicants may try to discuss the design of clinical trial with CMDE prior commencing it and submit the protocol. But, unless approval is required, they may not be able to obtain a meeting with the agency. Some requirements on clinical trial design for different types of devices may be found in the Medical Device Good Clinical Practice rules

[44] Sun (2021).

[45] National Medical Products Administration (2021d).

[46] National Medical Products Administration (2020c).

[47] State Council (2021), Article 27.

(GCP), discussed more below, and NMPA's various guidance documents. For example, 2015 (updated in 2022) guidance contains information on how to design trials for medical device software that is either a component of another device or standalone software. Unlike prior rules and guidelines, such as those on IVDs, current rules and guidelines stop short of providing minimum requirements for the number of subjects to prove the effectiveness of certain devices. For example, prior IVD guidelines, since abolished, provided that there should be "at least" 1000 subjects for Class III IVDs and 200 for Class II.[48]

Under the current regulations, once the ethics committee has approved the trial, the sponsor (or, if it is a company outside of China, its local agent) must submit a filing to the provincial-level medical products administration where the trial takes place. Once a clinical trial is approved it must be conducted in accordance with GCP rules, which were issued in 2016 and amended in 2022. These rules set out the requirements for sponsors, institutions, and ethics committees during the trial, including requirements to protect subject rights, to ensure appropriate design and data integrity, and to handle the investigation product (including requirements for production quality and proper handling).[49]

6.4.1.1 GCP on Conducting a Trial

The Medical Device GCP (last revised in 2022) set forth the roles of the parties involved in a clinical trial, including the sponsor, the investigator, and the institution, and they contain various other requirements, including for the content of the informed consent form and special requirements on consent for children and other vulnerable groups, and the roles of the sponsor, investigator, and ethics committee, which must approve the trial at a given medical institution. The GCP also regulate the content of the clinical trial agreement and provide guidance on the design of the clinical trial and the coordination of multi-site clinical trials. The GCP regulate the preparation and content of the clinical study report, which must be submitted with the registration application, and concurrent and subsequent record-keeping obligations, including a group of core documents known as the clinical trial master file. The GCP are increasingly supplemented by guidelines, such as those issued by NMPA on data submission requirements[50] and model clinical trial protocols.[51]

6.4.1.2 Adverse Event Reporting for Clinical Trials

The Medical Device GCP also discuss the reporting of adverse events during a clinical trial. These requirements are separate from those for post-market adverse events. An adverse event is defined as an untoward medical event that occurs during a clinical trial, whether or not it is related to the investigational device. A serious adverse event is an event that occurs during the course of a trial that causes or could

[48] State Food and Drug Administration (2007a).

[49] State Administration for Market Regulation (2022a).

[50] National Medical Products Administration (2021g).

[51] National Medical Products Administration (2022i).

cause death or serious harm. This includes life-threatening disease or harm; permanent impairment of bodily structure or function; harm requiring inpatient treatment, long term hospital stay, or medical treatment or surgical intervention in order to avoid permanent impairment to bodily structure or function; or an event that jeopardizes or causes the death of a fetus or results in defects or abnormalities.

Investigators must monitor for unexpected and serious adverse events. These adverse events must be immediately reported to the sponsor, the institution, and the ethics committee. The sponsor must report serious events to its own provincial-level medical product administration (or that of its agent, if the sponsor is outside China) and to the provincial-level medical products administration and health authorities in the locality where the clinical trial institution is located.[52] When adverse events occur during a trial, investigators are also responsible to secure treatment of the subjects. Unexpected adverse events may require revisions to the informed consent forms, and events that indicate a defect in the investigational device may require recalls or other measures to protect subjects, such as putting a hold on the trial.

If the trial utilizes an approved medical device, the R/F of that device would also be under obligation to report serious, mass, and in the case of innovative medical devices, non-serious adverse events as described in the section on post-market surveillance below.

6.4.1.3 Investigational Products

The Medical Device GCP provide some general rules on the disposal or return of investigational devices following the trial. These devices are required to be labeled as for investigational use only and represented as such when being imported into China. Investigational devices must be distributed through proper channels and stored according to instructions; they are generally not permitted to be diverted into clinical use after the trial, although in practice a hospital may request to continue using this equipment. In practice, there can also be circumstances in which devices are passed between hospitals without informing the relevant manufacturers or distributors.

Once NMPA approves a medical device for marketing, customs authorities may expect that devices imported and manufactured under the license are made *after* the date of approval. This can make it problematic to store devices in China prior to approval and market them thereafter, without the devices being refused entry and/or deemed to be a form of unapproved device.

Sometimes it is necessary to use another unapproved device alongside the device under investigation or to use a specific delivery device in a drug trial. In contrast to comparator drugs used in drug trials, China has not yet developed a clear pathway to import these unapproved devices for use in the clinical trials studying other devices or drugs. This can create choices about changing the design of the trial to study the device too or to use an existing device approved in China that may not be as effective.

[52] State Administration for Market Regulation (2022a), Article 44.

6.4.1.4 In Vitro Diagnostic Clinical Studies

IVDs were not subject to the 2016 version of the GCP; instead, NMPA enacted a separate set of guidelines, Technical Guidelines on Conducting Clinical Evaluations for In Vitro Diagnostics, to cover IVD clinical trials with similar requirements. These guidelines contained the core human-subject requirements, such as ethics committee approval for the study and informed consent for the subjects. Other prior guidelines also set minimum subject requirements for IVD trials.

The 2022 Medical Device GCP now cover IVDs, and the IVD Technical Guidelines were revised in 2021 and continue to provide more specific requirements for IVD trials. These include definitions and requirements for observational and interventional studies, as well as small-scale exploratory studies for new IVDs that are not similar to others on the market.[53] The Technical Guidelines also provide guidance on general design, including sample size. Unlike prior device or IVD clinical development guidelines (see above) there has been a move away from minimum case requirements and toward using statistical methods to determine the sufficiency of the number of samples. The Technical Guidelines contain general guidance and specific illustrative examples as to the power of the trial (i.e., the number of subjects).

6.4.2 Use of Foreign Data and Real-World Evidence

As with drugs, NMPA has accepted existing clinical data and literature to support registration of medical devices, although the guidelines and rules for accepting this data were not institutionalized until recently. In 2018, with the implementation of the Innovation Opinion, NMPA released guidelines on the acceptance of overseas clinical data on medical devices.[54] These guidelines include whether the device was investigated in a trial according to good clinical practice rules that are equivalent to China's or, if not, whether such differences can be explained and compensated for in other ways. The design of the trial must also be sufficient to demonstrate safety and performance of the device to the level that China would expect.

More recently, in 2020, NMPA and CMDE have developed guidelines on real-world data to permit its use in clinical evaluations throughout the lifecycle of the medical device. Real-world evidence is defined as "various data related to a patient's health status and/or routine medical care . . .collected from a variety of sources outside of traditional clinical trials."[55]

Perhaps one of the most prominent uses of real-world evidence has been data generated by special use of unapproved devices in China's medical tourism zones. As discussed in Chap. 5, China has developed or is developing medical tourism zones in Hainan Province, Guangdong Province, and other places. These zones

[53] National Medical Products Administration (2021v).

[54] National Medical Products Administration (2018a).

[55] National Medical Products Administration (2020b).

permit use of devices that are approved abroad but not approved in China if there is a clinical need to do so. These devices are usually imported on a small scale and used on a group of patients that hospitals (the applicants for permission to use the device) determine are eligible. CMDE will permit data from this use to support registration applications at the national level. It claims to have strengthened, through rules issued in 2022, its links to the Hainan Government to guide data collection on these devices.[56] As of December 2022, data from the Hainan medical tourism zone had supported clinical evaluations for four imported devices approved by NMPA,[57] including a laser eye-therapy system.[58] A Real World Data Institute in the Hainan Zone was jointly established by multiple agencies within the local government and endorsed by NMPA.[59]

6.4.3 Other Studies on Medical Devices

As with drugs, there is only a nascent legal framework for conducting research on medical devices outside of the registrational context. Sponsors of medical device clinical trials are defined as the manufacturer of the device, i.e., the marketing-authorization holder. Conceptually this leaves it in doubt as to whether a registrant can allow healthcare professionals or other collaborators to conduct post-market research on product without risk that agencies will view the product registrant as legally responsible for the study, although there is not clear publicly available evidence of enforcement.

These types of studies include investigator-initiated studies (IIS), which have only recently been the subject of experimental regulations issued by the NHC and NMPA, which were discussed in Chap. 5. Hospitals may have also developed their own individual rules for IIS. Regulations of IIS prohibit studies on devices for unapproved uses, absent special circumstances, and their acceptance by NMPA is not guaranteed. There remains a risk that these studies could be used improperly, e.g., as a way of evading NMPA scrutiny or as a way to establish a relationship between a healthcare professional and a company. However, they remain a popular method of post-market research and are sought out by researchers at hospitals in China, who are looking for legitimate collaborations with medical device companies.

6.4.4 In Vitro Diagnostics

Most IVDs are regulated under the medical device regime, with a few exceptions, such as IVDs for blood screening or that use radionuclide tracers. But in addition to

[56] Center for Medical Device Evaluation (2022c).

[57] Sun (2022).

[58] Hainan Real World Data Research Institute (2022a).

[59] Hainan Real World Data Research Institute (2022b).

the regulations on clinical development, there is also a separate regulation for IVD registration. IVDs are defined as "in vitro diagnostic reagents that are managed as medical devices, including those used in vitro for human samples in the process of disease prediction, prevention, diagnosis, treatment monitoring, prognosis observation, and health status evaluation. Reagents, test kits, calibrators, quality control products, and other products for detection can be used alone or in combination with instruments, appliances, equipment, or systems."[60]

IVD development became an increasing priority as China established its own precision medicine initiative. Specifically, China aimed to invest billions of Yuan into research and development of in vitro diagnostics that could be used gather genetic data to tailor treatments to individuals. This initiative followed the announcement of one in the United States.[61]

Despite the investment in precision medicine, CMDE was slower to develop a regulatory framework for companion diagnostics, or IVDs used in conjunction with drugs to determine whether a patient would respond to a specific medicine or therapy. China began to approve companion diagnostics before its regulatory framework was in place.

Beginning in 2020, CMDE began to issue guidelines for clinical research on and labeling of companion diagnostics for use with oncology drugs. The labeling of the companion diagnostic must name the specific drugs or, if there is evidence to support it, the specific class of drugs with which the diagnostic may be used. The data from the registrational drug clinical trial may be used to support this determination.[62] Guidelines issued in 2022 mapped out three scenarios for development: (1) the drug and the IVD are co-developed and registered together; (2) the clinical trial assay used to detect biomarkers during the drug trial is at an early stage but could be later developed into a companion diagnostic approved for use with the drug; and (3) there is already an approved companion diagnostic that can be used with the drug during development and data from the trial can be used for approval of the drug and a supplement to change the labeling of the IVD.[63]

6.4.5 Laboratory Developed Tests

Until the 2021 MDSAR, NMPA did not recognize what is known in other places, such as the United States, as laboratory developed tests (LDT), i.e., diagnostic tests that are medical devices but are not subject to device registration or record-filing requirements because they are made and used within one laboratory. The 2000 MDSAR contained a provision that allowed for medical institutions to develop and use medical devices necessary for their clinical needs under the guidance of

---

[60] State Administration for Market Regulation (2021a), Article 3.

[61] Cyranoski (2016).

[62] Center for Medical Device Evaluation (2021).

[63] Center for Medical Device Evaluation (2022d).

physicians, but that same provision required that these devices be reviewed and approved by either the provincial government (Class II) or the national government (Class III). However, from the 2014 MDSAR onward, all IVDs—regardless of whether they are developed in a laboratory or made for commercial production—were required to go through the applicable approval process. NMPA and its provincial counterparts have sometimes enforced these rules against unregistered IVDs.

Nevertheless, the practice of using IVDs that are unregistered (or pending approval) in laboratories and hospitals has continued, with different provinces taking different positions on enforcement. Some fine companies and laboratories for the use of unregistered IVDs.[64] As recently as 2022, Shanghai took action against laboratories that were using unregistered test kits.[65]

Furthermore, only certain IVDs, listed in a catalogue issued by the National Health Commission, may be used in hospitals. In 2016, the government appeared to move toward loosening this policy by encouraging quick evaluation and clinical use on new kinds of tests, although this notice did not permit the use of IVDs that are unapproved by NMPA.[66]

Interestingly, some device classification decisions by National Institute for Food and Drug Control, leading up to the 2021 MDSAR, exempted other devices that were made and used within medical institutions (i.e., not for sale or export, such as 3D medical models) from regulation, thus adopting LDT logic in specific cases.[67] The 2021 MDSAR now includes a provision that recognizes something similar to the concept of a laboratory developed test. Specifically, medical institutions are permitted to develop and use tests internally, provided there is no similar type in China and the test is used under physician guidance.[68] As of the time of this writing, no rules have been developed to implement this provision.

6.4.6 Research Use Only IVDs

NMPA's IVD registration regulations, enacted in 2007, recognized an exception to registration for IVDs that were only for research use and not for clinical diagnosis. The IVDs imported and marketed for this purpose had to be labeled as such, i.e., "for research use only," or RUO.[69]

In 2014, however, NMPA issued a notice related to genomics IVDs, indicating that the practice of using ostensible RUO IVDs for clinical diagnosis was unacceptable.[70] (For example, in some cases, laboratories used a RUO IVD to validate the

---

[64] Biodiscover.com (2016).

[65] Mian (2022).

[66] National Health and Family Planning Commission (2016).

[67] Center for Medical Device Standardization Administration (2019).

[68] State Council (2021), Article 53.

[69] State Food and Drug Administration (2007b).

[70] China Food and Drug Administration, National Health and Family Planning Commission (2014).

results of a separate registered test.) Later that same year, with the promulgation of the 2014 MDSAR and the subsequent amendment of its implementing regulations on IVDs, NMPA deleted the RUO exception and has not since replaced it.

6.5 Medical Device Registration Process

6.5.1 Division of Labor Within NMPA

Regarding the division of labor within NMPA related to registration, clinical trial approvals, and amendments to those approvals, device regulations divide responsibility based on the class of the medical device between central and provision authorities. The provincial food and drug administrations serve as the license and filing authorities for Class II and I domestically produced devices, whereas Class III domestic and all imported devices are handled by NMPA directly. In 2017, by a Document called Order 32, NMPA gave authority over approvals for clinical trial applications for high-risk Class III devices and the approval of amendments and renewals for Class III domestically produced devices and all licensed imported devices (i.e., Classes II and III) to the Center for Medical Device Evaluation. The goal, as stated in Order 32, was to make the amendment and renewal process more streamlined.

Table 6.2 illustrates this authority.

6.5.2 Registration Requirements

The classification of a device, whether it is made inside China, the clinical evaluation strategy, the product technical requirements ("PTR") (i.e., the product specifications), and the raw materials used are factors that can influence the registration or filing pathway and the content of the application, including the standards applied to the device. For some devices that are sufficiently similar to those on the market, the process can be tedious but doable. For other devices that are newer,

Table 6.2 Agency Authority over Medical Device Registration

Device pathway & class	Registration authority	Amendment and renewal authority
Imported Classes II and III	NMPA after a recommendation by CMDE	CMDE
Imported Class I devices	Filed with NMPA	Updated regularly through the electronic system
Domestic Class III	NMPA	CMDE
Domestic Class II	Provincial device regulatory authority	Same
Domestic Class I	Filing with the municipal level device regulatory authority	Same

without predicates on the market, working with NMPA on the evaluation of the design and functionality can take time and effort.

As with drugs, NMPA has deepened its process of inquiry into the design and functionality of devices, releasing new application requirements in 2021 that request, among other things, information about the design and manufacturing process for components.[71] And, similar to other spaces, NMPA can ask detailed questions about design history, manufacturing, and quality processes.

This section outlines some of the core parts of the registration (and record-filing) processes.

6.5.2.1 Registration Unit

We have already considered classification above; another important determination is the registration unit—meaning, whether medical devices of sufficiently similar technical principles and structure, performance indicators, and intended uses can be part of the same registration. NMPA released a guideline on registration units in 2017, which sets forth general principles for deciding whether devices can be part of one or more registration units, and it provides specific examples. For example, active (i.e., electric) devices and passive consumables used with them are typically included in separate registration units. Active accessories would likely be in the same registration unit as the host electronic unit that they connect into, but if accessories are managed as separate medical devices they are in separate units, each according to the intended use. Endoscopes with different scope attachments, such as arthroscope, sinuscope, hysteroscope, and laparoscope, may be in different registration units (i.e., subject to different registrations). Hemodialysis equipment is in a different registration unit from the consumables it uses, e.g., tubing, dialyzer, and perfusion devices.[72]

Considering the example of soft contact lenses, the registration review guidelines give seven scenarios in which lenses of different types should be divided into different registration units. This helps illustrate how differences in structure and design, performance, and intended use can separate these units:

- Aside from colors, soft contact lenses with different material formulations should generally be divided into different registration units.
- Soft contact lenses with different optical designs or uses should be divided into different registration units such as monofocal, multifocal, toric.
- Soft contact lenses with different dyeing purposes and methods should be divided into different registration units, such as tinted lenses for visibility, enhanced tinted lenses, and colored lenses for medical use.
- Soft contact lenses with and without UV resistance shall be divided into different registration units.

[71] National Medical Products Administration (2021l).

[72] China Food and Drug Administration (2017a).

- Soft contact lenses with different wearing cycles should be divided into different registration units: such as daily disposable, monthly disposable, and traditional (wearable for about 1 year).
- Soft contact lenses with different wearing methods due to different designs should be divided into different registration units such as day wear and night wear.
- Soft contact lenses with different surface treatments should be divided into different registration units.[73]

6.5.2.2 Select Components of a Registration Application

Once the classification and registration unit are set and preparation of various documentation is complete, including non-clinical and clinical materials and testing, an applicant may apply for registration of the medical device/in vitro diagnostic. If the applicant is offshore, the registration agent or another party that the foreign marketing-authorization holder designates would make this submission.

The first two sections of a device registration application are the supervision information and the summary materials. The supervision information includes company information, information about the registration agent in China (if it is an imported medical device) and an authorization from the overseas company to the agent outlining its duties, and certificates from the country of origin or manufacturing attesting to approval of the product for marketing as a device (if it is not a green channel application (a type of expedited review program) or a new-to-the-world device), and compliance of the product and the facility (if an imported device). There is also a declaration of compliance/conformity with China device rules.

The summary material section includes a description of the product, its scope of use (i.e., indications) and contraindications, and its marketing history. These materials are expanded on throughout the remainder of the application.

6.5.2.3 Non-Clinical Materials

The non-clinical materials section of the application includes the product technical requirements (PTR). The PTR are the product standard and are a core document throughout the life cycle of the device. The PTR contain the performance indicators for the product and its specifications. As discussed above, the PTR replaced the product standard as the blueprint for the device in NMPA's files and need to be formulated in advance of the development program; they serve as the map for the registration testing that needs to occur before the trial. The registration testing report is submitted to the ethics committee as evidence that the device can function as intended.

The PTR are critical because their creation helps an applicant evaluate how its device fits into NMPA's vast body of design and testing standards, which should be cited or otherwise incorporated into the PTR, if applicable to a given device. The medical device good manufacturing practice rules require that the device be released

[73] Ibid.

in accordance with the PTR, and compliance with this requirement can be a focus on inspections. In addition, as will be discussed below, the PTR will be a basis for testing the device once it is on the market in the actual distribution chain to ensure that the manufactured product conforms to the design that NMPA has approved. This can be particularly difficult for foreign applicants (as the discussion of the depth and complexity of the standards system above indicates), but it is important that the overseas registrant or filer understands what is in the Chinese PTR, including the implications of each standard mentioned, to ensure that the product they are making offshore (and its updates) can meet those requirements. Over the years there have been debates about the degree to which NMPA expects the applicant to carry out the PTR in practice (versus using equivalent international standards), but practice has shown that at least some NMPA inspectors and reviewers expect the PTR to be followed closely.

NMPA issued guidelines on writing PTR in 2014 and revised them significantly in 2022. The PTR must consist of basic product information, such as the product name and model; the performance indicators, which are objectively measurable elements of the device that reflect on its safety and effectiveness (e.g., degrees of suction or no leakage); test methods to determine whether the device complies with its requirements and performance indicators; and, for in vitro diagnostics, an appendix listing the primary raw materials. The guidelines also counsel against including the results of studies or specifications of raw materials in the PTR, as these are challenging to verify for the finished product.

PTR have also been controversial because of a requirement in a 2014 guideline that PTR performance indicators for each device be made public online.[74] This requirement was never fully implemented, although NMPA can consider the performance indicators public information when responding to open government information requests. A 2021 revision of the guideline removed this requirement.[75]

In a case in 2021, an individual filed a request with NMPA to disclose the PTR of a medical device. The NMPA refused disclosure on the grounds that the PTR contain commercial secrets and can only be disclosed with the consent of the holder, i.e., the medical device registrant, which NMPA had sought and had the holder had refused. The individual plaintiff provided evidence that the PTR should be considered public, including that certain provinces had pathways to make PTR publicly available, and materials to show that the plaintiff had a valid quality concern about the product, including a recall notice.

NMPA and the company provided materials to show that the PTR involved did contain non-public and valuable commercial secrets, including the company's efforts to keep the information private by limiting internal access and requiring employees to keep the information confidential. NMPA also submitted a public interest analysis to show that the non-disclosure of the PTR would not have a

[74] China Food and Drug Administration (2014d).

[75] National Medical Products Administration (2021s).

significant impact on the public interest. The court in its opinion did not discuss what that analysis entailed in detail.

The court rejected the evidence that local governments had rules concerning public nature of the PTR, stating that regulations and normative documents should not be used as evidence. It also rejected the evidence of the quality of the product as irrelevant to the actions of NMPA. The court went on to conclude:

> In a case in which the third party [holder of the rights to the information] expressly disagrees with the disclosure in writing, and there is no possibility that non-disclosure may cause damage to the public interest, and there are no other distinguishing features to the case, the defendant will not disclose [the requested information] after explaining the reasons to the plaintiff[.] [This approach is] in compliance with the provisions of the Regulations on Open Government Information[;] statutory notification obligations have been fulfilled. This court does not support the plaintiff's claim that the government information applied for does not constitute a trade secret. To sum up, the plaintiff's claim has [an] insufficient basis, and this court does not support it.[76]

Interestingly, the court did not consider whether the prior guidelines on PTR disclosure were evidence that the government could validly consider them to be public information, nor did it break down whether *parts* of the PTR could be considered public. Further, it seemed to fully defer to NMPA's determinations about confidentiality and public interest, although it did not expressly state that it was applying any type of administrative deference.

6.5.2.4 Other Non-Clinical Materials

The other components of non-clinical materials include product risk management materials, which includes the applicant's discussion of identified risks and hazards, determination of the need to minimize those hazards and risks, and the measures to control and reduce those risk to acceptable levels. They also determine the acceptability of those risks in light of the product's benefits.

Research materials and non-clinical literature are included in this section. Research materials can cover various areas, such as physio-chemical performance, electrical safety, radiation safety (China has separate sets of standards governing radiation-emitting drugs and devices), software research (including cybersecurity risks, artificial intelligence, and compatibility), biological properties, the safety of biological materials, sterilization and sanitation, and animal studies.

The applicant must also present copies of the product label (i.e., the label affixed to the package) and instructions for use, which is the longer explanatory document that must accompany the package. China issued special rules on device labeling in 2014 and has not changed those rules throughout multiple iterations of the MDSAR.[77] For imported products, the application must include a copy of the original foreign language labeling. As with imported drugs, the product imported

[76] Li Lei versus National Medical Products Administration (2020).

[77] China Food and Drug Administration (2014a).

into China is generally supposed to match the features of its foreign predicate, although in practice there can be changes in the labeling and even the statement of the scope of use for the China product that occur as part of the registration process.

As discussed above, the application must also include a clinical evaluation report or documentation to request an exemption from the clinical evaluation requirement.

6.5.2.5 Components and Masterfiles

Components of medical devices are not themselves considered devices, but applicants must submit information about the supply, composition, and manufacturing of components during the registration process. In 2021, NMPA released guidance on the creation of a masterfile system that allow components to be registered separately from the medical device.[78] This online system allows for the component maker of a medical device to submit information about the design and quality control of the component to an online platform. If the component is used by a registrant in its application, the registrant can reference the code assigned to the active masterful instead of submitting the technical documentation related to the component with the registration application. The system was designed to allow component manufacturers to confidential trade-secrets and know-how related to their raw materials. In addition to more traditional components, the device masterfile system can be used to register software contained in the medical device (source code of this software is often something companies consider sensitive).

6.5.2.6 Manufacturing and Quality Dossier

This is a portion of the dossier that can cause stress and confusion for an applicant, precisely because the requirements are phrased in a general, ambiguous manner. This can lead to an interpretation of "more is better" to meet regulatory expectations and ensure the approval process is efficient. But the preparation and transfer of such information can generate risk of disclosure (e.g., inadvertent disclosure or theft by third parties). NMPA's requirements for this section of the application can also be refined through a process of supplemental requests by the agency or when the agency arranges meetings or communications with the applicant to ask additional questions.

The manufacturing and quality section consists of general information about the manufacturing process, including the names and addresses of manufacturers, manufacturing information of important components, and raw material information. It also covers the quality system and its structure and function, including quality procedures, management producers, raw materials, and various product-related procedures, including for design and procurement, among others.

[78] National Medical Products Administration (2021e).

6.6 Technical Evaluation Process

Including the time for clinical and non-clinical development and testing, the medical device registration process can take several years. On paper, the time from registration application to approval or rejection is as short as approximately 70 business days for Class II devices and 100 business days for Class III. Class I devices may be marketed nearly immediately after a complete set of materials is accepted through NMPA's online system, although the review of Class I documentation can continue after submission and regulators may raise questions during that review.[79]

For Class II and III devices, the official process goes through three steps prior to approval: (1) acceptance for filing, (2) technical review, and (3) approval of registration by NMPA's registration division. In the first step, NMPA's administrative filing center reviews the materials to ensure that they are complete. If the materials are not complete, it has 5 business days to make a one-time request for the applicant to supplement those materials. Once the materials are complete, the filing center has 3 business days to pass the application to CMDE for a technical review.

CMDE must organize and conduct a technical review within 60 business days for Class II devices and within 90 business days for Class III.[80] During this time, CMDE's reviewers conduct an evaluation of the full application to determine whether they should make a recommendation for approval to NMPA's medical device registration division. Once the technical review by CMDE is complete, the recommendation on the application is passed to NMPA. It has 20 days to make its decision on the application and additional 10 days to notify the applicant.[81] Taken together, this process can become quite lengthy even if all these deadlines are met, considering that all these time frames are in business days.[82]

In reality the registration process can take longer. Applications for registration can take anywhere from 9 months to 2 years; if a clinical trial is required the process could take 3 or more years. When a device is truly new to China, and there are no clear standards or practical experience, the process can get stuck and not move for several (e.g., 10) years.

Some of this delay is due to issues contemplated under the regulations. For example, the 60/90-business-day estimate does not include the time that CMDE may need to assemble an expert panel. In addition, many (if not most) applications are met with a deficiency letter requesting supplemental materials, which gives the applicant possibly a year to assemble additional materials and data and then gives CMDE an additional 60 business days of review time after receiving those additional materials. NMPA may also decide to return the application to CMDE for further review.

---

[79] National Medical Products Administration (2022b).

[80] State Administration for Market Regulation (2021a), Article 92.

[81] Ibid. Articles 94–95.

[82] Ibid. Article 99.

Table 6.3 Drug/Device Combination Product Examples

Combination products for medical device licenses	Combination products for drug approval numbers, marketing authorizations, and import licenses
Drug-coated stent	Wound paste containing antibacterial and anti-inflammatory drugs
Catheter with antibacterial coating	External application of traditional Chinese medicine
Medicated condom	
Medicated birth control ring	

Combination products may proceed through a separate preliminary process, if the applicant cannot determine which is the dominant mechanism of action. A combination product is one that has drug and device components manufactured as a single unit. Therefore, it is different from a companion diagnostic, in which the drug and the diagnostic components are approved and labeled separately but rely upon one another.

Pursuant to a notice issued in 2009 and then an amended set of procedures issued in 2021, NMPA indicated that it would review these applications as either drugs or devices, depending on their features. In other words, the product will receive either a drug marketing permission (see above) or a device license. Applicants can submit applications for drug-device combination products to the Medical Device Standards Management Center of NIFDC. The application should describe the attributes of the product, which the Center will review to determine the dominant mechanism of action. If the dominant mechanism of action is the pharmacological one, then CDE will coordinate the review of the application, seeking assistance from CMDE on device data if needed.[83]

In certain cases, prior decisions or notices will define which is the dominant mechanism of action. For example, the 2009 notice contained the division as shown in Table 6.3.

Since 2017, NMPA has regularly released the Notice on Determination of Drug-Device Combination Products.[84] These notices show the applicant, the product, the determination that it is a combination product, and according to which product it will be licensed for marketing. These catalogues can be another useful source for determining product classification.

[83] National Medical Products Administration (2021r).

[84] See, e.g., National Medical Products Administration (2019e).

6.7 Naming and Labeling

6.7.1 Medical Device Naming

As part of a medical device registration or filing application, the applicant must state the generic name of the medical device. Devices are not registered by their brand names, although a brand name may be included in the proposed label and package insert samples, which are reviewed by the agency. NMPA issued a set of naming principles for medical devices. Generally the name consists of a "core" term and three or less "characteristic terms."

"Core" words are those that modify the core words and describe medical devices with the same technical operation, structural composition or intended use, such as "hearing aids," "medical rehabilitation equipment" or "wheelchair."

Characteristic words describe the site of use, the structural features, technical characteristics or material composition. The Naming Principles state:

> The site of use refers to the place where the product acts on the human body, such as the system, organ, tissue, or cell. Structural features describe the specific structure and appearance of the product. Technical characteristics are the description or limitation of the special action principle, mechanism or particular performance of the product. Material composition describes the main material or main component(s) of the product.[85]

Characteristic words might be "behind the ear," "in the ear" or "cassette" hearing aids.[86] These names can be aided by standards, guidelines, and the Medical Device Classification Catalogue discussed herein. IVDs have separate naming principles described in the IVD registration measures. They have three parts: Part I, the tested substance, Part II, the application, and Part III, the method or principle. A tested substance would be, for example, glycosylated serum protein. An application would be an "assay test kit" and a method or principle would be "fluorescent PCR."[87] As in other spaces, naming rules also forbid certain misleading or promotional language in medical device names, such as names or people or places, registered trademarks, and/or words like "best," "unique," "precise" and "quick effect."[88]

6.7.2 Labels/Instructions for Use

Medical devices must have labels and "explanation books," which are often translated as "instructions for use" (IFU). An IFU is technical documentation produced by the medical device registrant or filer that covers the basic information on its safety and effectiveness, is used to guide its correct installation, debugging, operation, use,

[85] China Food and Drug Administration (2015a), Article 6.

[86] National Medical Products Administration (2021q).

[87] National Medical Products Administration (2021i), Article 111.

[88] China Food and Drug Administration (2015a), Article 7.

and maintenance, and has the distinguishing characteristic of being *provided to the user along with the product.*[89] Labels are "text, graphics and symbols *attached to medical devices or their packages* to identify product features and indicate safety warnings and other information." (Emphasis added to indicate the importance of each item's location.)

Copies of Chinese (and foreign) labels and IFUs must be submitted for review with applications. One approved they must be used in that form; failure to do so may incur penalties or cause recalls. Labels and IFUs may be updated concurrently with other amendments to the registration. For example, if the registrant's name changes, so too will the label and the IFU. New labels and IFUs are presented with the amendment application and used after its approval. There is also a notification procedure for amending IFUs, separate from a change to the registration. Under this procedure, a comparison table and other materials are presented to NMPA and if no objection is presented within 20 business days, then the change may be implemented.

The items on a label must include:

- The product name, model, and specification;
- The name, address, and contact information of the registrant or filer. For imported medical devices, the name, address, and contact information of the registered agent shall also be stated;
- Medical device registration certificate number or record filing certificate number;
- The name, production address, and production license number or production record-filing certificate number of the entrusted enterprise;
- Production date, service life, or expiration date;
- Power connection conditions and input;
- Labeled graphics, symbols, and other relevant content;
- Required warnings and precautions;
- Special storage and operating conditions or instructions;
- Warning signs or warning instructions in Chinese;
- For radiation medical devices, applicable warning signs or warning instructions in Chinese.

The items in an IFU must include:

- The product name, model, and specification;
- The name, address, and contact information of the registrant or filer. For imported medical devices, the name, address, and contact information of the registered agent shall also be stated;
- Medical device registration certificate number or record filing certificate number;
- Product technical requirements number;
- Product performance, main structural composition, scope of application;

[89] China Food and Drug Administration (2014c).

- Contraindications, precautions, warnings, and reminders;
- Installation and use instructions or illustrations, and in the case of medical devices used by consumers, special instructions for safe use;
- Product maintenance methods, special storage and transportation conditions and methods;
- Production date, service life, or expiration date;
- List of accessories, including instructions on the replacement cycle of accessories, accessories, consumables, and replacement methods;
- Explanation of the graphics, symbols, abbreviations, and other such content used in the device label;
- Date of preparation or revision of the instructions;
- Other required content.

IVD labeling is also governed by other labeling guidelines and documents. For example, NMPA has released guidelines on the compilation of IVD labels.[90] In 2016, NMPA released a notice that contained various points on labeling, including that certain labeling changes could be made by applicants on their own, such as if they are adjusting the wording in labeling (without more substantive changes associated with a registration amendment) to better comply with YY/T 0466.1 (2016) covering Symbols Used in Medical Device Labels and Instructions for Use—Part 1 General Requirements.[91] CMDE has emphasized that this principle is still applicable, but has not provided further specific guidelines as to when an applicant may change labeling on its own versus applying for a notification in accordance with more general labeling rules.[92]

6.8 Device License Renewals

Medical device registrations for Class II and III medical devices must be renewed every 5 years (the period under the 2000 MDSAR was four). Filers must update or refile Class I device filings when there is new information. The renewal process involves an application submitted to CMDE, and it can require new testing for products to ensure compliance and potentially a supplemental request for materials. There is a rush to complete renewal prior to the expiry of the registration in order to avoid delays in importation and/or distribution.

In the 2014 revision of the MDSAR, the State Council changed the renewal provisions to state that if the renewal application is not approved prior to the expiration of the license, then the application would be deemed to be renewed. In practice, however, the newly issued license with the new valid period is required to permit importation and distribution of products, so delays in issuance of that license

[90] China Food and Drug Administration (2014g).

[91] China Food and Drug Administration (2016a).

[92] *See* Center for Medical Device Evaluation (2023).

can still cause delays for the registrant. As with similar licenses in China, it is not clear why this renewal is necessary, given the system for amendments of licenses and other periodic reports, such as the good manufacturing practice self-inspection reports (see below).

Yearly reports from the Center for Medical Device Evaluation indicate that the majority of registration applications that it reviews every year are renewal applications for existing devices, with many fewer applications for newly registered devices. In 2021, there were 5875 renewal applications, more than twice the number of first-time registration applications (2750).[93]

NMPA and CMDE will receive new information through the requirement for amendments to the existing licenses, and if there are safety issues it can require that applicants undertake a variety of actions, including conducting new testing or completely re-evaluating the safety and effectiveness of their devices (see section on device re-evaluation below).

6.9 Amendments to Pending Applications

It should be noted that the medical device regulations only allow for amendments to approved licenses. In other words, there is no pathway under the regulations for amendments to pending applications. If information on a pending application changes, the applicant has to refile the application and start from the beginning. Sometimes it is possible to introduce new information when the registration applicant makes a supplemental submission in response to a deficiency letter during the technical review process, but the agency always has the discretion to make the applicant start again if the change is very significant.

6.9.1 Change of License Holder Name Amendment

Amendment applications to change the name of the registrant deserve special mention here. These amendments can serve to support not only when the name of the company that holds the registration actually changes, but they can also be used when there is also a change of control of the license to a new registrant company, such as after an asset acquisition.

As discussed, in any sale of the entity the license will transfer with that entity. In the case of a merger or a dissolution, a new company may have to apply for an entirely new license or have the prior one transferred to it depending on the circumstances. The medical device regulations themselves do not contemplate another circumstance in which a registration can be transferred as an asset. Indeed, the device regulations expressly forbid the sale, alternation, lease, or sharing of medical device licenses.

[93] Center for Medical Device Evaluation (2022b).

These applications typically require ministerial documentation and an explanation of the change, and for an imported medical device, the transferee must present proof of an overseas approval in its name, e.g., through a certificate of foreign government issued by US FDA showing the medical device can be legally marketed in the US.

In 2020, NMPA also created a pathway to facilitate the establishment of domestic manufacturing affiliated with the imported device registrant. The pathway allows for a domestic company applicant, which can be the subsidiary of a foreign company, to rely on an imported device dossier to apply for a domestic device registration via a simplified process, once a manufacturing site, with an equivalent quality system, to produce the same device in China using the same manufacturing process as that of the imported product is in place.[94] The design of the domestic device must also be the same as the imported device.

6.9.2 New Amendment Regime Under the 2021 MDSAR

Prior MDSARs had two types of minor amendments: registration amendments (e.g., changes of address or company names) and licensure amendments. The latter require a technical review, because arguably those changes touch upon the safety and effectiveness of the device.

Under the 2021 MDSAR, there are major, moderate, and quality changes. The major changes are material changes to aspects of the device, such as design, raw materials, production processes, and scope of use (e.g., indications) that substantially affect safety and effectiveness. Changes to the following aspects of the device are deemed to meet this standard: the name, the model, specifications, structure, scope of use, the PTR, and production site of an imported drug.

There is not a standard as to whether a certain level or degree of change to PTR content requires an amendment to the PTR; there is no *di minimis* change exception. This is challenging because a technical reading of the rules can lead to the conclusion that no product can be released until a final, approved PTR is in place.

Other changes that do not affect safety and effectiveness may be notified to the agency through a record-filing process. Finally, the implementation rules for the 2021 MDSAR also now allow for changes to the quality system that do not fit into the above categories to be changed according to a company's internal quality procedures. Those types of changes may also need to be reported in GMP self-inspection reports, which are now due every March for both overseas and domestic registrants and filers (see discussion below).[95]

In practice, the amended registration-change system is an improvement, as it allows for a record-filing process for certain changes. But it can still be challenging to keep up with changes to registrations and the determination as to what constitutes

[94] National Medical Products Administration (2020d).

[95] State Administration for Market Regulation (2021d), Article 79.

safety and effectiveness remains unclear in certain cases. Change applications can entail significant preparation, such as formal documentation related to the overseas changes, e.g., documentation to reflect the overseas government approval status of the device or the manufacturing facility involved, and possibly new testing data (completed at laboratories in China) to evaluate the performance of the device under the new PTR. This preparation can take months to compile, particularly if formalities, such as wet signatures and legalization of documents, are required.

6.9.3 Special Pathways

In addition to the standard registration pathways, there are various special pathways that can expedite an approval. These include the innovative device pathway, which has existed since 2014, and the emergency device pathway, which gained popularity during the pandemic and is under revision at the time of this writing, and a pathway for devices that meet unmet medical needs. Certain devices may also be conditionally approved based on early phase data.

A popular special pathway is the innovative medical device pathway. This pathway can reduce the review time somewhat by giving an applicant greater access to consultation with CMDE reviewers. Another point of attractiveness may be in the fact that, if admitted to the innovative device pathway, an applicant avoids the need to submit proof of foreign approval thereby reducing time needed to wait for a foreign approval with the appropriate scope for China.

As discussed above, imported devices must receive a foreign approval at some point. If the device is an innovative device (e.g., has never been approved for marketing), then an applicant will not need to submit this proof until it receives such approval, possibly at the renewal stage.

There are three broad qualifying criteria for entry into the innovative medical device pathway. First, the applicant for the device registration must hold in its name the intellectual property on the "core technology" of the device in China. This means either holding a China invention patent, a license of an invention patent or use rights (filed with the China National Intellectual Property Administration), or having an invention patent application that has been published for public comment in China. A search report showing that the invention is novel is also required. Second, the applicant must have completed the preliminary research, the research must be controlled and traceable, and have a finished product. Third, the principles of operation or the mechanism of action must be the first in China or a significant improvement over other products, as well as be leading internationally and have significant clinical value.[96]

According to published statistics, NMPA admitted 39 devices to the pathway in 2019, 53 in 2020, and 60 in 2021. Since 2014, it has ultimately admitted

[96] National Medical Products Administration (2018c).

357 innovative devices to the review pathway, both domestically produced and imported,[97] and approved 190.[98] Once approved NMPA may publish a high level description of the technical review of the device on its website.

The medical device emergency use pathway only applies in the event of a public health emergency declared by the government. If China has not approved the necessary devices or there is a short supply of necessary devices, makers of imported and domestically produced Class II and III devices may apply to NMPA (for any Class III and Class II imported) or provincial authorities (Class II domestic) for emergency approval. These authorities will organize a panel of experts to review and approve the applications. The process of approval, including laboratory testing (organized in 24 h), quality verification (organized in 3 days), and technical review (10 days) are all completed on a compressed timeline. Some of these devices are conditionally approved for a period generally not to exceed a year and must reapply for approval at that time, having met relevant conditions (e.g., a study). The reapplication can extend the device license for the full 5-year period.[99] NMPA approved over 30 IVD COVID-19 tests through this pathway during the pandemic.

In 2022, NMPA proposed rules on urgent use of unapproved medical devices during particularly severe public health emergencies. The NHC and its Center for Disease Prevention and Control can propose a list of devices for urgent use if there is no similar device in China or applicable devices are in short supply. NMPA then organizes expert panels to review data submitted by manufacturers. The result is permission to use the device, but not a registration, although the data from use can be used to apply for registration. NMPA proposed the potential exemption from liability for injury caused by design or production issues with urgent use devices if the violation by the manufacturer is unintentional.[100]

NMPA also administers a priority approval pathway to approve devices used to treat rare diseases or diseases prevalent amongst the elderly or children or those in clinical demand in certain areas,[101] and a conditional approval pathway for urgently needed devices to treat life-threatening illnesses.

6.10 Post-Market Obligations

This section covers post-market obligations, in terms of ordinary and special reporting obligations, reporting evaluations and risk assessments, and control measures to handle safety and efficacy concerns. Control measures can include all manner of recalls, including repairs and software updates, and medical device re-evaluation, which is a process in which the safety and effectiveness of the device

[97] Zhongcheng Medical Device Network (2022).

[98] National Medical Products Administration (2023a).

[99] National Medical Products Administration (2021j).

[100] National Medical Products Administration (2023b).

[101] China Food and Drug Administration (2016c).

is examined de novo. It also includes an emerging set of responsibilities for marketing-authorization holders to conduct continued studies on their medical devices. As discussed above, clinical evaluation of the device is now a life-cycle concept, continuing into the post-market environment, as is NMPA's overall approach to monitoring the safety of the medical device.

6.10.1 Adverse Event Regime

China's post-market medical device adverse event ("AE") monitoring, reporting, and evaluation regime has grown increasingly complex over the last 10 years. The latest significant revision of the regulations in this area became effective on January 1, 2019.

An adverse event refers to the various types of harmful events that cause *or could cause* harm to humans under conditions of normal use. A serious adverse event, which is the primary reported incident, must include the possibility of a serious injury that results in: (1) loss of life, (2) permanent damage to body function or structure, or (3) the need for medical measures to prevent permanent damage.[102]

In prior regulations, this definition included an additional criterion that the event involved a device that meets quality standards. This criterion has now been removed, meaning that quality issues may cause adverse events. The prior definition applicants to report quality-related issues according to manufacturing and quality rules, which was not a useful way to gather safety data considering that one-time or recurring quality issues can be the cause of serious safety concerns.

This AE regulation covers the obligations to monitor for adverse events, report them to the government (and in some cases other stakeholders), evaluate those events and the products. These evaluations may occur after an individual serious event, or on a yearly or other periodic basis, and take measures to reduce risk, referred to in the regulations as "risk-control measures." There are three subjects of these obligations: marketing-authorization holders (MAHs) and their representative agents (i.e., device manufacturers, as they were called in earlier regulations), device distributors, and user entities (i.e., hospitals and clinics that use the devices). Individuals are *encouraged* to report adverse events under this framework, but they are not required to do so.

6.10.2 Who Is Required to Report Adverse Events?

MAHs, device distributors, and user entities are required to report adverse events that rise to a certain level, including serious adverse events and some non-serious events. These reporting obligations are across all classes of medical devices. Both distributors and user entities have obligations to report these adverse events

[102] State Administration for Market Regulation (2018a), Article 4.

involving devices that they either distribute or use on individuals, respectively. Both types of entities also have an obligation to report those events to the MAH itself. MAHs (both domestic and foreign) have obligations to report adverse events related to their devices, including those that involve similar devices that they market abroad.

The practical reality in China is that user entities and distributors report the vast majority of AEs. The Chinese government releases reports on the number of medical device AEs each year and the percentage of events reported by user entities outstrips the other categories. In 2020, registrants only reported 2.09% of the overall number of reported events, compared to user entities (over 85%) and distributors (over 12%).[103]

6.10.3 What Is a Reportable Event?

The mandatory reporters above have legal obligations to report serious adverse events. A serious adverse event is generally one that causes or that someone with professional clinical judgment would consider possible to cause death or serious injury if the event were to recur.

Events that result in death must be reported within 7 calendar days of awareness or discovery of the event. Events that result or that could result in serious injury must be reported within 20 calendar days of awareness or discovery of the event.

Mandatory reporters are also required to report group or mass adverse events. Group events occur when multiple—but not necessarily serious—events happen close in time and in geographic proximity. The regulations are vague on whether a mass event in another country needs to be reported, but any ex-China serious adverse events associated with the mass event are reportable. Finally, the adverse event regulations require that a MAH report all events associated with innovative devices, whether serious or not, for the first 5-year license term. Innovative devices would be designated as such during the registration stage. This is not unlike the prior requirement that all imported drugs and all domestic drugs be under a monitoring period for the first 5-year license term or duration of the period, respectively. In addition to reporting the event (and providing updates), the registrant must conduct a risk evaluation for the event. Depending on the seriousness of the event, they may need to engage with authorities and provide materials as to the relationship of the event to the device and the risk to patients.[104]

A registrant must also report serious AEs that occur in other countries if they are related to the same device registered in China. This reporting must take place within 30 days. If the overseas event results in control measures, the registrant must report within 24 h. The 24-h timeframe presents tight timeline for reporting in the event of an overseas recall.[105]

[103] National Medical Products Administration (2021n).

[104] State Administration for Market Regulation (2018b), Articles 29–30.

[105] Ibid., Articles 27 and 53.

Periodic reporting and risk assessments are also required. For Class II and III medical devices, the registrant must submit a report within 60 days of the anniversary of approval of the registration. For Class I devices, this report is only retained for inspection. The provincial-level AE monitoring institutions and medical products administrations then compile these reports for NMPA.

6.10.4 Targeted Monitoring of Medical Devices

For some time, CFDA and now NMPA has run a targeted monitoring program (sometimes referred to as "key" or "priority" monitoring). Targeted monitoring is defined in current regulations as active monitoring of medical device adverse events and safety information during a defined phase of time to study the risk profile, features, and severity and rate of occurrence of safety signals of a particular medical device or type of device post-market. Under this program, the government generates a list of medical devices for heightened monitoring each year. The heightened monitoring refers to greater scrutiny of AEs and an annual report to local regulators on these AEs and company's subsequent evaluations during the year.

NMPA confers with the NHC to compile the annual list of targeted devices based on their risk profiles and the extent of distribution, and the provincial-level adverse event monitoring institutions provide guidance and training to the individual companies and receive and review the initial reports. The data from the program is passed on to NMPA for a higher-level review.[106]

Targeted monitoring requires that individual companies submit a monitoring plan for the period during which they will undergo heightened monitoring. This sets out how the company will regularly monitor a specific device for reportable adverse events and abnormalities (some government documents have used this term without defining it). The company is required to submit periodic reports of its monitoring and evaluation of events, its risk assessment of the results, and any proposed control measures, such as amendments to the device license that it proposes to take as a result of the monitoring.

6.10.5 Medical Device Re-Evaluation

Re-evaluation is required when either the government or a registrant or filer discovers a significant safety signal or change in scientific research that calls into question the original evaluation for marketing of the device. Re-evaluation is defined as "the process of re-evaluating the safety and effectiveness of medical devices that have been registered or filed, and marketed, and taking corresponding measures."[107] When this occurs the registrant must conduct a risk assessment to determine whether

[106] Ibid., Articles 42–47.

[107] Ibid., Article 4.

it should make changes to the device and thus its registration or remove the device from market.[108]

6.10.6 Recalls

When adverse events or other quality or safety signals indicate actual or potential issues, companies must conduct an evaluation to determine whether a control measure is required. A control measure can be any number of things, including internal quality investigations, but when a control measure involves label or software changes or other changes to devices in the field it is referred to as a recall in China.

Under the recall rules, a recall is defined as any of the following measures: warning, inspection, repair, re-labeling, modifications to instructions for use, software updates, device or part replacement, return of devices, or destruction of devices.[109] There is no category of field actions other than a recall in China, meaning there is no option to conduct a recall at the distributor level (if the devices are in the hands of consumers) or to simply withdraw stock or send hospitals letters if the risk associated with the change is determined to be low.

Rather, when a safety signal is received that may trigger a recall, the registrant or filer must conduct an investigation and evaluation to determine the risk, draft a recall notice to inform the regulator, create a recall plan (to communicate the timing and details to the regulator), and draft a recall notification (for the regulator's website) and communications to user entities and distributors (e.g. letters to instruct them on how to carry out the recall). These documents are filed with the local regulator for review. Essentially, the recall does not begin until the regulator has agreed upon the plan.

The triggers for a recall can be quite broad. A recall should be conducted either proactively or, if the registrant or filer refuses upon order of the government, when:

- There is an unreasonable risk to human safety or life;
- The product(s) do not meet applicable mandatory standards or the registered PTR;
- Non-compliances with good supply or manufacturing practice rules could (i.e., possibly) result in unreasonable risks; or
- Other circumstances in which products need to be recalled.[110]

Although a there are a set of criteria for determining the extent of the risk—for example, injuries and malfunctions (i.e., device failures that could have caused injury) and the scope of the area and characteristics of potentially affected individuals—recalls may also be required if the standard non-compliance is minor, non-safety related, or the result of an error in the PTR. In some cases, other

[108] Ibid., Article 4.

[109] China Food and Drug Administration (2017d), Article 2.

[110] Ibid.

inconsistencies with the registration (beyond the PTR) can result in a recall. The rules are not written to avoid comparatively small issues causing recalls.

Each recall is classified as 1, 2, or 3, based on risk. That classification affects the timeline according to which documentation must be submitted and may affect the timeline that the regulator may expect to see when handling the recall. The three classes are:

- Class 1: the use of the medical device may or has caused serious health hazards (report within 1 day of decision);
- Class 2: the use of the medical device may or has caused temporary or reversible health hazards (report within 3 days of decision);
- Class 3: the use of the medical device is less likely to cause harm but still needs to be recalled (report within 7 days).

The registrant or filer must monitor and report on the results of the recall according to the recall plan. In serious cases, the provincial regulator may ask for more frequent reports and subject the registrant to greater scrutiny. At the end of the recall, the registrant must submit a final report and account for the distributed devices and other information in the recall plan.

In a recent case, a court articulated two important points about medical device recalls. The case involved a sales contract for medical devices and a dispute over a sales contract for medical devices. In revoking the first instance judgment in favor of the plaintiff, the appeals court noted that first instance court had not properly recognized the determination by a local market supervision bureau that the devices had to be recalled because their English and Chinese labels were inconsistent. The first instance court had concluded that a label inconsistency did not amount to a quality problem.[111]

The appeals court made two points relevant to medical device regulation. First, "labels and instructions [for use] are directly related to the method, scope and limitations of product use. If there are defects, serious consequences will result. . ." Thus, label defect can be a quality defect. Second, the court of first instance failed to defer to the decision of the local medical products administration, which had declared the products illegal for sale. The appeals court did not question this basis, but instead recognized that if the local medical products administration had declared the products illegal, then the object of the contract too should have been declared illegal on that basis.

The case demonstrates the wide discretion given to local medical products administration in terms of quality and safety issues. Inconsistencies between English and Chinese labels are not necessarily a clearly defined ground for a recall in the medical device recall rules—the medical device labeling rules do not regulate

[111] X Medical Sales Co. and Y Medical Device Company Civil Contract Dispute in the Second Instance (2020).

foreign languages[112]—and yet, the appeals court did not question this ground. Perhaps this is because a challenge to any administrative actions surrounding the recall were not at issue in the case.

6.11 Medical Device Distribution

6.11.1 Definition of Distribution

The government in China regulates device manufacturing, distribution, and use as separate systems subject to different sets of standards. While these are three distinct areas, unfortunately the regulations do not contain very clear definitions of them. For example, it is not clear the extent to which distributors can engage in repair activities or simple labeling of devices, versus when the repair is so significant that the device must be shipped to a manufacturer to complete work. Similarly, it is not clear the extent to which distributors in the free trade zones (special zones in the ports where full importation has not yet occurred and therefore all requirements do not apply) can assist with fixes to mislabeled devices versus shipping those devices back (at enormous cost) or destroyed (substantial waste).

Distribution is vaguely defined in the medical device distribution rules. The clearest definition is in the 2017 version of the Medical Device Distribution Supervision Measures, where distribution is fined as: providing medical devices by means of purchase and sale, including procurement, acceptance, storage, sale, transportation, and after-sale services.[113] The 2021 rules remove this definition and rely on definitions of medical device wholesale, i.e., *sale* to medical device distributors, manufacturers, user entities, and other entities with a reasonable need to use medical devices, and medical device retail, i.e., the *sale* of medical devices directly to individual users.[114]

This definition means that a company must be a distributor in order to sell and book sales of medical devices. A registrant/manufacturer can also sell its own devices.[115] Although this definition refers mainly to sale, in practice provision of samples, including their physical transportation, should also require that a company has distribution credentials.

The earlier definition also refers to procurement by public hospitals, which is run at the local levels and supervised nationally by the National Healthcare Security Administration (see Chap. 2).[116]

---

[112] China Food and Drug Administration (2014c).

[113] China Food and Drug Administration (2017c), Article 62.

[114] State Administration for Market Regulation (2021b), Article 71.

[115] Ibid., Article 3.

[116] U.S. Department of State (2022).

A distribution license or filing (see below) allows an entity to apply for permission for a medical device advertisement. Registrants and manufacturers can also apply for medical device advertisement approvals.

6.11.2 Distribution License System

If an entity can be said to distribute a medical device, then it is subject to a pre-market license (Class III) or record-filing (Class II), observance of a set of good supply practices (which can also be translated as good distribution practices and are required for all devices), and inspections and enforcement for compliance with these systems. The license or filing must have the appropriate scope and the entity must have the appropriate capacity to distribute the applicable range of devices. A distributor of software, for example, may have fewer requirements (e.g., in terms of storage) than does an IVD distributor. Although a bit counterintuitive, an entity that seeks to distribute both Class III and II devices may need to have both a license and a record-filing, although the submissions and inspections may be joined for efficiency.[117]

During investigation of any complaint, authorities will typically first ask for relevant licenses. For that reason, it can be helpful to err on the side of caution and qualify an affiliate as a device distribution enterprise, even if an argument can be made that it is not engaging in those activities. Authorities have such discretion that they could easily decide that any range of activities could fall under that rubric and take the opportunity to penalize an entity accordingly.

It should be noted that a distribution license is not required for companies that serve in the role of an agent to the registrant. Some agents are small and do not have such distribution licenses or record-filings, although having the license or filing can allow the agent to engage in a broader range of activities, including assisting with importation. The distribution license is required to handle certain testing at the border.

Device distribution licenses are specific to the entity and its premises, so separate and independent entities or branches in other cities will require other licenses—although a license permits the company to distribute devices nationwide.

Device distribution licenses must be renewed every 5 years, like other product and facility licenses. Changes to the information on the license require amendments. Class II filings do not require this renewal, but new information must be submitted to ensure that the filing remains current.

Devices may also be sold online. These websites require their own permits or record-filings, including from telecom authorities, and are governed by special rules that provide rules for transactions and divide the responsibility for ensuring safety and quality and other aspects of legal compliance between the operator of the website (i.e., the platform operator) and the actual distributor/seller of the medical

[117] China Food and Drug Administration (2017g).

device.[118] A similar division of responsibility exists for cross-border e-commerce which allows certain, foreign labeled, unregistered medical devices to enter China from abroad and be sold directly to consumers (see also Chaps. 8 and 9 regarding how this pathway is used for food and cosmetics).

6.11.3 Basic Requirements for Becoming a Licensed Device Distribution Enterprise

Each local municipal government runs the process for obtaining a device distribution license according to its own rules, but the basic parameters are set at the national level. Each device distribution enterprise must have an appropriate quality control system and personnel to execute it, storage space (including any necessary cold storage), procedures for receiving and tracking sales, a training program, and the capacity to conduct after-sale services (e.g., maintenance and repair activities). In some cases, device distributors will contract with third parties to store their products and conduct after-sale repairs. The regulations permit this practice of entrusting these legal obligations to a third party, provided that third party is accredited and the device distribution enterprise itself verifies those qualifications.

6.11.4 Medical Device Good Supply Practice

Like distributors of drugs, device distributors must comply with a set of good supply practices (GSP). These are regulations that set requirements for how device distributors are run, e.g., the required systems and procedures, the rules for receiving products and vetting suppliers, and provisions on distributor obligations for post-market surveillance.[119] Device distributors are inspected for compliance with the device GSP when they obtain their device distribution license and thereafter periodically.

When inspected for GSP compliance to issue or renew a license for a Class III medical device distributor, the local authorities inspect the company in accordance with a checklist that contains certain "core items" and "standard items."[120] If all the items are present, the facility will receive a passing grade. If a core item is missing or more than 10% of the standard items are missing, then the entity will receive a failing grade and will not receive its medical device distribution license. If less than 10% of the standard items are found to be non-compliant or missing, then the company will receive a "notice for correction by a certain deadline." The company must submit a report to the local agency correcting all the items. If the company can't complete

[118] National Medical Products Administration (2021k).

[119] China Food and Drug Administration (2014b).

[120] China Food and Drug Administration (2015e).

corrective actions by the deadline or those corrective actions are unacceptable to the local authority, the company will not receive a device distribution license.

In the case of a Class II device, the inspection will occur after the filing is accepted. If all the applicable items are presented during this type of inspection (or a post-market supervisory inspection), then the company will receive a passing score. If some of the items are missing, then the company will receive a notice for correction within a certain deadline.

As called for in the 2021 MDSAR, NMPA has exempted the distributors of certain Class II devices from the distribution record filing requirement based on lower risk of those devices as determined by NMPA, but distributors must still comply with distribution requirements. In 2021, NMPA issued a catalogue of devices that would qualify a distributor for an exemption. The catalogue includes 13 devices, including certain electronic blood pressure monitors, certain birth control devices, and electronic and manual wheelchairs. Although there was a comment period, the final catalogue did not differ significantly from the draft, and it is unclear how these devices were selected for inclusion.[121]

In 2023 NMPA proposed and then finalized a revised set of Medical Device Good Supply Practices that will take effect in July 2024.[122] The revised rule, among other things, increases requirements for device quality systems and the obligation to consistently improve them to ensure that stringents quality standards are applied. It also adds provisions encouraging the use of informational technology to transmit and store records and recognizes the validity of electronic registration documents (distributors may not sell unregistered devices and must check for valid registrations). Also incorporated is the concept to entrustment in different contexts, e.g., the entrustment of distribution enterprises by the registrant itself and the obligations of the registrant to sign a distribution agreement and ensure that those distributors have appropriate credentials and are property trained.[123] The revised rule also incorporates the concept of entrustment of device transportation, and in 2022 NMPA released a specialized appendix to the Good Supply Practice rule covering specialized provisions for medical device storage and transportation.[124] The effect of these new rules remains to be seen.

6.11.5 Traceability and Emerging Unique Device Identifier (UDI) System

China's medical device regulations contain several references to medical device traceability. In general, pursuant to their respective GxPs, manufacturers and distributors are required to keep records of sales and records of repairs in order to

[121] National Medical Products Administration (2021a).

[122] National Medical Products Administration (2023c).

[123] National Medical Products Administration (2023d).

[124] National Medical Products Administration (2022h).

effectively implement recalls when necessary. In 2018, the government also developed a proposal for a unique device identifier (UDI) system and issued a short set of rules in 2019 to define the procedures for coding each medical device.

The UDI code is divided into one part that provides information on production (batch number, production date, expiration date) and another part that provides information on the product (model, specification). The registrant and filer must apply to local coding institutions for procedures for assigning UDIs to the device at the time of production. Different devices may have different data carriers: one- or two-dimensional barcodes or radio-frequency carriers. Once the UDI is assigned, the information must be uploaded to NMPA's database and amended as needed.

The system has been implemented in phases, each one requiring certain devices to be assigned UDIs. NMPA maintains a searchable online database of identifiers that includes the number, the product name, and the enterprise name.[125] The first phase, in 2019, consisted of certain Class III devices, including certain active and passive implantable devices.[126] The second was released in 2021,[127] and the third was released for public comment in November of 2022 and finalized in 2023, including a variety of Class II devices, such as opthamalic instructions and assisted reproduction and contraceptive devices, and respiratory and anesthesia equipment.[128]

6.11.6 Two-Invoice System

As indicated in the chapter on drugs, China established a two-invoice system to set constraints on the distribution of drugs in 2016. Pursuant to this system and subsequent provincial rules, there can only be two tax invoices between the manufacturer and the user entity (i.e., the hospital procuring the drug). This would slim down corrupt distribution chains and encourage a class of larger, more sophisticated drug distributors. NMPA and its predecessors greatly curtailed the issuance of drug distribution licenses for similar reasons.

The two-invoice system was established for drug regulatory purposes, with aspirations to expand to medical devices at a later point. As of this writing there has been no expansion to devices on the national level, but certain provinces have begun to enforce the system against medical devices. For example, as of mid-2018, approximately four provinces and a number of localities within other provinces had adopted some measure to apply the two-invoice system to medical supplies.[129]

---

[125] National Medical Products Administration (2019c).

[126] National Medical Products Administration (2019d).

[127] National Medical Products Administration (2021p).

[128] National Medical Products Administration (2022d) and National Medical Products Administration, National Health Commission, et al. (2023).

[129] Health Pink (2019).

6.11.7 Distributor Self-Inspections

Device distribution enterprises must comply with yearly self-inspection requirements. These self-inspections have been an important tool for NMPA and the provincial MPAs as they require enterprises to evaluate their own compliance and then submit reports to local authorities; enforcement, including loss of the license, may follow if these reports are not submitted or are incomplete or inaccurate.[130]

Although inspections are now annual (due March 31st of every year), distributors were previously given smaller windows. One required self-inspection in 2016 gave only a few weeks to investigate and complete evaluation reports.[131] As with manufacturing self-inspections, companies must continually prepare for these self-inspections and the information in the report must be consistent with other enforcement or inspections, such as routine or for-cause inspections, AE reporting, or penalties as the result of non-compliance.

6.11.8 Distributor Risk-Tiered Inspections

As the sections above show, NMPA continually embraces a risk-based approach to medical devices, slimming down the approvals and requirements for different devices based on a determined risk level; this risk determined is what goes into the class determination and into exemptions from clinical evaluations and distribution permissions. In 2014, shortly after the enactment of the heavily revised MDSAR, NMPA issued a regulation that divided both device manufacturers and distributors into risk-based tiers, taking into account the features of the device (class, risk level) and the record of the distributor. Those with a higher risk rating would be subject to more frequent inspections.[132]

In 2022, NMPA issued an opinion cancelling these rules as of January 1, 2023 and calling on local governments to issue their own regulations to determine the risk of different device distribution and manufacturing enterprises and to implement inspections accordingly. This approach seems appropriate given that these local governments issue the manufacturing and distribution permissions and conduct inspections. This opinion embraced four risk-based tiers, similar to the prior regulations.

- Level 4: Highest risk, including distributors that provide storage and transportation services for registrants and filers, and other production and distribution companies determined by considering the risks. These enterprises should be inspected at least once a year.

[130] State Administration for Market Regulation (2022c), Article 44.

[131] China Food and Drug Administration (2016b).

[132] China Food and Drug Administration (2014f).

- Level 3: Still high risk, including wholesalers of devices selected for targeted supervision and deemed higher risk as determined by the local governments, and distributors that have been subject to penalties or lower credit scores within the past year. Inspected once every year, with (and this is a bit confusing) all inspection items covered every 2 years.
- Level 2: Moderate risk, including other (i.e., not level 4 or 3) distributors of Class II and III devices and those in NMPA's catalogue. Inspected once every 2 years, except contact lens and [medical] protective product retailers, for which local authorities can determine the appropriate frequency.
- Level 1: Other devices that are not in the prior categories. For these, local authorities should randomly select 25% for inspection every year and cover all of them every 4 years.[133]

These inspections can be combined with the inspections required prior to registration or with significant changes to the registrations, in order to ensure efficiency. Given the increased attention to device GxP enforcement in recent years, companies are increasingly focused on inspection readiness and implementing internal procedures to that effect.

6.11.9 In-Market Sampling Inspections

Sampling inspections—testing of medical devices on the market—determine whether they meet their registered standards. These inspections occur at the local level for a specific set of devices determined on an annual basis by NMPA and the provinces; individual devices selected from a given lot for testing are supposed to be chosen at random and failure to meet registered PTRs or standards can result in recalls. Sampling inspections are a potent form of enforcement because they can reveal gaps between the device and its PTR. In 2020, NMPA issued revised rules for these inspections, which added procedural protections, including those to ensure random selection of samples and prohibit tampering with inspections.[134]

6.12 Medical Device Manufacturing

As with drugs, the policy has been to increasingly improve quality, including compliance of all device makers with applicable quality rules, and to move to larger, more technologically advanced manufacturing operations for medical devices. The move to a marketing-authorization holder system in which one registrant/filer assumes responsibility for the device, including supervision of separate third-party manufacturers, has led to changes in the enforcement of manufacturing rules, with

[133] National Medical Products Administration (2022f).

[134] National Medical Products Administration (2020a).

new rules established in 2022 that articulate the specific responsibilities of an "entrustment" arrangement.[135]

China has also issued quality-system guidance that discusses the role of the registrant in design, implementation of approved specifications through release, and requirements and supervision of suppliers.

As discussed, under prior device regulation the manufacturer was the sole responsible entity. Current regulation moves from a regime in which all design and quality responsibilities fall to one entity to one in which other entities, whether affiliates in a corporate group or third parties, are permitted to play a role in design, supply, production, marketing, and distribution, with the appropriate integration and supervision. With this regulator change, NMPA is heavily focused on the registrant/filer entity itself (arguably overly so), sometimes expecting core employees and systems all to reside in one entity, which can present challenges for corporate groups, which have these personnel spread out across different entities.

6.12.1 Manufacturing License and Filing System

NMPA controls medical device manufacturing in ways similar to manufacturing regulation in the drug space. There is a manufacturing license system for domestically made devices; imported devices must submit information about the manufacturing process with their applications, but they are not licensed or registered as facilities. As with drugs, the trend has been to require increasingly more information related to overseas manufacturing and quality processes. The new medical device manufacturing rules make it clear that they apply to imported medical devices, and the appointed agents of overseas registrants and filers should be assisting those companies with understanding their obligations under production rules.[136]

A few points are different from the drug system, however. First, licenses are not required for all medical device manufacturing facilities in China. Under the 2014 MDSAR, only facilities that make Class II and III devices and IVDs must have medical device manufacturing licenses. These manufacturing licenses are associated with specific devices, meaning the facility is linked with a registration and vice versa. One innovation in the 2014 MDSAR was that the facility does not need to be licensed to permit the submission of a product registration application. Like with drugs, parties can submit manufacturing license applications and have good manufacturing practice qualification inspections completed prior to the marketing of products. Such a GMP inspection might also be required if the facility already has a manufacturing license but is the subject of an amendment to expand or change its scope to bring on new products. The 2021 MDSAR now also adds pre-approval manufacturing verifications, which can occur at the same time as the GMP

[135] State Administration for Market Regulation (2022b), Chapter 3.

[136] Ibid., Articles 61–62.

inspections. The verifications are focused on determining whether the facility has the appropriate quality system for the candidate device under review, and these verifications can be conducted onsite through inspections or through the collection and review of documentation.

The provinces oversee issuing medical device manufacturing licenses, making amendments to them when product or other core license information changes, and renewing them every 5 years. Each manufacturing license has a specific scope of products associated with it, e.g., IVDs or other types of devices.

In contrast to Class II and III devices, facilities that produce only Class I medical devices do not require medical device manufacturing licenses. Those facilities must still comply with medical device GMP, but they submit materials to the local regulator in the form of a record-filing (see Chap. 3).

The second difference from drugs is that the registrant/filer itself does not need a supervision manufacturing license in addition to its product registration in order to supervise the manufacturer. The registrant/filer obtains a product permission, and the manufacturer obtains its own license/filing.

6.12.2 Medical Device GMP Implementation

Currently all device registrants, filers, and actual manufacturers (i.e., manufacturers entrusted to production by registrants or filers) of finished devices must follow China's medical device good manufacturing practice (GMP) rules. The implementation of the medical device GMP has been gradual. NMPA's predecessor released a set of "trial implementation" GMPs in 2009. Implementation of the "trial" medical device GMP remained uneven and was therefore a concern of regulators in 2014 when the MDSAR was amended. Regulators originally focused primarily on conducting inspections of facilities within China and upgrading the quality of domestically produced devices. The GMPs are currently under revision again as of the time of this writing, and the role of the "registrant" versus the "actual manufacturer" in quality compliance is an important topic in that discussion.

As discussed, the medical device GMP rules apply to all devices (without qualification), but the government has implemented GMP in stages. Upon issuance of the new GMPs in 2014, NMPA published a notice requiring Class III device manufacturers to comply by January 1, 2016, and from that time to conduct self-inspections to ensure they met GMP requirements.[137]

In 2017, the government issued another notice requiring that all Class I and II manufacturers (and thus all manufacturers) to comply by January 1, 2018 (see below). This notice also required that these manufacturers conduct self-inspections to determine compliance and either shut down or remediate issues.

All manufacturers that import products qualifying as medical devices into China must comply with the medical device GMP and the inspectors from China that go

[137] *See* China Food and Drug Administration (2016d).

abroad to conduct overseas inspections will likely expect familiarity with the various standards and guidelines the relate to the GMP.

The medical device GMP consists of 13 chapters, including general principles and miscellaneous provisions. Arguably most important are the 11 chapters governing the core systems required for device manufacturers: departments and personnel, facilities, equipment, document management, design and development, procurement, manufacturing, quality control, sales and after-sale (i.e., repair), and adverse event monitoring, analysis, and correction. The rather general requirements in these provisions have been further interpreted in guidance documents.

Another regulation on device GMP site inspections further breaks the requirements down into core and non-core items that inspectors must examine. The failure to pass on a core item can lead to a failing grade on the inspection. The manufacturer is permitted the opportunity to correct failure on non-core items to the satisfaction of NMPA. If the issues are successfully remediated, there will be no interruption in the manufacturing or importation of the product into China.

In addition to these general requirements, NMPA has enacted several appendices to the GMP. These appendices cover specialized areas, such as dental implants, sterile devices, implantable devices, and in vitro diagnostics. In 2020, NMPA finalized a similar appendix on GMPs for standalone software, which contains various design and testing requirements. The push for NMPA and CMDE to upgrade regulations and guidelines to effectively regulate China's vigorously expanding digital health sector is discussed below.

In addition to the GMP, its specialized appendices, and inspection checklists, all of which are critical sources of requirements, there is a vast body of government-issued standards that should be considered. There are, as discussed, product standards, but also various design, testing, and manufacturing standards. For example, the China Pharmacopeia contains various testing standards, such as those on sterility testing. And China has adopted (with some modifications) various International Organization for Standardization (ISO) standards, such as those for device quality systems. ISO 13485 Medical Device Quality Management System Requirements for Regulatory Purposes has been adopted as YY/T 0287. (As mentioned previously, the YY code for standards in China means "medical standards" and the T stands for *tuijian* or "recommended" standard.[138])

There are also guidelines that may be applicable, some of which may come from provincial regulators. For example, NMPA has a set of guidelines on water used in the medical device production process, and Beijing has a set of guidelines on sample retention.

The challenge, particularly for international manufacturers, is to understand this vast and expanding set of materials (all of which are in Chinese) and show inspectors that come to their facilities either in-person or virtually (through cameras and video call technology) that they have considered these rules and concepts and incorporated them, either with specific reference to the China guidance or through equivalent

[138] China Food and Drug Administration (2017b).

international methods that those manufacturers can prove are equivalent through data or gap-analyses.

6.12.3 PTR and Release

At least a few words should be said about the role of PTR in manufacturing. China's medical device rules require that production is organized around the PTR. China inspectors will expect that the product is released from the manufacturing facility in accordance with its registered PTR. Experience has proven that registrants and filers should structure the PTR with product (not component) specifications and test methods that they can realistically implement, to the extent possible. This may not always be possible, though, as China standards may dictate relevant test methods for certain products and therefore, certain compromises or adjustments may have to be made. Both NMPA and its Center for Food and Drug Inspections are focused on having products produced and released in accordance with the most recent version of the PTR that has been approved by the agency. This is also evidenced by the sampling program discussed above.

The manufacturing rules state:

> Registrants and filers shall be responsible for product market release and establish a product market release procedure, clear release standards and conditions, and evaluate and verify records of the production process and the results of quality testing. If [the product] meets standards and conditions, the product may enter the market after authorized release personnel have signed [off]. If manufacturing is entrusted, registrants and filers still shall evaluate and verify the release documents issued by the entrusted manufacturer.[139]

Therefore, the responsibility for releasing compliant product sits with the registrant and filer, even if they entrust manufacturing to other entities. CFDI focuses on release and the registrant's role in implementing or at least supervising this process for all products.

6.12.4 Contract Manufacturing

Increasingly, as a consequence of the implementation of the marketing authorization holder system, medical device registrants and filers may now use one or more contract manufacturers (referred to under the rules as "entrusted manufacturers") to make all or part of the product. Contract manufacturing was permitted prior to this system but the contract manufacturing rules have developed considerably under it.

The registrant or filer must have a quality agreement in place with each contract manufacturer, including those entities that are part of the corporate group (the rules do not distinguish between third parties and manufacturers within the same corporate

[139] State Administration for Market Regulation (2022b), Article 34.

family), and it must exercise supervision over those entities. When intermediates of the device are produced at other facilities, those entities may be treated like suppliers and are subject to NMPA's medical device supplier guidelines,[140] which set forth rules on contracting, supervising, and auditing suppliers.[141]

Supervision consists of an initial vet to ensure the manufacturer is qualified, provision of documentation and instruction as to manufacturing and product technology, and the conduct of audits by the registrant or filer to ensure initial and continued compliance with their instructions and the general rules on manufacturing and quality. The device manufacturing rules contain various provisions on supervision by registrants and filers over contract manufactures.[142]

Further, NMPA has articulated a list of devices for which contract manufacturing is not permitted. These include pacemakers, implantable cardiac assistance devices, Dural and intracranial stents, among others.[143] For these devices, at least domestically, the registrant must be the manufacturer. For imported devices, it is not clear how strictly these prohibitions on contract manufacturing are applied.

6.12.5 Refurbished Devices

China has no clear policy on refurbishing domestically produced devices or their components. In 2005, NMPA issued a draft regulation on refurbished medical devices[144] that would have standardized the practice of device refurbishment and permitted manufacturers to obtain licenses for and to refurbish all or part of certain devices, provided it is done in accordance with manufacturing regulations. Refurbished devices were defined as used, recovered, or refurbished devices that are used or sold again, and device refurbishment is repairing, replacement of parts, calibration, cleaning, disinfection, cosmetic repair, testing, upgrades, or software upgrades.[145] NMPA never finalized that draft, and NMPA has no official prohibition on the use of refurbished devices or parts.

However, NMPA has enforced against the practice of manufacturers refurbishing devices without holding a marketing certificate or the appropriate manufacturing certificate for those devices. That enforcement may be under the prohibited acts of marketing of an unlicensed device or impermissible manufacturing.

Customs has been clearer about its prohibition on refurbished devices (issuing a warning notice on importation of "old" medical devices),[146] and in the 2021 MDSAR it states that devices (not components) that are expired, invalid, outdated,

[140] National Medical Products Administration (2022c).

[141] China Food and Drug Administration (2015b).

[142] State Administration for Market Regulation (2022b), Chapter 3.

[143] National Medical Products Administration (2022e).

[144] China Food and Drug Administration (2005).

[145] Pharmaceutical Economic News (2011).

[146] General Administration for Quality Supervision, Inspection, and Quarantine (2016).

or have been used may not be imported.[147] However, there is still no definition as to what constitutes a used device and whether certain parts can be still be refurbished. Officials do not make clear distinctions between products that are essentially discarded items or trash and products that have been properly refurbished as new or were never used and have only been cleaned up and perhaps upgraded to permit use after sitting dormant. Whether a device for importation that is refurbished in whole or in part violates this rule remains a case-by-case, risk-based determination.

6.12.6 Manufacturing Inspections

GMP are enforced through routine and for-cause GMP inspections, both announced and unannounced. This includes government-led domestic and overseas inspections and self-inspections.

6.12.6.1 Manufacturing Self-Inspections

As with distributors, manufacturers are subject to self-inspection requirements. NMPA at first used them periodically or on a select basis and now on a regular basis for all devices. For example, as noted previously, NMPA required that manufacturers of all classes of devices be compliant with device GMPs by January 1, 2018, and thereafter they were required to conduct a comprehensive self-inspection to determine GMP compliance; if they failed, they were to shut down and report to the local government.[148] These self-inspections were coupled with government-led inspections.

Now, device registrants and filers are required to submit a yearly self-audit of their quality-control system. Under the 2014 MDSAR, these self-inspections were required for domestically produced devices.[149] As of 2021, all manufacturers, whether foreign or domestic, are required to submit a self-inspection report on their quality system by the end of March, reporting on activities for the prior year (December through January). The registrant or filer and agent of an overseas registrant or filer submits the report to the province where they are located.[150]

The annual self-inspection report for manufacturers includes basic information about the company, any changes during the year (e.g., to design, production area, testing equipment, or primary suppliers, such as those that provide key ingredients or components or provide sterilization, shipping, or testing services), and information about the operation of the quality system (e.g., personnel, trainings, customer feedback, traceability, penalties, internal reviews, and audits). The report also includes post-market surveillance (e.g., serious AEs) and enforcement information,

[147] State Administration for Market Regulation (2021b), Article 57.

[148] China Food and Drug Administration (2017f).

[149] State Administration for Market Regulation (2014), Article 24.

[150] National Medical Products Administration (2021h).

such as non-compliances, inspections, and penalties. Failure to submit this report can result in warnings and fines.

6.12.6.2 Domestic Inspections

Domestic device manufacturers are subject to various types of inspections. To obtain the manufacturing license, there must be a qualifying good manufacturing practice inspection. For each device manufactured at the facility, NMPA can conduct a verification inspection onsite (or by paper) to ensure that the manufacturer has the appropriate quality system for the product.

Thereafter, device manufacturers are subject to regular compliance inspections, which can be announced or unannounced (referred to as "flying," or *feixing*) inspections. As with distributors, NMPA released a rule in 2014 dividing the risk level of manufacturers into four tiers, four being those with the highest risk and one the lowest. The risk tier is determined based on a variety of factors, including the risk of the products, the market share, the results of inspections and sampling, adverse event data, recalls, and quality complaints from consumers. Level four manufactures may be inspected once a year, level three once every 2 years, level two once every 4 years, and level one subject to spot checks.[151] As discussed above, these regulations were terminated in 2022 and provinces tasked with creating their own rules for similar risk-based systems.

6.12.6.3 Overseas Inspections

Overseas inspections for medical device manufactures have been occurring since 2015, approximately 3–4 years after similar inspections began for drugs. The number of products covered by such inspections have regularly increased. These inspections ceased during the COVID-19 pandemic but have resumed virtually since 2020. There were approximately nine virtual inspections for drugs and devices in 2022 and announced publicly, but more may have occurred and were just not publicized. In many cases, a brief report of the findings and any importation ban, penalties, and/or resumption of importation is posted publicly. The level of detail released by NMPA and the CFDI has varied over time.

Like drug inspections, these overseas inspections are in accordance with China good manufacturing practices. Three to five inspectors from CFDI and some perhaps from other provincial-level medical product administrations or other related local governmental entities either in person or virtually assess compliance according to the on-site inspection checklist, which exists for the general GMP and each of its appendices. After the inspection concludes, the inspectors give the preliminary results at a meeting. They have 10 days to give the findings in writing. The inspected entity has 10 days to object to the findings or make factual corrections, although at times much more communication occurs during this time to resolve debates that arose during the inspection. The extent of this activity and exchange depends on the inspection team at CFDI. Once the company receives the inspection report, it has

[151] China Food and Drug Administration (2014e).

50 business days to file a response with all the remediation measures that it completes and those that it starts but will require additional time to complete. During the writing of this report, the company may be asked to update CFDI on its progress.

Once the registrant's response is complete, CFDI reviews and issues a recommendation to NMPA, which may result in the commencement of a penalty investigation, depending on the extent of the non-compliance. CFDI scores inspections by determining compliance with critical items and general items on the inspection checklist. Failure on a critical item can result in failure of the inspection, as can failure on a number of general items that affect safety and quality. CFDI has the power to report more serious violations of the GMP earlier, which can result in recalls just after or in parallel to the inspection, and blocks on the importation of the product, among other control measures. Importation cannot resume until the registrant submits a corrective action and preventive action (CAPA) report to NMPA's satisfaction and, possibly, another inspection is performed. Once NMPA is satisfied that the violation is remediated, it will grant a letter application by the registrant company to resume importation and issue a public notice to that effect.

6.13 Medical Device Use Responsibilities

There are also rules that govern the user entity environment for medical devices. User entities are institutions that provide treatment and technical services with medical devices, including medical institutions, family planning institutions, blood stations, blood plasma collection stations, and institutions that deal in the adaption of medical devices for rehabilitation.[152] Both NMPA and NHC have issued rules on this area. NMPA has issued Medical Device Good Use Practice rules that require that user entities ensure that the devices they are receiving, whether through purchase or donation, are registered or record-filed, that they comply with applicable standards, that they are used in accordance with their instructions for use, and that those devices are appropriately maintained by distributors and manufacturers. These rules also regulate transfers of medical devices between medical institutions, including by donation. Each year a user entity is required to conduct a self-inspection of its system for managing the quality of medical devices and keep the report on file; local medical products authorities conduct spot checks of these reports, in addition to other inspections.[153]

In 2021, the NHC released rules on the management of "clinical use" of medical devices. These rules overlap with the NMPA rules from 2015 in requiring that user entities have the appropriate personnel and system for using medical devices, and they include provisions on assessing and reporting on the safety of devices.[154]

[152] State Administration for Market Regulation (2021b), Article 103.

[153] State Administration for Market Regulation (2015).

[154] National Health Commission (2021).

6.14 Export of Medical Devices

Unlike devices for import and use in China, devices for export are not subject to the same requirements as in China. Generally speaking, a device for export needs to be compliant with the requirements of the receiving country. When a company is manufacturing devices for export, it must obtain a certification of its quality system by a third party, or it must obtain the device manufacturing or record filing that would be needed to produce the devices for China.[155] Provincial governments are authorized to issue certificates of free sale (referred to as export sales certificate) to prove the marketable status for devices that have obtained registration and manufacturing permission for China.[156]

There are some exceptions to this flexibility, however. For example, under the 2007 U.S. Food and Drug Administration and China State Food and Drug Administration Agreement on the Safety of Drugs and Medical Devices, China agreed condoms and glucose test strips for export would also need to be registered for marketing in China, adding another layer of regulation.[157]

During the pandemic, the requirement that certain devices be registered in China before they could be exported was originally extended to personal protective equipment (PPE), such as masks. NMPA also reinforced that only licensed or record-filed distributors should handle such devices, and Customs, the Ministry of Commerce, and the State Administration for Market Regulation issued a notice that jointly required that exporters sign an undertaking prior to the export of such devices listing the manufacturing license and registration number, although a more flexible notice later permitted PPE if it complied with the standards of the receiving country.[158] China customs also tested devices at the border to ensure that they complied with relevant standards.

Masks for industrial use (KN masks) are not subject to medical device requirements but must meet the requirements of the receiving country. However, China also maintains standards for industrial masks, including GB 2626 2019 "Respiratory Protection—Non-Powered Air Purifying Particle Respirator," and they could be approved in the receiving country, such as NIOSH approved in the United States. China maintains a list of masks accepted for export either because they were medical devices or industrial masks with approval in the receiving country.[159] China also maintains a list of companies known to counterfeit masks.

There were various issues with PPE during the pandemic. Distributors in China were in demand during the early stages of the pandemic when there were PPE shortages, but they could be cagey about what masks they could export, citing unpublished restrictions on the exportation of more trusted brands. Despite the

[155] State Administration for Market Regulation (2021c), Article 70.

[156] China Food and Drug Administration (2015d).

[157] U.S. Food and Drug Administration (2007).

[158] Covington (2020) and Ministry of Commerce (2020).

[159] China Chamber of Commerce for Import and Export of Medicines and Health Products (2021).

aforementioned regulatory measures, some sources indicate that more than half of the 13 million counterfeit masks that authorities seized in 2020 came from China.[160] Similar problems were noted with respect to test kits and other medical devices coming from China.[161]

6.15 Digital Health and Medical Devices

China is undergoing a digital health revolution, as technology is increasingly incorporated into health care. This includes everything from general use apps registered as standalone software to more complex artificial intelligence medical devices. Pursuant to the 14th Five Year Plan on the Development of the Digital Economy, China continues to make the advancement of digital health a national priority by accelerating the development of mechanisms to provide health care services through the Internet and smart health equipment.[162]

Digital health has been a policy priority for China for various reasons, including the goal (discussed throughout this book) of providing greater quality health care throughout China. In many respects this has been government-led, meaning that the government has promoted use of digital technology to expand the reach of the state healthcare system. However, medical device and other technology companies have participated both by supporting the government's efforts and by introducing "software as a medical device" with various accessories (e.g., wearables to track bodily functions) into the market.

In 2014, the NHC issued a regulatory document in the form of an opinion allowing for hospital-to-hospital digital health exchanges so that physicians throughout China could obtain second opinions from key experts and medical institutions abroad. This institutionalized an already occurring practice. The opinion defined telemedicine activities as those "to provide technical support for the diagnosis and treatment of patients in this medical institution by using communication, computer, and network technologies" and required a cooperation agreement between the two parties and informed consent from individual patients.[163] The opinion forbade non-medical institutions (e.g., those without such a medical institution license) from providing such services. Shortly thereafter the NHC released lengthy technical guidelines that provided parameters for obtaining the necessary communications equipment to make such exchanges possible.[164]

Later, China established rules, unique globally, for virtual internet hospitals and e-diagnosis services for bricks and mortar hospitals. These remote online operations and services can provide follow-up visits and certain prescriptions for patients

[160] Hayashi (2021).

[161] Morgenson (2020).

[162] Li (2022).

[163] National Health and Medical Development (2014).

[164] National Health and Family Planning (2014).

around China.[165] By 2020, China was reporting that approximately 600 internet hospitals were online, and during the heart of the COVID-19 pandemic remote care provided through the internet increased significantly, according to government reports.[166] Among the government's other stated goals, it has encouraged "medical institutions [to] make full use of the Internet, big data and other information technologies to expand service ... content, actively provide patients with online convenient and efficient services, as well as follow-up management and remote guidance, and gradually [support] patients' home rehabilitation."[167]

There are technology and software applications underlying China's unique telemedicine and internet hospital regulatory regime that may not have been registered as medical devices (although those applications function in that way), and NMPA has not provided specific guidelines that articulate the line between health and wellness software, unregulated healthcare technology, and regulated medical devices.

That said, CMDE has embarked on a significant digital health initiative and produced several basic guidelines that have advanced the structure for registering digital health solutions as medical devices. One of the starting points, the 2015 Guiding Technical Principles on the Registration Review of Software, articulate the requirements for registering software, including classification, specification (i.e., PTR) development, clinical development, and labeling. They also include important provisions on how to determine whether version changes and software updates are major amendments that affect the safety and effectiveness of the device and therefore require an amendment to the device license itself or more minor changes that do not affect safety and effectiveness and can be reported in annual reports to the agency. The 2015 guidelines were amended in 2022.

There are also several important specialized guidelines. For example, in 2017 CMDE issued guidelines on the registration of mobile medical applications. An important feature was CMDE's statement that any "app" that meets the definition of a medical device will be regulated as such, meaning NMPA and CMDE will not exercise enforcement discretion to allow lower risk apps that might otherwise qualify as medical devices to go on the market without registration as medical devices (similar to other jurisdictions).[168] Only apps that serve an information gathering (but not an analytical) function for fitness purposes would be considered non-devices. Indeed, technology companies have been moving to register apps and wearable technology for preventing, assisting with diagnosis of, and monitoring diseases or conditions.[169]

To accompany the newly promulgated Cybersecurity Law, CMDE also released a guideline on medical devices and cybersecurity. This guideline provides a roadmap

---

[165] National Health Commission, Chinese Medicine Administration (2018a, b, c, d).

[166] Qu et al. (2020).

[167] National Health Commission (2020).

[168] National Medical Products Administration (2017b).

[169] *See, e.g.,* Huawei (2022/2023).

for evaluating the cybersecurity risks of medical devices,[170] and NMPA's 2021 medical device registration application requirements name cybersecurity research as one of the types of pre-clinical research that a registrant may need to conduct to support an application. The NHC has also worked to establish requirements for an information infrastructure for medical institutions, including the establishment of data security measures and a system for conducting security assessments. Similarly, the NHC has raised concerns with the protection of the data generated by the operation of medical devices and ensuring its appropriate storage and non-disclosure.[171]

Similar to the discussions in the area of cybersecurity in China generally, the digital health regulatory trend has included discussions about data localization. For example, an amendment to a cybersecurity guideline that CMDE proposed in 2020 included a provision that would require that "population health information," a broadly defined concept of health data about the Chinese population, be kept on servers in China, potentially blocking data from being stored using technology that is accessible beyond China's borders.[172] The final version stated that big data should generally be stored on servers in China and only transferred overseas in accordance with applicable laws and regulations. It defined medical data as important data that is subject to additional scrutiny and must follow applicable rules, including the Human Genetic Resource Regulations, when being exported.[173]

A central area of China's digital health efforts has been in artificial intelligence. China has arguably been a leader in this area of regulation regionally, releasing several detailed regulatory guidelines for the classification, development, registration, and amendment of AI medical devices, including AI standalone software (SaMD) and embedded AI software (SiMD). This regulatory work represents the culmination of years of various guidelines from CMDE that are supportive of AI development.

By 2020, the body of regulatory documents had grown sufficiently substantial, and NMPA and CMDE approved several AI devices, including:

- Fractional Coronary Flow Reserve Software
- ECG Analysis Software
- Intracranial Tumor MRI Assisted Diagnosis Software
- Fundus Image Aided Diagnosis Software for Diabetic Retinopathy
- CT Angiography Image Vascular Stenosis Assisted Diagnosis Software
- Pulmonary Nodule CT Image Aided Detection Software
- CT Facture Intelligent Analysis System
- Pulmonary Nodule CT Image Aided Detection Software[174]

[170] Center for Medical Device Evaluation (2017).

[171] National Health Commission (2022).

[172] Center for Medical Device Evaluation (2020).

[173] Center for Medical Device Evaluation (2022a).

[174] Arterial Network (2021).

In 2022, NMPA released a guideline on classification of AI software that is regulated as a medical device. In that guideline, "medical use AI software" is defined as standalone software that is "based on medical data and uses AI technology to implement its medical treatment use." The core function of the software must be data processing, measurement, model calculation, or analysis, among other such functions. "Medical data" is defined as "objective data" that is produced by a medical device and is used for a medical treatment use.

Classification of the AI software requires a collective determination of factors such as the intended use and the maturing of the calculation method. Devices with a low level of maturity, meaning they have not previously been marketed or their safety and effectiveness has not been fully confirmed, and with a function of assisting in clinical diagnosis and treatment decision making, are regulated as Class III. Those devices that do not assist in decision-making, such as those that process and measure data for purposes of providing a clinical reference can be regulated as Class II.

For those devices with a high level of maturity, i.e., their safety and effectiveness has been fully confirmed, such as through clinical evaluation or market practice, can be classified according to the Classification Catalogue and related classification documentation.[175] But, even under the Catalogue certain medical software that assists in medical decision-making or with treatment planning is still Class III, whereas other types of medical use software can be Class II.

The guideline applies to standalone software, but component software can use the guidelines as a reference.

During the pandemic, China used AI to collect personal data about its citizens and transferred it to a national platform to analyze risk and control movement, and it was also considered a useful tool to assist already.[176] Each citizen received a health code based on this system of big data that determined when they had green, yellow, or red statuses for travel and even entry into certain buildings. Like its efforts with precision medicine, China's digital health regulation can be used in various ways, allowing for the provision of high-quality healthcare and preventive medicine to a broad range of its citizens beyond the major eastern cities, but it can also dovetail with its efforts to gather and deploy data for biosecurity, including national security, purposes.[177]

6.16 Penalties and Enforcement

The penalties for device regulatory violations are like those for drugs. They are primarily administrative penalties, including fines, disgorgement of profits, seizure of products and illegal implements, orders to cease production and distribution, and/or revocation of product or facility licenses. The MDSAR is the source of strict

[175] National Medical Products Administration (2021b).

[176] Li (2022), Zhang (2022).

[177] Bernot et al. (2021).

liability prohibited acts (e.g., importation of an unregistered device) that trigger these penalties. As with the Drug Administration Law (discussed in Chap. 5), the 2021 MDSAR increased the fines for various prohibited acts, from 20 to 30 times the value of the illegal product.[178]

The 2021 MDSAR also sharpened the penalties against individual employees, including individual fines and debarment in serious cases (see Chap. 3). In some serious cases, such as very egregious cases of producing or distributing unregistered medical devices, debarment can be for life.[179] As noted above, it is unusual for authorities to pursue such extreme punishments on an individual, but their existence can provide substantial leverage in obtaining cooperation from regulated entities in discussions surrounding the remediation of legal violations.

As with drugs, all of the violations are phrased in terms of strict liability, i.e., any person or entity or any registrant and filer that commits a certain act may be liable. Committing an act willfully or with the intent to break the law or engage in illegal conduct can be an aggravating circumstance that makes the violation serious and increases the penalty (see Chap. 3).

Violations of the MDSAR can also be criminal or civil in nature. Article 102 of the MDSAR states that individuals can be liable to others for compensation for the harm caused by violations of its provisions, and those individuals can be prosecuted criminally for violations of the device rules. The details of when compensation is available and when criminal prosecution is appropriate would be handled under the provisions of other laws, such as the Tort Law within the Civil Code and the Criminal Code. As with drugs, the Criminal Code contains provisions punishing distribution of substandard and fake medical devices with substantial fines, imprisonment, and in cases in which those devices cause particularly serious health consequences, life in prison.[180]

References

Arterial Network (2021) The drastic changes in medical AI approval. Jiemian. https://www.jiemian.com/article/6348212.html.

Bernot A, Trauth-Goik A, Trevaskes S (2021) China's 'surveillance creep': how big data COVID monitoring could be used to control people post-pandemic. The Conversation. https://theconversation.com/chinas-surveillance-creep-how-big-data-covid-monitoring-could-be-used-to-control-people-post-pandemic-164788.

Biodiscover.com (2016, May 10) Seven violations of domestic IVDs. https://www.biodiscover.com/interview/121859.html.

Center for Medical Device Evaluation (2017) Medical device cybersecurity technical review guidelines. https://www.cmde.org.cn//flfg/zdyz/zdyzwbk/20170328091100413.html.

[178] *See, e.g.,* State Administration for Market Regulation (2021b), Article 83.

[179] Ibid., Article 81.

[180] National People's Congress (2020), Article 145.

Center for Medical Device Evaluation (2020) Medical device cybersecurity technical review guidelines (draft for comment). https://www.cmde.org.cn//zhuanti/zqyj/20200908164412146.html.
Center for Medical Device Evaluation (2021) Guiding principles for the updates of package inserts and technical review of companion diagnostics for oncology drugs. https://www.cmde.org.cn/flfg/zdyz/zdyzwbk/20210416143648925.html.
Center for Medical Device Evaluation (2022a) Medical device cybersecurity registration review guidelines. https://www.cmde.org.cn//flfg/zdyz/zdyzwbk/20220309085900737.html.
Center for Medical Device Evaluation (2022b) 2021 Medical device registration work report. https://www.cmde.org.cn//xwdt/zxyw/20220128091800763.html.
Center for Medical Device Evaluation (2022c) Communication procedures (trial) for the pilot varieties of clinical real world data application of medical devices in Lecheng International Medical Tourism Pilot Zone. https://www.cmde.org.cn//xwdt/shpgzgg/gztg/20220429134511125.html.
Center for Medical Device Evaluation (2022d) Guiding principles for registration review of clinical trials of original companion diagnostic reagents developed simultaneously with antineoplastic drugs. https://www.cmde.org.cn//flfg/zdyz/zdyzwbk/20220711100756101.html.
Center for Medical Device Evaluation (2023) Guidelines for examining changes in the registration of in vitro diagnostic reagents. https://www.cmde.org.cn//zhuanti/zqyj/20230531152938141.html.
Center for Medical Device Standardization Administration (2019) 2018 Consolidated medical device classification decisions. https://www.nifdc.org.cn/nifdc/bshff/ylqxbzhgl/qxxxgk/fljd/20190218152201546.html.
China Chamber of Commerce for Import and Export of Medicines and Health Products (2021) White list of makers of medical masks and medical suppliers for export. https://www.guanwuxiaoer.com/thread-1262-1-1.html.
China Food and Drug Administration (2005) Administrative provisions on the supervision of refurbished medical devices (draft for comment). https://www.cpa.org.cn/index.php?do=info&cid=70788.
China Food and Drug Administration (2014a) Administrative provisions on medical device labels and instructions for use. https://www.nmpa.gov.cn/xxgk/fgwj/bmgzh/20140730180001248.html?type=pc&m=.
China Food and Drug Administration (2014b) Medical device good supply practice. https://www.nmpa.gov.cn/ylqx/ylqxggtg/ylqxqtgg/20141212120001470.html.
China Food and Drug Administration (2014c) Medical device label and instruction for use provisions. https://www.nmpa.gov.cn/xxgk/fgwj/bmgzh/20140730180001248.html?type=pc&m=.
China Food and Drug Administration (2014d) Operational guidelines on the examination and approval of registrations of Class III domestic and imported medical devices. https://baike.baidu.com/item/%E5%A2%83%E5%86%85%E7%AC%AC%E4%B8%89%E7%B1%BB%E5%92%8C%E8%BF%9B%E5%8F%A3%E5%8C%BB%E7%96%97%E5%99%A8%E6%A2%B0%E6%B3%A8%E5%86%8C%E5%AE%A1%E6%89%B9%E6%93%8D%E4%BD%9C%E8%A7%84%E8%8C%83/15817936.
China Food and Drug Administration (2014e) Regulations on the classification and grading supervision and administration of medical device manufacturing enterprises. https://www.nmpa.gov.cn/directory/web/nmpa/xxgk/fgwj/gzwj/gzwjylqx/20140930120001554.html.
China Food and Drug Administration (2014f) Regulations on the supervision and management of classification of medical device distribution enterprises. https://www.nmpa.gov.cn/directory/web/nmpa/xxgk/fgwj/gzwj/gzwjylqx/20150817120001189.html.
China Food and Drug Administration (2014g) Guiding principles on the compilation of in vitro diagnostic reagent instructions for use. https://www.nmpa.gov.cn/ylqx/ylqxggtg/ylqxqtgg/20140911120001763.html?type=pc&m=.

China Food and Drug Administration (2015a) Medical device naming principles. https://gkml.samr.gov.cn/nsjg/bgt/202106/t20210629_331730.html.
China Food and Drug Administration (2015b) Medical device supplier evaluation and verification guideline. https://www.nmpa.gov.cn/directory/web/nmpa/ylqx/ylqxggtg/ylqxqtgg/20150119120001343.html.
China Food and Drug Administration (2015c) Medical device classification principles. https://www.nmpa.gov.cn/ylqx/ylqxfgwj/ylqxbmgzh/20150714120001554.html?type=pc&m=.
China Food and Drug Administration (2015d) Administrative measures on export sales certificates for medical devices. https://www.nmpa.gov.cn/xxgk/ggtg/ylqxggtg/ylqxqtggtg/20150601120001159.html?type=pc&m=.
China Food and Drug Administration (2015e) Notice on guidelines on site inspections for medical device good supply practices. http://www.nmpa.gov.cn/WS04/CL2197/324866.html.
China Food and Drug Administration (2016a) Notice on textual changes in the instructions for use of in vitro diagnostic reagents. https://www.nmpa.gov.cn/directory/web/nmpa/xxgk/fgwj/gzwj/gzwjylqx/20160805162701333.html.
China Food and Drug Administration (2016b) Announcement of the general administration on rectifying illegal operation in the field of medical device circulation. No. 112. https://www.nmpa.gov.cn/directory/web/nmpa/ylqx/ylqxggtg/ylqxqtgg/20160607180201128.html.
China Food and Drug Administration (2016c) Explanation of priority approval for medical devices. https://www.nmpa.gov.cn/directory/web/nmpa/xxgk/zhcjd/zhcjdylqx/20161026164301389.html.
China Food and Drug Administration (2016d) Notice on several matters related to the implementation of the medical device GMP by Class III device makers. https://www.ccfdie.org/zryyxxw/zxdt/webinfo/2017/01/1485585738337592.htm.
China Food and Drug Administration (2017a) Guiding principles for the division of medical device registration units. https://www.nmpa.gov.cn/xxgk/ggtg/qtggtg/20171123171101393.html?type=pc&m=.
China Food and Drug Administration (2017b) Interpretation of YY/T 0287-2017. https://www.nmpa.gov.cn/directory/web/nmpa/xxgk/zhcjd/zhcjdylqx/20170204102501630.html.
China Food and Drug Administration (2017c) Medical device distribution supervision administrative measures. https://www.nmpa.gov.cn/directory/web/nmpa/ylqx/ylqxfgwj/ylqxbmgzh/20171121171601668.html.
China Food and Drug Administration (2017d) Medical device recall measures. http://www.gov.cn/xinwen/2017-02/08/content_5166508.htm#2.
China Food and Drug Administration (2017e) Notice on medical device classification work. https://www.nmpa.gov.cn/directory/web/nmpa/xxgk/fgwj/gzwj/gzwjylqx/20170926173301389.html.
China Food and Drug Administration (2017f) Notice regarding the implementation of medical device good manufacturing practices for Class I and II medical devices. https://www.nmpa.gov.cn/directory/web/nmpa/xxgk/fgwj/gzwj/gzwjylqx/20170901161101542.html.
China Food and Drug Administration (2017g) Notice on matters related to medical device distribution registration. https://www.gov.cn/xinwen/2017-11/03/content_5236985.htm.
China Food and Drug Administration, National Health and Family Planning Commission (2014) Notice on strengthening the management of genome sequencing products and technology. http://www.nhc.gov.cn/yzygj/s3593/201402/c395a20d3815430d8b1a54313ce23b2b.shtml.
Covington (2020) China and the United States adjust export/import policies on PPE. https://www.cov.com/-/media/files/corporate/publications/2020/04/china-and-the-united-states-adjust-export-import-policies-on-ppe.pdf.
Cyranoski D (2016) China embraces precision medicine on a massive scale. Nature 529:9–10. https://doi.org/10.1038/529009a.
Deloitte (2021) China medical device industry: how to thrive in an increasingly competitive market? https://www2.deloitte.com/content/dam/Deloitte/cn/Documents/life-sciences-health-care/deloitte-cn-lshc-medical-device-white-paper-en-210301.pdf.

General Administration for Quality Supervision, Inspection and Quarantine (2016) Warning notice on importation of old medical devices, no. 13. http://fuzhou.customs.gov.cn/changsha_customs/zfxxgk76/2962331/crjjyjyygd/tzgg38/3950467/index.html.

Hainan Real World Data Research Institute (2022a, March 9) Johnson & Johnson Catalys Femtosecond Laser Eye Therapy System - approved for introduction using real-world data. https://hnrws.cn/zhenyanchengguo/1357/1419.html.

Hainan Real World Data Research Institute (2022b) Organization profile. https://hnrws.cn/about/1396/.

Hayashi Y (2021, February 4) Millions of counterfeit masks flooded U.S. customs facilities last year. The Wall Street Journal. https://www.wsj.com/articles/millions-of-counterfeit-masks-flooded-u-s-customs-facilities-last-year-11612436403.

Health Pink (2019) Summary of China's medical device reform policy. https://www.sohu.com/a/338547652_648361.

Huawei (December 2022/January 2023) From lifestyle accessory to medical device. Transform. https://www.huawei.com/en/media-center/transform/06/from-lifestyle-accessory-to-medical-device.

International Trade Administration (2022) China country commercial guide: healthcare. https://www.trade.gov/country-commercial-guides/china-healthcare.

Li J (2022, March 11) E-prescription, remote diagnosis and treatment, mobile phone bill payment—online service keeps patients from going to the hospital. People's Daily Overseas Edition. http://www.gov.cn/xinwen/2022-03/11/content_5678414.htm.

Li Lei versus National Medical Products Administration (2020) Beijing No. 1 Intermediate Court, Jing, First Adjudication No. 755

Mian Q (2022, February 14) Well known laboratory fined, why is a medical device (IVD) registration certificate necessary? IIVD.net. https://www.iivd.net/article-28153-1.html.

Ministry of Commerce (2020) Notice number 12 increasing regulation of non-medical masks for export. http://www.mofcom.gov.cn/article/b/g/202005/20200502967516.shtml.

Ministry of Health (2004) Administrative measures for the configuration and use of large medical equipment. http://www.nhc.gov.cn/guihuaxxs/s3585/200905/5c4a6b4a07af44978b854a981ed4e32f.shtml.

Ministry of Health (2011) Medical device recall rules (trial implementation). https://www.nmpa.gov.cn/directory/web/nmpa/xxgk/fgwj/bmgzh/20110520104201832.html.

Morgenson G (2020) Unapproved Chinese coronavirus antibody tests being used in at least 2 states. NBC News. https://www.nbcnews.com/health/health-news/unapproved-chinese-coronavirus-antibody-tests-being-used-least-2-states-n1185131.

National Health and Family Planning (2014) Technical guidelines for the construction of a telemedicine information system. http://www.nhc.gov.cn/mohwsbwstjxxzx/s8555/201501/120c912de1c4474c95b345d001271368/files/92f278e879d743de8e2ba5f38c1ed3ae.pdf.

National Health and Family Planning Commission (2016) Notice on issues related to clinical testing items. http://www.nhc.gov.cn/yzygj/s7659/201603/4949119e603448bbb17db1db7ab6d363.shtml.

National Health and Medical Development (2014) Opinions of the Health and Family Planning Commission on promoting telemedicine services in medical institutions. http://www.gov.cn/gongbao/content/2014/content_2792664.htm.

National Health Commission (2018) Catalogue of large-scale medical devices. http://www.nirp.cn/userfiles/file/16WJWWJ.pdf

National Health Commission (2020) Notice on further promoting "Internet + Medical Health" and "Five Ones" service actions. http://www.gov.cn/zhengce/zhengceku/2020-12/10/content_5568777.htm.

National Health Commission (2021) Administrative measures on the clinical use of medical devices. http://www.nhc.gov.cn/fzs/s3576/202102/db876b0a2d2f4910bd1cf8009dfbe074.shtml.

National Health Commission (2022) Measures on the administration of network security for medical institutions. http://www.gov.cn/zhengce/zhengceku/2022-08/30/content_5707404.htm.
National Health Commission, Chinese Medicine Administration (2018a) Measures for the administration of internet diagnosis and treatment (for trial implementation)
National Health Commission, Chinese Medicine Administration (2018b) Internet hospital management measures (trial)
National Health Commission, Chinese Medicine Administration (2018c) Internet hospital basic standard (trial)
National Health Commission, Chinese Medicine Administration (2018d) Telemedicine service management standard (trial). http://www.gov.cn/gongbao/content/2019/content_5358684.htm.
National Medical Products Administration (2017a) Notice on the release of the medical device classification catalogue. https://www.nmpa.gov.cn/directory/web/nmpa/ylqx/ylqxggtg/ylqxqtgg/20170904150301537.html.
National Medical Products Administration (2017b) Guidelines on the technical review of mobile medical device registration. https://www.nmpa.gov.cn/directory/web/nmpa/ylqx/ylqxggtg/ylqxqtgg/20171229134501226.html.
National Medical Products Administration (2018a) Technical guidelines on the acceptance of overseas medical device clinical trial data. https://www.nmpa.gov.cn/directory/web/nmpa/zhuanti/ypqxgg/ggzhcfg/20180111175801167.html.
National Medical Products Administration (2018b) Guidelines for the technical review of soft contact lenses. https://www.nmpa.gov.cn/directory/web/nmpa/xxgk/ggtg/qtggtg/20180426172301719.html.
National Medical Products Administration (2018c) Innovative medical device special review procedures. https://www.nmpa.gov.cn/zhuanti/cxylqx/cxylqxzyxx/20181105160001106.html?type=pc&m=.
National Medical Products Administration (2019a) Interpretation of medical device UDI system rules. http://mpa.gd.gov.cn/ztzl/wybssd/wjjs/content/post_3918743.html.
National Medical Products Administration (2019b) Consolidated medical device clinical trial exemption catalogue. https://www.nmpa.gov.cn/directory/web/nmpa/xxgk/zhcjd/zhcjdylqx/20191223093301810.html.
National Medical Products Administration (2019c) Medical device unique identifier system. https://udi.nmpa.gov.cn/toDetail.html?infoId=37&CatalogId=2.
National Medical Products Administration (2019d) Notice on completing the work on the first batch of UDI medical devices. https://www.nmpa.gov.cn/directory/web/nmpa/xxgk/ggtg/qtggtg/20191015160401461.html.
National Medical Products Administration (2019e) Notice on determination of drug device combination products. https://www.nmpa.gov.cn/zwfw/zwfwgggs/tzgg/20190718140001703.html?type=pc&m.
National Medical Products Administration (2019f) Our countries medical device standard system continues to improve. http://www.gov.cn/xinwen/2019-01/14/content_5357901.htm.
National Medical Products Administration (2020a) Administrative measures on the quality sampling inspection of medical devices. http://www.gov.cn/zhengce/zhengceku/2020-03/25/content_5495451.htm.
National Medical Products Administration (2020b) Interpretation of technical guidelines for clinical evaluation of medical devices using real-world data (for trial implementation). https://www.nmpa.gov.cn/directory/web/nmpa/xxgk/zhcjd/zhcjdylqx/20201130144911110.html.
National Medical Products Administration (2020c) List of Class III medical devices requiring clinical trial examination and approval. https://www.nmpa.gov.cn/directory/web/nmpa/zhuanti/ypqxgg/ggzhcfg/20200918103742111.html.
National Medical Products Administration (2020d) Notice on several matters concerning the domestic manufacture of imported medical devices. https://www.nmpa.gov.cn/directory/web/nmpa/ylqx/ylqxggtg/ylqxqtgg/20200925152407171.html.

National Medical Products Administration (2021a) Catalogue of Class II devices exempt from distribution record filings. https://www.nmpa.gov.cn/directory/web/nmpa/xxgk/ggtg/qtggtg/20210630150227194.html.
National Medical Products Administration (2021b) Artificial intelligence medical use software classification guiding principles. https://www.nmpa.gov.cn/xxgk/ggtg/qtggtg/20210708111147171.html?type=pc&m=.
National Medical Products Administration (2021c) Catalogue of medical devices exempt from clinical evaluation. https://www.nmpa.gov.cn/ylqx/ylqxggtg/20210918150708171.html?type=pc&m=.
National Medical Products Administration (2021d) Guidelines on determining when to conduct a medical device clinical trial. https://www.nmpa.gov.cn/xxgk/ggtg/qtggtg/20210928170338138.html?type=pc&m=.
National Medical Products Administration (2021e) Notice on registration of medical device masterfiles. https://www.nmpa.gov.cn/xxgk/ggtg/ylqxggtg/ylqxqtggtg/20210312150925118.html.
National Medical Products Administration (2021f) The General Department of the State Food and Drug Administration publicly solicits opinions on the optimization of mandatory standards for medical devices. https://www.nmpa.gov.cn/directory/web/nmpa/xxgk/zhqyj/zhqyjylqx/20210714162921101.html.
National Medical Products Administration (2021g) Guiding principles for the registration review of medical device clinical data submission requirements. https://www.nmpa.gov.cn/ylqx/ylqxggtg/20211125173931195.html.
National Medical Products Administration (2021h) Guidelines on annual quality system self-inspection reports. https://www.nmpa.gov.cn/directory/web/nmpa/ylqx/ylqxggtg/20220324100322197.html.
National Medical Products Administration (2021i) IVD registration and filing measures. Article 111. https://gkml.samr.gov.cn/nsjg/fgs/202108/t20210831_334232.html.
National Medical Products Administration (2021j) Medical device emergency evaluation and approval procedures. https://www.nmpa.gov.cn/directory/web/nmpa/images/1640855587877082585.docx.
National Medical Products Administration (2021k) Medical device internet sales supervisory measures. https://gkml.samr.gov.cn/nsjg/bgt/202106/t20210628_331663.html.
National Medical Products Administration (2021l) Medical device registration application requirements and approval documentation forms. https://www.nmpa.gov.cn/ylqx/ylqxggtg/20210930155134148.html.
National Medical Products Administration (2021m) Medical device standard catalogue. https://www.nmpa.gov.cn/ylqx/ylqxjgdt/20210531143615123.html?type=pc&m=.
National Medical Products Administration (2021n) National medical device adverse event surveillance annual report. https://www.nmpa.gov.cn/directory/web/nmpa/xxgk/fgwj/gzwj/gzwjylqx/20210329170305102.html.
National Medical Products Administration (2021o) NMPA issued medical device registration technical review guiding principles catalogue. https://www.nmpa.gov.cn/ylqx/ylqxjgdt/20210531143442106.html?type=pc&m=.
National Medical Products Administration (2021p) Notice on completing the implementation work on the second batch of UDI medical devices. https://www.nmpa.gov.cn/directory/web/nmpa/ylqx/ylqxggtg/20210917171201171.html.
National Medical Products Administration (2021q) Notice on the issuance of six sets of guiding principles for the naming of medical devices including those on the naming of medical rehabilitation equipment. https://www.ccfdie.org/cn/yjxx/ylqx/webinfo/2021/08/1622499299051709.htm.
National Medical Products Administration (2021r) Notice on several issues related to drug-device combination products. https://www.nmpa.gov.cn/xxgk/fgwj/xzhgfxwj/20210727154135199.html.

National Medical Products Administration (2021s) Operational guidelines on the examination and approval of registrations of Class III domestic and imported medical devices. https://www.nmpa.gov.cn/directory/web/nmpa/xxgk/fgwj/gzwj/gzwjylqx/20211104141907123.html.
National Medical Products Administration (2021t) Release of five technical guidelines including the guidelines for technical evaluation of medical devices. https://www.nmpa.gov.cn/directory/web/nmpa/ylqx/ylqxggtg/20210928170338138.html.
National Medical Products Administration (2021u) Second revision of medical device clinical trial exemption catalogue. https://www.nmpa.gov.cn/zhuanti/ypqxgg/ggzhcfg/20210119105058137.html?type=pc&m=.
National Medical Products Administration (2021v) Technical guidelines on the clinical trials of in vitro diagnostics. https://www.nmpa.gov.cn/ylqx/ylqxggtg/20210927152837140.html?type=pc&m=.
National Medical Products Administration (2021w) Technical guidelines on medical device clinical evaluation. https://www.nmpa.gov.cn/directory/web/nmpa/ylqx/ylqxggtg/20210928170338138.html.
National Medical Products Administration (2022a) Adjustment of part of the content of the medical device classification catalogue. https://www.nmpa.gov.cn/directory/web/nmpa/ylqx/ylqxggtg/20220324153614110.html.
National Medical Products Administration (2022b) Class I medical device filing operation standard. https://www.nmpa.gov.cn/xxgk/ggtg/qtggtg/20220811172649189.html.
National Medical Products Administration (2022c) Medical device manufacturing entrustment agreement guideline. https://www.nmpa.gov.cn/ylqx/ylqxggtg/20220324170135111.html?type=pc&m=.
National Medical Products Administration (2022d) Notice on completing the implementation work on the third batch of UDI medical devices (draft for comment). https://www.nmpa.gov.cn/xxgk/zhqyj/zhqyjylqx/20221130144240106.html?type=pc&m=.
National Medical Products Administration (2022e) Notice of the State Food and Drug Administration on issuing the catalogue of medical devices prohibited from entrusted production, no. 17. https://www.nmpa.gov.cn/xxgk/ggtg/qtggtg/20220324101855146.html?type=pc&m=.
National Medical Products Administration (2022f) Opinion on increasing categorized supervision of medical device manufacturers and distributors. https://www.nmpa.gov.cn/xxgk/fgwj/gzwj/gzwjylqx/20220909171207137.html?type=pc&m=.
National Medical Products Administration (2022g) Statistical report on drug administration (3rd quarter 2021). https://www.nmpa.gov.cn/zwgk/tjxx/tjnb/20220318160852122.html.
National Medical Products Administration (2022h) Medical device good supply practice appendix on quality management for enterprises specializing in medical device transportation and storage. https://www.nmpa.gov.cn/xxgk/fgwj/xzhgfxwj/20221101154122176.html.
National Medical Products Administration (2022i) Medical device clinical trial protocol sample https://www.nmpa.gov.cn/directory/web/nmpa/images/1648714444800058874.docx.
National Medical Products Administration (2023a) Catalogue of approved innovative devices. https://www.nmpa.gov.cn/zhuanti/cxylqx/ylqxxgmlhz/20210531163512106.html?type=pc&m.
National Medical Products Administration (2023b) Urgent use medical device procedures (trial implementation) (draft). https://www.nmpa.gov.cn/xxgk/zhqyj/zhqyjylqx/20230119161948141.html?type=pc&m=.
National Medical Products Administration (2023c) Notice releasing medical device good supply practice rules. https://www.nmpa.gov.cn/xxgk/ggtg/ylqxggtg/ylqxqtggtg/20231207173141173.html.
National Medical Products Administration (2023d) Interpretation I of medical device good supply practice rules. https://www.nmpa.gov.cn/xxgk/zhcjd/zhcjdylqx/20231208112259137.html?type=pc&m=.

National Medical Products Administration, National Health Commission et al (2023) Notice on implementation of third batch of unique identifiers for medical devices. https://www.nmpa.gov.cn/ylqx/ylqxggtg/20230217152350198.html?type=pc&m=.
National People's Congress (2020) Criminal code of the People's Republic of China, Article 145. http://www.chnlawyer.net/law/subs/xingfa.html.
Pharmaceutical Economic News (2011) Don't let refurbished medical devices be equated with fake and inferior devices. HC3I.cn. https://www.hc3i.cn/php/marticle.php?id=15369.
Qu T et al (2020) 600 internet hospitals online, has the spring of China. Xinhua News Agency. http://www.xinhuanet.com/politics/2020-11/24/c_1126778767.htm.
Shanghai Medical Products Administration (2020) Draft guidelines for the technical review of contrast medium injection devices. https://www.ciopharma.com/supervise/10681.
State Administration for Market Regulation (2014) Medical Device Supervision and Administration Regulation
State Administration for Market Regulation (2015) Medical device good use practices. https://gkml.samr.gov.cn/nsjg/bgt/202106/t20210629_331731.html.
State Administration for Market Regulation (2017) Medical device administrative measures. https://gkml.samr.gov.cn/nsjg/bgt/202106/t20210628_331673.html.
State Administration for Market Regulation (2018a) Medical device adverse event monitoring and re-evaluation measures. Article 4. https://gkml.samr.gov.cn/nsjg/bgt/202106/t20210624_331375.html.
State Administration for Market Regulation (2018b) Supervisory administrative measures on agents of imported medical devices (draft for comment). http://www.mofcom.gov.cn/article/b/g/201901/20190102831736.shtml
State Administration for Market Regulation (2021a) In vitro diagnostic registration and record-filing measures. https://gkml.samr.gov.cn/nsjg/fgs/202108/t20210831_334232.html#:~:text=%E5%A2%83%E5%86%85%E7%AC%AC%E4%B8%80%E7%B1%BB%E4%BD%93%E5%A4%96%E8%AF%8A%E6%96%AD%E8%AF%95%E5%89%82%E5%A4%87%E6%A1%88%EF%BC%8C%E5%A4%87%E6%A1%88%E4%BA%BA,%E7%AE%A1%E7%90%86%E5%B1%80%E6%8F%90%E4%BA%A4%E5%A4%87%E6%A1%88%E8%B5%84%E6%96%99%E3%80%82.
State Administration for Market Regulation (2021b) Medical device distribution supervision administrative measures. https://gkml.samr.gov.cn/nsjg/fgs/202203/t20220322_340682.html
State Administration for Market Regulation (2021c) Medical device manufacturing measures. https://gkml.samr.gov.cn/nsjg/bgt/202106/t20210629_331736.html.
State Administration for Market Regulation (2021d) Medical device registration and recordation administrative regulations
State Administration for Market Regulation (2022a) Medical device good clinical practice rules. https://www.samr.gov.cn/zw/gz/lhzd/202206/P020220629544294696316.pdf.
State Administration for Market Regulation (2022b) Medical device manufacturing supervision measures. https://gkml.samr.gov.cn/nsjg/fgs/202203/t20220322_340674.html.
State Administration for Market Regulation (2022c) Medical device distribution supervision and administration rules. https://gkml.samr.gov.cn/nsjg/fgs/202203/t20220322_340682.html.
State Administration of Medicine (1996) Medical device registration rules. http://fgcx.bjcourt.gov.cn:4601/law?fn=chl075s045.txt&term=5.
State Council (2021) Medical Device Supervision and Administration Regulation. http://www.gov.cn/zhengce/content/2021-03/18/content_5593739.htm.
State Council of the People's Republic of China (2000) Medical Device Supervision and Administration Regulation (no. 276). http://www.gov.cn/gongbao/content/2000/content_60576.htm.
State Council of the People's Republic of China (2014) Medical Device Supervision and Administration Regulation
State Council of the People's Republic of China (2017) Medical Device Supervision and Administration Regulation. http://www.nhc.gov.cn/fzs/s3576/201808/c8a195d01cfe4b8da8908e920e637180.shtml.

State Council of the People's Republic of China (2018) Administrative measures on the configuration and use method of large-scale medical devices. http://www.gov.cn/gongbao/content/2018/content_5338246.htm.
State Drug Administration (1991) Interim administrative measures on medical devices. https://www.srlz.gov.cn/m/view.php?aid=4623.
State Drug Administration (2002) Medical device classification catalogue. https://www.nmpa.gov.cn/directory/web/nmpa/xxgk/fgwj/gzwj/gzwjyp/20020828010101509.html.
State Food and Drug Administration (2004) Provisions on medical device clinical trials. http://www.nhc.gov.cn/zwgk/wlwl/201105/f23ec06b5acb4dcc870776df0dc367f9.shtml.
State Food and Drug Administration (2007a) Notice on the issuance of the "Technical Guiding Principles for Clinical Research of In Vitro Diagnostic Reagents" and the "Guiding Principles for Writing Instructions for In Vitro Diagnostic Reagents," [Canceled]. https://www.nmpa.gov.cn/directory/web/nmpa/xxgk/fgwj/gzwj/gzwjylqx/20070428010101666_2.html.
State Food and Drug Administration (2007b) In vitro diagnostic reagent registration measures (trial implementation). http://fgcx.bjcourt.gov.cn:4601/law?fn=chl356s183.txt&term=63.
State Food and Drug Administration (2008) Medical device adverse event monitoring and re-evaluation measures (trial implementation). https://www.nmpa.gov.cn/directory/web/nmpa/xxgk/fgwj/gzwj/gzwjylqx/20081229134901842.html.
State Food and Drug Administration (2011) Notice on questions regarding the implementation of the medical device good manufacturing practices (trial implementation) and related documents. https://www.nmpa.gov.cn/directory/web/nmpa/xxgk/fgwj/gzwj/gzwjylqx/20110127095401382.html.
Sun J (2021) Important content in the technical guidelines for the clinical evaluation of medical devices. China Food and Drug Network. http://mpa.jl.gov.cn/zxfw_84842/xwlm/gndt/202110/t20211025_8256554.html.
Sun L (2022, December 12) Real-world data helps medical device innovation and development. China Medical News Release. https://www.cmde.org.cn//xwdt/zxyw/20230103085521172.html.
U.S. Department of State (2022) Investment climate statements: China. https://www.state.gov/reports/2022-investment-climate-statements/china/.
U.S. Food and Drug Administration (2007) Agreement on the safety of drugs and medical devices. https://www.fda.gov/international-programs/cooperative-arrangements/fda-sfda-china-agreement-safety-drugs-and-medical-devices.
X Medical Sales Co. and Y Medical Device Company Civil Contract Dispute in the Second Instance (2020) Civil Judgment No. 11167. Guangdong
Zhang C (2022) Application of AI technology in the diagnosis and treatment of Covid-19. Center for Medical Device Evaluation device review. https://www.cmde.org.cn//splt/ltwz/ltyy/20220812092027172.html.
Zhongcheng Medical Device Network (2022) Inventory of 2021 innovative medical device data. IIVD.net. https://www.iivd.net/article-28166-1.html.

Biosecurity and FDD Research and Development

7

7.1 Introduction to Biosecurity

China's legislative efforts have increasingly focused on the connections between data, health, and national security. Key examples include the Cybersecurity Law, the Data Security Law, the Personal Information Protection Law, and the National Security Law. In addition, there is the Biosecurity Law (2021), which covers a broad range of topics from bioterrorism to the protection of biological resources stemming from animals, plants, and (importantly for our discussion here) human beings, to combating infectious disease and antimicrobial resistance.[1]

For many years, China has had a concern about foreign exploitation of its natural resources and people. This is reflected in its approach to clinical trials and its safeguards against using Chinese people as a testing ground for drugs that have not already been tested in other countries. For example, until 2017 international multi-center clinical trials were not allowed to begin in China until *after* they had entered a more advanced stage abroad. But since the late 1990s, and more intensely since 2015, there also has been concern about misuse or abuse of biotechnology and the use of the genetic code of Chinese citizens.

This concern was spurred on by various incidents, including reports that the scientist Jiankui He, in China, had performed gene editing on embryos, and his subsequent condemnation and criminal sentence.[2] As Bai Chunli, a deputy to the National People's Congress, stated, "the formulation of the law requires a good grasp of the 'degree' between scientific norms, avoiding misuse and abuse, encouraging scientific research and exploration, and not giving up food because of choking."[3]

[1] Standing Committee of the National People's Congress (2021), Article 2.

[2] Bai (March 11, 2019).

[3] *Ibid.*

J. C. Balzano, *China Food and Drug Law*, Law for Professionals,
https://doi.org/10.1007/978-3-031-61901-4_7

Various policies reflect concerns that biotechnology could allow China human genetic resources to be misused or exploited, with potential ramifications for China's national security. The dangers of uncontrolled collection and exportation of such resources by foreign parties, including both samples and data, became an area of focus and preventive measures deemed necessary by the government, particularly as the number of foreign-sponsored trials for innovative drugs began to grow due to drug reforms, as discussed in Chap. 5.

China has joined various international conventions aimed at protecting plant and animal resources, such as the United Nations Convention on Biological Diversity, Nagoya Protocol on Access and Benefit Sharing, and Cartagena Protocol on Biosafety. While China has been slower to implement the letter of these documents with detailed legislation, their principles about the need to protect natural resources and requirement of foreign developers to share the benefits with local stakeholders have, in part, driven other laws and regulations.

7.2 China's 1998 "Interim" Human Genetic Resource Regime

The "Human Genetic Resource Regulations (Trial Implementation)" ("HGR Regulations") were established in 1998, with the basic premise that foreign companies, including companies with foreign investment, were not permitted to collect or utilize samples containing the genetic material of Chinese individuals in China on their own. These rules were established in response to "leakage" incidents in which samples taken from humans in China were transported abroad with little constraint.[4] Because of public health and national security concerns, both these samples and their associated data needed to be collected by a Chinese entity, and the foreign company was then permitted to use those samples and data in the context of an approved international collaboration between it and the chosen Chinese entity.

The Ministry of Science and Technology (MOST), under which an Office of Human Genetic Resources (OHGRA) used to sit, must approve any international collaboration, and a separate approval was required to export human genetic resources, which included samples and possibly included data (the regime regarding data is now different). At some points, the OHGRA suggested that data could be exported if properly declared under an international collaboration. Finally, the 1998 HGR Regulations contained provisions that required co-ownership between the foreign and Chinese parties of patent rights arising from the use of human genetic resources.

The regime has been under MOST (with some early and more recent 2024 transfer of HGR administration to what is now the National Health Commission (NHC)). MOST plays a role in directing human biological research in China through its use of national research grants. MOST also administered laws governing science and technology and the promotion of government-led or funded research. As of this

[4] Xin and Wang (2023).

writing, the National People's Congress in 2023 passed a government reorganization plan that placed the China National Center for Biotechnology Development (under which the Office of Human Genetic Resource Administration (OHGRA) sat) under the NHC.[5] The State Council implemented this change to NHC jurisdiction via an amendment to the HGR Regulations that took effect in May 2024 (State Council of the People's Republic of China 2024). This change represents one of the many significant changes to the HGR regime that occurred over 2023–2024 and that are still ongoing as of the time of this writing.

Between 1998 and 2015, the HGR Regulations were not regularly observed in practice, perhaps because they were interpreted as inapplicable if the clinical research did not focus on genetics and the human genome, nor is there evidence that MOST vigorously enforced them. The government increasingly emphasized the need for research involving foreign participation in the collection of human genetic resources to be approved, but the scope of research projects subject to this approval remained unclear.[6] The government admitted in 2013 that "[s]ince the implementation of the measures, the protection and management of human genetic resources in China has achieved good results, but the phenomenon of loss of human genetic resources still occurs from time to time." It also noted that human genetic resources needed to be protected in the context of "promoting new drug innovation."[7]

7.2.1 Emergence of the Current Regime

In 2015, the scope became clearer and arguably broader. MOST and OHGRA released an extensive guideline that year that more clearly defined the scope of the HGR regime, including applying it to all research involving a foreign party, and only providing limited exceptions from the rule for funerals, police investigations, blood donation and, perhaps most broadly, clinical treatment. But not drug and device research. The new guideline revamped the administrative approval process and renamed the approval "Human Genetic Resources collection, receipt, trading, export, exit approval." Applications under this new system were accepted starting on October 1, 2015.[8]

The 2015 guideline and a subsequent grace period for unapproved clinical trials to obtain approval led to a scramble for existing clinical trials. To obtain approval, the foreign sponsors had to work with hospitals, which at that time were required to submit the applications. Hospitals were not ready for this work and needed support from trial sponsors to put together the necessary application information, which included information about the trial and related parties supporting the sponsor. Over the next four years, HGR approval would become a regular feature of clinical

[5]National People's Congress (2023).

[6]Lifei and Xin (June 19, 2018).

[7]Ministry of Science and Technology (2013).

[8]Ministry of Science and Technology (2015).

research in China. Sponsors needed to obtain this approval after the clinical trial approval from NMPA and the ethics committee but before the clinical trial agreement could be signed. As time went by, OHGRA seemed to be expanding its requirements, including requiring that an increasing array of entities associated with the trial, including laboratories, contract research organizations, bioanalytics firms, and waste disposal companies, be reported and present agreements with the sponsor and credentials.

But, because the original HGR Regulations were quite general, few other guidelines existed (besides those in 2015), and OHGRA officials did not have extensive experience with medical device or drug development, many questions remained. For example, it was unclear what constituted a foreign party or a Chinese party, including what level of investment was required, as well as what data was covered and what was needed to export covered data.

The renewed emphasis on enforcement of the HGR rules and the release of an even more stringents amendment for comment in 2016 raised concerns that the HGR Regulations could become a serious impediment to clinical development in China. Other countries that China seeks to match in terms of research and development, such as those in the EU, Japan, and the U.S., at that time did not currently limit the collection of samples by foreign companies in the same way, i.e., in a way that goes beyond rules to prevent public-health issues and human-subject protections. The HGR regime was an increasing burden, at a time when NMPA's process was becoming more efficient. And penalty notices from 2015 through 2018 showed that MOST was serious about enforcement of the letter of the rules, with increasing fines and debarment for companies and hospitals.[9] This was something of a shock because NMPA had not often pursued debarment penalties available under its laws and regulations against foreign industry.

Another important change occurred in 2019. In late May, the State Council passed a significant amendment to the HGR Regulations. That amendment set more detailed rules for the structure of the collaboration between the foreign and Chinese parties and established a stringents regime in terms of the requirements of an application, rules governing data transfer, and increased penalties and enforcement powers for MOST and OHGRA. The Biosecurity Law, which was enacted in 2021, further enhanced penalties and, although its provisions were only extremely high level, signaled a continued commitment to the HGR regime by writing it into a national-level law and signaled some changes that were ultimately reflected in HGR implementing rules, such as changes to the data transfer regime from a record-filing to a "prior report".

In 2023, MOST released new HGR implementing rules (HGRIR).[10] These rules significantly reduced the scope of the HGR regime in many ways and made three major changes affecting drug and device companies doing clinical research in China

[9] A list of penalty notices, including names of those penalized, is available at https://fuwu.most.gov.cn/html/jcxtml/20181201/2837.html?tab=xzcf.

[10] MOST (2022c, 2023a) Implementing Rules of the Human Genetic Resource Regulations.

(although at the time of this writing the effects of the new rules are still emerging). First, the implementing rules narrowed the definition of a foreign entity subject to the international collaboration and other sample collection and data transfer approval or notification requirements to one that had 50% or more ownership by foreign shareholders or is otherwise controlled by foreign shareholders (e.g., through contractual arrangements). One or a few foreign investors holding a minority of the shares of stock will no longer deem the company to be a foreign company, unless there is evidence of other significant control. This appears to allow Chinese companies that have some form of foreign investment to be the China party in their collaborations with partners from other major markets. The HGIR also allows for the Hong Kong-based entities of Chinese companies to be considered Chinese parties. Previously, Hong Kong entities were considered foreign parties.

Second, the HGRIR narrowed the definition of human genetic resource information (i.e., the data covered by the rules) to human gene and genome data and other such information generated by utilizing human genetic resource materials. This definition was further narrowed under guidelines released shortly after the HGRIR: data on genes, genomes, transcriptomes, epitomes, ctDNA, and other nuclei acid bio markers, and related disease and race information.[11] The HGRIR expressly exclude the following types of data from coverage: clinical data, imaging data, protein data, and metabolic data. Although each type of data is not defined, the guidelines and application materials released online to date seem to indicate that covered data is limited to general or targeted genome sequencing data and a limited set of biomarker data, i.e., nucleic acid biomarkers.

Under the HGRIR, HGR materials (i.e., human samples covered by the rules) remains broadly defined. But, a question and answer document issued later narrows the definition of samples subject to HGR rules, i.e., plasma, urine, feces or samples with little to no shed cells are no longer covered.[12] This changes represent a significant narrowing of the scope of the rules.

Third, the HGRIR allow for broader use of existing pathways to obtain approval for studies through notification procedures, streamline the mechanisms for data transfer during an international collaboration (e.g., by allowing all entities on an approval to receive data contemplated in the relevant agreements without the need for additional approvals), and create faster procedures for amendments to existing approvals by creating a clearer structure for minor amendments via notification and major approvals via pre-approval. Importantly, the HGRIR allow for applicants to obtain approval for their international collaborations through a record-filing procedure (see Chap. 3), while still using central laboratories outside of the clinical trial sites to do complex sample analysis. The remaining sections will set forth the

[11] MOST (2023b) Notice on Updating Service Guidelines for Administrative Licensing Matters and Record-filing and Prior Report Scope and Procedures.

[12] MOST (2023c) Notice on the Release of Frequently Asked Questions about Human Genetic Resource Management, https://www.most.gov.cn/tztg/202309/t20230912_187931.html [Last Visited December 30, 2023].

different requirements under the HGR Regulations in more detail and noting how these different areas have evolved between the different stages of development of this regulatory regime. However, while regulatory change in China is constant and arguably faster than in other peer countries in the FDD area, the HGR regime is one of the more fluid regulatory regimes in China. Therefore, by the time this is published more change could occur. The sections presented here are a snapshot as of the end of 2023.

7.3 Collaborators and Collaboration

The HGR Regulations (2019) state that foreign companies or companies that are actually controlled by foreign companies cannot collect either HGR materials or the data associated with them on their own. Foreign-controlled standards are still applied very broadly. If a company has a majority or a minority with actual control of investors or shareholders that is not established in or citizens of mainland China, then it may be considered foreign. This includes investment from those in Hong Kong, Macao, and Taiwan—although the affiliates of Mainland Companies in Hong Kong or Macao may be considered Chinese parties for purposes of HGR.

Under the HGRIR, foreign companies are defined as having a majority foreign share-owner. But as noted it could also be a company in which foreign investors or shareholders are in the minority but have a significant impact over voting or operations or control the company through contracts or other means. In practice, more Chinese companies can now claim this status. The implementing rules also allow the Hong Kong and Macao affiliates of these companies to claim Chinese party status.

Foreign entities may only "utilize" HGR in collaboration with one or more Chinese parties ("international collaboration"). The HGR regulators must approve these international collaborations prior to the study commencing after reviewing various materials, including the protocol for the study and the informed consent form, and evaluating whether the international collaboration research will harm China's public health and national security and whether the purpose, content, and length of time of the research is "reasonable." Such research must also have been evaluated by an ethics committee in China and, as will be discussed below, there must be a reasonable and specific distribution of the benefits of the research plan between the foreign and Chinese parties.

To understand the research plan, the HGR regulator evaluates the protocol of the research, the agreement between the Chinese and foreign collaborators, agreements between the foreign sponsor and other entities that will receive HGR, and information about the type and quantity of samples and associated data set forth in the application. The HGR regulator also reviews the proposed informed consent form to ensure that subjects will give proper consent to the collection of their samples and that the scope of the project in the consent is consistent with that proposed in the application. The international collaboration application requires a description of the different types and amounts of samples they will take (e.g., whole blood, tumor,) per subject, which laboratories will test them, and how they will test (i.e., what type of analysis) and store or dispose of them. Similarly for data, the applicant must state how much of different types of data (i.e., biomarker and genetic) it will collect on both individuals that are screened and enrolled subjects over the course of the study. The application must list the entities that will analyze these data and where and for how long those data will ultimately be stored (typically permitted at least with the sponsor).

The foreign and Chinese collaborators are responsible for the implementation of the collaboration in accordance with the approval. The HGR Regime only recognizes four types of entities as substantive collaborators in an international collaboration: the Chinese entity (typically a hospital), the sponsor(s) of the trial (typically the overseas and local entities listed on the clinical trial approval), one main contract research organization (CRO), and the central laboratory. The flexibility to list more collaborators than this or entities that are not officially approved as sponsors has decreased over time. Since 2019, the MOST rarely made exceptions to allow, for example, an applicant to add more sponsors than are listed on the clinical trial approval from NMPA. Previously there were cases in which companies were allowed to list multiple foreign sponsors to allow better access to data.

In an earlier period of enforcement, the collaborators also had stronger rights to the data generated by the study. The 2019 HGR Regulations make it clear that collaborators should share "data and records" amongst each other. MOST also allowed data from the trial to be passed between these parties without further approval, whereas other non-collaborator parties must go through another record-filing step to transfer specific packets of certain covered data. Currently all entities, whether collaborators or Other Entities (see below) can receive data from the study without further approval. This data for transfer must be contemplated.

As will be discussed in the next section, the earlier constraints on data transfers to Other Entities previously set substantial constraints on the sponsor and the third parties assisting it in running a clinical trial and analyzing the data.

7.3.1 Other Entities

If a foreign entity does not fit into one of these three categories (sponsor, main CRO, central laboratory), then it cannot be a collaborator. Now, all non-collaborators—including entities within the same corporate family as the sponsor—must be listed as "Other Entities" on the HGR application. This can include electronic data capture vendors, data analytics firms, and possibly CROs that perform specialized roles. In some cases, the listing of "other entities" has been stretched to extreme lengths, including companies that play no role in analyzing samples or data for research purposes, such as companies that ship samples or translate or make electronic copies of data. However, under the HGRIR and its subsequent Q&A, an other entity should be listed if it substantively participates in the collaboration, including receipt of HGR information or trial results or samples. If a foreign company merely provides drug or partial funding for a clinical study (such as might be the case in an investigator initiated study, see Chap. 5) then it would not be considered to be substantively participating and may not be listed on the application.

The requirements for an application go further that simply listing "Other Entities" and require information about their responsible and contact persons and a contract, with a collaborator such as a service or partnership agreement to show the relationship and the HGR data that may be received. The requirement for an agreement causes additional burden, delay, and confidentiality concerns, as some of the service agreements contain business confidential information (and the MOST has sometimes insisted on seeing the full agreement as opposed to a scope of work).

International collaboration (and other types of HGR applications) must be submitted electronically. After the promulgation of the HGRIR, MOST established a new platform for this purpose.[13] Notably, the new platform requires that all entities on an application (including overseas entities) directly register with the HGR platform. This process includes naming an individual who is legally responsible for entity (legal representative) and a contact person. The registrant must provide an identification document, i.e., a passport. This direct registration by foreign entities appears different from the approach NMPA has taken, wherein a local agent holds the account to the online platform and does the registration work on behalf of the overseas entity.

7.4 Data Transfer

The HGR Regulations contain procedures for providing HGR information to the "outside," meaning to foreign parties, including foreign-invested companies in China, foreigners, and the public. In other words, these procedures are necessary whenever certain covered HGR information goes to any party other than a participant in an international collaboration or another Chinese party. There are two types of data filings: those to transmit data to the "outside," when non-public data covered under the HGR Regulation are published in a journal outside of China, and those to transmit data to a specific non-collaborator or other foreign entity.

Prior to the 2019 amendment of the HGR Regulations and its implementation, there were a number of questions about what data from a trial could be considered HGR Information and therefore subject to restrictions on transfer, including the aforementioned record filings to send the data to foreign parties. Under the 2019 HGR Regulations, at first, a policy was adopted where there were virtually no limits on the data covered. It could include any clinical data, imaging data, biomarker data, protein data, metabolic data, and genetic data. It seemed no data was excluded from coverage.

[13] MOST (2023d) Announcement on Adjustment to Matters Related to the Management of Human Genetic Resources, https://www.most.gov.cn/tztg/202306/t20230627_186788.html [Last Visited December 30, 2023].

In 2022, OHGRA began to liberalize this policy. It released question-and-answer documents that made it clear that there is no need to obtain approval to collect data that are already within the public domain[14] and certain routine, clinical, non-genetic data. It listed examples of data excluded from coverage: B-ultrasound, CT, PET-CT, nuclear magnetic resonance, X-ray and other image data, interventional, ophthalmoscopy, endoscopy, dermatoscopy, pathological diagnosis and other image data, and blood routine, urine routine, liver and kidney function, blood biochemistry and other general laboratory examination information, growth and development indicators, such as height and weight, questionnaire information, and other picture-result data.[15]

As noted, in 2023, the HGRIR went further. It categorically excluded clinical data, imaging data, protein data, and metabolic data. Later Guidelines made it clear that only genetic data (see definition above) and certain biomarker data would be covered. This shrunk the universe of covered data significantly.

If a transfer of covered HGR information to the "outside" is necessary, a Chinese party (i.e., the China side of the collaboration) must engage in a two-step application process in order to obtain permission from the the HGR regulator, in the form of a record-filing number. Under the HGRIR, this process is now called a prior-report.

In the first step, the Chinese party uploads a copy of the data to the HGR platform website. The HGR regulator will then keep this information on file as a record of the data that has gone to foreign parties. Once this upload is successful, the platform will generate a back-up number (*beifenhao*)—essentially a confirmation number—indicating as much.

In the second step, a form is filed identifying the party (or the public) to which the HGR information will be transferred. If transferred to a foreign company, this form will indicate responsible person for the transferee entity, the reason fo transfer, a description of the cybersecurity measures that the transferee will take to keep the data secure, its storage location, and an accounting of types and quantity of data for transfer. Previously, the OHGRA required, as a matter of policy, that the entity receiving the data must be an "other entity" on the original international collaboration approval. That means that any underlying international collaboration approval had to be amended as a first step before commencing the two-step process. Once that form is accepted and a second *beianhao*—this one essentially a filing number—is issued, the data can be legally transferred. The entire process could take several weeks to a month to complete, and possibly more if OHGRA determined that a "security review" is required or if the filing was rejected one or more times. Fortunately, it is no longer the case that a transferee must be reflected on an international collaboration approval. Indeed, data transfer record filings—which are now referred to as "prior reports" under the HGRIR—can occur regardless of

[14] Ministry of Science and Technology (2022a) Notice on Updating Frequently Asked Questions on Human Genetic Resources Management.

[15] Ministry of Science and Technology (2022b) Notice on Updating Frequently Asked Questions on Human Genetic Resources Management (Q&A Series 2).

whether they are associated with an international collaboration or not, and in a complete reversal, their purpose is now to transfer data to entities when such a transfer is not covered by an international collaboration, either because the entity is not on the approval, there data is outside the scope of the parties agreement, or there is no approval.

The security review may still be required if the HGR regulator determines that the data transfer affects China's public health, national security, or public interest. The HGRIR state that the following circumstances could trigger such a review during the data transfer process: (1) the information relates to Important Genetic Pedigrees, (2) the information relates to isolated Groups Specific Regions, (3) the information involves exome or genome sequencing for 500 or more cases, or (4).other circumstances that affect national security, public health or the public interest. The latter two terms "important genetic pedigrees" and HGR from "Specific Regions" are further defined in the HGRIR, reflecting the trend of narrowing the scope of the rules to focus resources on "genetic resources:"

- The former relates to a blood related group with a hereditary disease or a specific constitution or physiological makeup that persists for three generations.
- The latter refers to groups (not necessarily ethnic minorities) that have special physical characteristics or adaptive traits because they have lived in isolation or a special environment for a long-time.

The data-transfer process and restrictions, which were not truly implemented until 2020, were perhaps the most difficult and complex burden under the 2019 HGR Regulations. A company could tie itself in knots determining whether (a) data are covered by the HGR Regulations and (b) how to ensure that the correct permissions are in place prior to its transfer. The restriction on non-collaborators receiving data until a data transfer record-filing was complete restricted the real-time, free flow of data during a clinical trial, including adverse event data (depending on how broadly a company interpreted the term HGR information). Further, the fact that companies were required to have copies made of this data and upload them to a government platform raised concerns about data integrity (if data could be lost in the process or unblinded by asking the site to submit the back-up) or security.

Despite the difficulties with compliance, regulators did not focus enforcement on punishing companies or a hospitals purely because of violation of data-transfer rules, although there is a prohibited act in the HGR Regulations covering transfer of any human genetic resources, including materials and/or information, overseas without approval—this provision is phrased in a strict liability fashion (intent is not required). But penalties have been levied against companies for other reasons (including illegal sample transfer).

7.5 HGR Collection Permits

An additional permit to collect human biospecimens may be required in addition to the international collaboration. This is required if the collection involves Important Pedigrees of HGR— as defined above—or if the HGR comes from populations that have lived in Specific Regions. Finally, permits used to be required if the study involved 500 or more subjects.[16] Under the HGRIR, the first two circumstances continue to apply but the 500 subject limit has been raised to 3000 subjects. Also, under the HGRIR the special collection permit is not required if the HGR are collected as part of a clinical trial to support a marketing authorization of a drug or medical device in China.

7.6 Intellectual Property Sharing

Since the 1998 HGR rule, the HGR Regulations have contained provisions that require companies to share patents arising from the international collaboration. With the revision to the rule in 2019 and subsequent practice, the IP-sharing requirements became more complex. First, the HGR Regulations still require that any the rights to any patents arising from the international collaboration will be co-owned by the parties. This requirement does not refer to patents over the drug or background intellectual property, but rather patents for exploratory research that are unrelated to applications to obtain approval to market the drug. Although this idea of "exploratory research" is an undefined formulation, the point is that, while companies can own the drugs they develop in China, should something more emerge from the clinical trial (e.g., an unexpected byproduct of the research), then that patent right may have to be co-owned.

The HGR Regulations state that, other than these patent rights arising from exploratory research, parties can decide according to an agreement how to allocate ownership over other types of intellectual property, including (if applicable) data and know-how. If they do not do so, the IP is co-owned and the collaborators must ask permission of each other in order to transfer their rights to another. Benefits arising from the IP are allocated pro rata depending on the relative contributions of the collaborators, or equally if the contribution cannot be determined.

In practice, the MOST has at times gone further than only requiring sharing of patent rights over inventions arising from exploratory research. It has required that the parties co-own and share equally the rights not just to these patents but also to the exploratory research data, i.e., data that are unrelated to drug registration. Although not applied more recently, this unwritten policy has presented considerable difficulty because it is very difficult to determine which data are and are not related to the development of the drug for registration. Sometimes that does not become clear until later in the development process.

[16]Ministry of Science and Technology (2019).

For these reasons, the IP-sharing requirement over "exploratory research" has proven to be a deterrent to innovation in the country, particularly early-stage research. Arguably, exploratory research is a natural feature of the very early drug-development process, which entails the screening of numerous molecules with respect to a specific target in the body, or in research on biologics, as researchers attempt to engineer a solution based on a defined disease pathway (e.g., a mutated gene). Either way, the process inherently encompasses many successes and failures. The requirement to co-own patents over exploratory efforts means that there would be a risk that drug-company sponsors would have to co-own intellectual property with hospitals or other partners they work with in China. Although no court case wherein a hospital has sued to obtain the benefits of such research has tested enforcement of this rule, it introduces uncertainty into the process.

The allocation of IP rights among the collaborators is reflected both in the collaboration agreement and in the IP section of certain international collaboration-approval applications. The application instructions state that the language of both should be consistent on the sharing of ownership and dispute resolution mechanisms. For this reason, companies may need to carefully define with their partners what constitutes exploratory research and the ownership rights in the event of such co-ownership. Commitments from the sites or other Chinese parties to obtain the necessary rights to data and patents may be important to continuing the research.

In October 2022, the National Intellectual Property Administration made the need to comply with HGR Regulations even more important. NIPA issued a third draft of Patent Examination Guidelines that propose to deny patentability in China should genetic information (i.e., HGR information) not be properly received, pursuant to a record-filing or approval.[17] The final Patent Examination Guidelines were released in late 2023 and require that when an invention patent application relies entirely on human genetic information must be compliant with the rules on data transfer to foreign organizations, and if that is not the case, then the patent cannot be granted.[18] Given that the scope of human genetic information has been significantly narrowed, this provision should be much narrowerer than before when human genetic resource information covered all data from a trial in China,, although the impact of these changes, which are only just effective as of January 20, 2024 is not yet clear.

7.7 Application Delay and Other Difficulties Impeding Development

Under the HGR Regulations, regulators have two significant powers: the power to approve studies and the power to police compliance through inspections. Starting in 2017, the OHGRA exercised considerable control over studies on the front-end, putting in place a series of increasingly stringents policies that reached an apex in

[17] China National Intellectual Property Administration (October 31, 2022).

[18] China National Intellectual Property Administration (2023) Patent Examination Guidelines.

2019 and 2020, which made it more difficult to obtain approvals of international collaborations. If they go through an approval pathway, applications are first submitted electronically and reviewed for format and, if they pass, are reviewed by experts (e.g., professors or scientists) selected at random by the HGR regulator. Expert committees only meet once or twice per month to consider and approval applications. The HGR regulator does not permit formal meetings in advance of applications, but questions can be directed via phone or email to the National Biotechnology Development Center.[19]

Applications can be rejected for any number of reasons, such as inconsistencies between sections in the application setting forth the quantity and types of data samples for collection and other application materials, the "reasonableness" of the post-study storage arrangement, the "reasonableness" of the IP arrangement, and the consistency of the collaborators on the HGR application with those entities listed on the clinical trial approval. Some of these issues could cause an application to be rejected multiple times. Each rejection can add 2–3 months to the clinical trial preparation process, and some studies have been delayed by years or abandoned altogether, although the process as of this writing proceeds more smoothly compared to 2019–2021.

In 2019, the revised HGR Regulations also established a record-filing procedure to approve an international collaboration that operates like a notification, allowing the study to proceed once all the materials were received and considered complete. Use of the record-filing procedure is subject to certain conditions under the regulations: the study has to be for the registration of a drug or medical device in China and no samples exported during the study. Although not immediately apparent from the regulations, the OHGRA also interpreted the language in the HGR Regulations to condition use of the record-filing pathway on applicants promising that samples would be analyzed within the clinical trial site laboratories. As most studies use a commercial central laboratory in China due to the complexity of testing, this final condition was one of the reasons that this record-filing pathway was rarely used.

It is not clear why this restriction was necessary, given the realities of clinical trial practice and the fact that registrational trials (as opposed to exploratory research) should be the focus of less, not more, scrutiny. And, as mentioned, no samples would leave China (another controversial aspect). Certainly, the central laboratories used in these trials often have foreign investment or are branches of foreign companies, but under the HGR Regulations these laboratories are only permitted to store samples temporarily before and after testing, as called for in the protocol. Therefore, they do not pose a risk of banking them and using them for other purposes. The OHGRA did not specifically limit the timing of temporary storage, but it expects that after a reasonable amount of time the samples will be transferred to a biobanking facility, which are all required to be Chinese entities that are separately licensed by the OHGRA as preservation facilities.

---

[19] China National Center for Biotechnology Development. https://www.cncbd.org.cn/Index/.

In the 2023, the HGRIR allow for registrational studies that use central laboratories, contracted with by the sponsor, to proceed with a record-filing instead of an approval, provided that the conditions above are met *and* the sample testing is done within the clinical trial sits *or* laboratories designated in the protocol. This change loosens the restriction above and, potentially, allows for more liberal use of the record-filing pathway to approve of international collaboration. As this pathway is only for registration of drugs and devices, the HGRIR state that exploratory research will require a separate approval. The record-filing applications do not require IP sharing sections because the research is not supposed to be exploratory.

7.8 Sample Transport

Human samples may be transported within Mainland China without a separate approval by shipping companies and couriers, provided that those couriers comply with applicable rules on safe transport. As discussed above, the laboratories that handle and test these samples and the companies that dispose of waste must be listed on the international collaboration application forms as collaborators or other entities. Some of the waste-disposal companies may also have their own shipping/transport services. Entities that biobank samples long-term must be domestic entities that are licensed under the HGR Regulations.

If the samples will leave China, then that plan should be noted on original approval application form, if possible, and a separate approval to export samples is also required. Or, sample transport will need to be applied for at a later time. Unfortunately, experience has shown that even when requested, approval for sample export is not frequently granted, particularly during the height of the pandemic. During that time there was increased concern about unapproved transport of COVID-related blood or tissue samples.

While difficult to pin down the facts, stories circulated online of doctors or others who attempted to remove samples from China without permission. Such conduct was considered threatening to national security and at the heart of what the HGR regime is designed to protect against. Criminal charges against such individuals and debarment (sometimes permanent) from HGR application processes are possible in such circumstances.

7.9 Site Inspections, Self-Inspections and Surveys

Besides running approval and amendment processes, the MOST's local science and technology bureaus monitored compliance through investigation of complaints and different types of inspections. On complaints, the HGR regulations and the subsequent Biosecurity Law allow for anyone to report a suspected violation of the HGR Regulations. Once such a report has been lodged, the local science and technology bureaus, which are largely responsible for inspections and investigations, are obligated to investigate the issue.

In addition to these for-cause investigations, the MOST and the local science and technology bureaus that enforced the HGR regulations conducted surveys and other routine inspections of China-based entities. Although HGR regulators retain the right to conduct an overseas inspection, as of this writing it is not clear they have ever done so.

Self-inspections and HGR-related surveys are usually conducted on a more routine basis. Self-inspections have been done yearly over some periods, but the HGRIR require national surveys of HGR every five years.[20] The content of the self-inspection form may vary, but the various forms to date have required a listing of approved studies that the subject of the inspection has conducted in China, over a specific reporting period. The company may need to identify any non-compliances associated with its research activities, such as studies that did not obtain the necessary international collaboration approval or went beyond the scope of the approval, or transfers of data that were not covered by the applicable record filing.

Onsite inspections can be routine, specially focused, or for-cause. The HGRIR indicate that a record of administrative penalties for a company within the last three years can result in increased surveillance, and the HGR regulators and their provincial-level counterparts may arrange for a specialized inspection if needed to contain serious illegal actions or urgent tasks.[21]

The HGRIR also contain more detailed procedures for collecting evidence during inspections and investigations and ensuring that official investigators are objective (i.e., recusal procedures) than are in the HGR Regulations themselves. Inspections and enforcement since 2015 have been fairly varied in their content and the level of compliance required. Therefore, these provisions on investigation and penalty procedures may be an attempt to ensure a more standardized and predictable process for stakeholders.

7.10 Enforcement and Penalties

As discussed above, there was no clear evidence of significant enforcement of the HGR Regulations until recently but, beginning in 2018, OHGRA began releasing records online of the penalties that it had issued to companies for violations of rules requiring approval for studies and for sample transfer. These published penalty decisions dated back only to 2015. The penalties covered both foreign-invested and Chinese companies, and were a mix of fines and debarment, meaning that certain of these entities and hospitals were prohibited from obtaining approval for HGR studies for six months to a year. The companies were also sometimes asked to divest from illegal gains, such as the value of the samples or the amount of money invested

[20] Ministry of Science and Technology (2023) Human Genetic Resource Regulation Implementing Rules, Article 22.

[21] *Ibid* Articles 58, 60.

in them, and/or destroy the samples and halt the research. An example decision is as follows:

> Unit name: Huashan Hospital Affiliated to Fudan University
>
> Address: []
>
> Legal representative: []
>
> According to the "Interim Measures for the Administration of Human Genetic Resources" (Guobanfa [1998] No. 36), the "Administrative Penalty Law of the People's Republic of China" and other relevant regulations, the Human Genetic Resources Administration of China (hereinafter referred to as [OHGRA]) has . . .conducted an investigation of the international scientific research collaboration of the "Large-sample Case-Control Study of Unipolar Depression in Chinese Women," and the investigation has now been concluded. After this investigation, the following violations were found:
>
> Huashan Hospital and Shenzhen Huada Gene Technology Service Co., Ltd. carried out international cooperative research on Chinese human genetic resources with Oxford University in the United Kingdom without permission. Huashan Hospital transmitted some human genetic resources information from the Internet without permission.
>
> The above acts violated the provisions of Articles 4 [prohibition on collection, export, and otherwise providing HGR to the outside without the appropriate permit], [Article] 11 on the [requirement for approval of an application for international collaboration submitted by the Chinese party in advance of signing the cooperation agreement] and [Article] 16 [providing that Customs will rely on HGR approval prior to releasing samples for export] of the Interim Measures for the Administration of Human Genetic Resources. According to Article 21 of the "Interim Measures for the Administration of Human Genetic Resources" and the relevant provisions of the "Administrative Penalty Law of the People's Republic of China," the decision to punish [your Unit] is as follows:
>
> 1. Your unit shall immediately stop the implementation of the research work from the date of receipt of this decision.
> 2. [Your unit shall] destroy all genetic resource materials and related research data that have not been exported in the research work.
> 3. From the date of delivery of this decision, Huashan Hospital's international cooperation involving human genetic resources in China will be stopped, and it will only be carried out after the rectification and acceptance are passed.
>
> If your unit refuses to accept this decision, you may apply to the Ministry of Science and Technology for administrative reconsideration within 60 days from the date of receipt of this decision, or you may file an administrative lawsuit with

the competent people's court within six months. The implementation of this decision will not be suspended during the period of reconsideration and litigation. If the decision is not fulfilled within the time limit, the punishment organ will apply to the people's court for compulsory execution.

Ministry of Science and Technology of the People's Republic of China

September 7, 2015

***

This decision is similar to those for other violations of the HGR Regulations. First, the sensitivity is to the mishandling of human samples. Second, the more sensitive activity is arguably the export of such samples from China, as opposed to sending them to laboratories in China that may not be listed on the approval. The HGR Regulations were originally premised on leakage of HGR overseas. Third, unlike other administrative enforcement, the more devastating effects of an HGR penalty are on research, because HGR regulators can stop research in its tracks, seize important materials for analysis, and/or state that the research results are illegal and may not be used. In addition, there is a debarment component to the penalty, which the MOST employed considerably more often than NMPA.

The 2019 revision of the HGR Regulations, the Biosecurity Law, and, in 2021, the Criminal Law have given the HGR Regulators increasingly stronger enforcement tools. If the penalties are administrative, the worst-case scenario would be over a million dollars in fines for a given violation and permanent debarment imposed on both the individuals and entities involved. The HGRIR provide more detailed provisions on seizing evidence (i.e., to prevent its dispersal), establishing a case, holding hearings (when requested by the applicants that are entitled to them), and review of the facts of the case and determination of the extent of the penalty, although they do not clearly delineate how to determine the length of the debarment.

The Criminal Law was amended in 2021 to include HGR-specific penalties:

> Criminal Law Article 334-1: Crime of Illegally Collecting Human Genetic Resources and Smuggling Personal Genetic Resource Materials: Those who illegally collect China human genetic resources in violation of national provisions [of law or regulations] or illegally transport, mail, or carry China human genetic resource materials out of the country, endangering public health or public interests, if the circumstances are serious, shall be sentenced to fix term imprisonment three years or less, criminal detention or public surveillance, and currently or solely fined. If the circumstances are particularly serious, then they shall be [sentenced to] a fixed term imprisonment of three to seven years and a concurrent fine.

This provision includes several elements and a fair amount of room for the exercise of official discretion. For example, illegal collection could include unapproved studies generally, and it is not entirely clear if this activity must also be deemed to "endanger public health and public interests." As each study approved by the HGR regulator requires an assessment of public health and national security, not seeking that approval could presumably be deemed to endanger those interests to some degree.

Certain high-level prosecutions have made these types of provisions feared in the industry. For example, in 2020, the CEO of a promising biotech company was arrested for allegedly illegal exporting material related to development activities.[22] The individual was ultimately cleared but the case underscores the sensitivity of any type of HGR leaving the country.[23] Unapproved transfer of COVID-19 related samples was also a subject of scrutiny under the HGR regime. These and other similar cases demonstrated that HGR regulatory requirements were to be strictly enforced, even compared to enforcement of the drug and device regulatory regimes governing safety and quality.

7.11 The Biosecurity Law and Surveillance of National Human and Biological Resources

In 2021, the National People's Congress enacted the Biosecurity Law. This is a broad-based law covering everything from bioterrorism to infectious disease control. It also includes the protection of human genetic and plant and animal resources. However, a couple years after its passage, there were still neither a set of specialized implementing rules nor a clear plan for implementation, other than the HGRIR and existing regulations on laboratories.

The regime on protection of plant and animal resources has also been slow to emerge. Since ratifying The Nagoya Protocol on Access to Genetic Resources and Fair and Equitable Sharing of Benefits Arising from their Utilization to the Convention on Biological Diversity in 2016,[24] China has been working on legislation to protect animal and plant resources in the form of regulations. It released some high-level administrative documents and the agency that is now the Ministry of Ecology and the Environment released a draft implementing rule in 2017.[25] Some provinces like Guangxi have rules that touch on access and benefit sharing of plant and animal resources, establishing collaboration and sharing provisions similar to those in the HGR Regulations.[26] Like the HGR Regulations, this draft would have required that foreign parties partner with a Chinese party to conduct research using protected plant and animal resources, enter into a collaboration agreement, obtain approval, and ensure that there is access and benefit sharing from both parties. This draft was never finalized, and at the date of this writing there has still not been further nationwide implementation of the provision on the protection of plant and animal resources in the Biosecurity Law.

[22] Patchen (March 6, 2022).

[23] Keown (November 24, 2020).

[24] Conference of the Parties to the Convention on Biological Diversity (2010).

[25] Ministry of Ecology and Environment (2017).

[26] Guangxi Zhuang Autonomous Region Depart of the Ecology and the Environment (2021).

The Biosecurity Law established a governmental data platform, consisting of collected data from various biosecurity activities. This presumably includes those data generated by human genetic resource regulatory activities and state-funded research. A similar HGR big data platform is mentioned in the HGRIR, presumably consisting of data that has been uploaded or shared as part of HGR activities. This includes HGR information, structured for "big data" and open to "scientific research institutions, medical institutions, and enterprises in accordance with [unnamed] laws and regulations"[27] and it should not, at least under the HGR regime, include data generated during development in other countries because those data would not need to be shared with the HGR regulator—the HGR Regulations do cover foreign individuals—although such data may be submitted to NMPA. It is not clear what China does with these various types of data, but its focus on the use of large and complex data sets, known as "big data," particularly to conduct surveillance and warning activities during COVID and related to other social surveillance in China using biometric data, raises questions about the potentially broad intended use of these aggregated HGR and other data.[28]

References

Bai C (March 11, 2019) Research and development of gene editing technology cannot be abandoned because of choking. Science and Technology Daily. http://www.ce.cn/xwzx/gnsz/gdxw/201903/11/t20190311_31650113.shtml

China National Center for Biotechnology Development. https://www.cncbd.org.cn/Index/

China National Intellectual Property Administration (October 31, 2022) Notice seeking comment on an additional draft of the patent examination guidelines. https://www.cnipa.gov.cn/art/2022/10/31/art_75_180016.html

China National Intellectual Property Administration (2023) Patent examination guidelines. https://www.cnipa.gov.cn/module/download/downfile.jsp?classid=0&showname=%E4%B8%93%E5%88%A9%E5%AE%A1%E6%9F%A5%E6%8C%87%E5%8D%97.pdf&filename=8d8282d3c26b4900b82dbadfbdfc7aa9.pdf

Conference of the Parties to the Convention on Biological Diversity (2010) Nagoya Protocol on access to genetic resources and the fair and equitable sharing of benefits arising from their utilization. https://www.cbd.int/abs/nagoya-protocol/signatories/

Guangxi Zhuang Autonomous Region Depart of the Ecology and the Environment (2021) Guangxi Zhuang Autonomous Region Biogenetic Resources and Related Traditional Knowledge Acquisition and Benefit Sharing Management Measures (Trial). http://sthjt.gxzf.gov.cn/zfxxgk/zfxxgkgl/fdzdgknr/zcfg/gfxwj/t10236044.shtml

Keown A (November 24, 2020) Chinese government arrests GenScript and legend biotech founder for smuggling. Biospace. https://www.biospace.com/article/genscript-and-legend-biotech-founder-frank-zhang-arrested-by-chinese-government/

Lifei F, Xin H (June 19, 2018) Summary of the 20-year implementation of the interim measures on human genetic resources. Chin J Sci. https://news.sciencenet.cn/htmlnews/2018/6/414556.shtm

Ministry of Ecology and Environment (2017) Letter on Public Solicitation of Comments on the Regulations on the Administration of Access to and Benefit-sharing of Biological Genetic

[27] MOST (2023) Human Genetic Resource Regulation Implementing Rules, Article 16.

[28] Wu et al. (2020).

Resources (Draft for Comments). https://www.mee.gov.cn/ywgz/zrstbh/swdyxbh/201703/t20170323_408704.shtml
Ministry of Science and Technology (2013) Notice of the Ministry of Science and Technology on Further Strengthening the Management of Human Genetic Resources. https://www.most.gov.cn/xxgk/xinxifenlei/fdzdgknr/fgzc/gfxwj//gfxwj2013/201310/t20131015_109772.html
Ministry of Science and Technology (2015) Notice of MOST on the Implementation of Administrative Approval for Collection, Receipt, Trading, Export, and Exit of Human Genetic Resources. https://www.most.gov.cn/xxgk/xinxifenlei/fdzdgknr/fgzc/gfxwj/gfxwj2015/201509/t20150930_121850.html
Ministry of Science and Technology (2019) Collection Permit Evaluation and Approval Guideline. https://fuwu.most.gov.cn/r/cms/zwpt/web/assets/pdf/1.rlyczycjspfwzn.pdf
Ministry of Science and Technology (2022a) Notice on Updating Frequently Asked Questions on Human Genetic Resources Management. https://www.most.gov.cn/tztg/202203/t20220304_179634.html
Ministry of Science and Technology (2022b) Notice on Updating Frequently Asked Questions on Human Genetic Resources Management (Q&A Series 2). https://www.most.gov.cn/tztg/202204/t20220415_180263.html
Ministry of Science and Technology (2022c) Draft Human Genetic Resource Regulation Implementing Rules. https://www.most.gov.cn/tztg/202203/W020220322405257162084.docx
Ministry of Science and Technology (2023a) Implementing Rules of the Human Genetic Resource Regulations, https://www.most.gov.cn/xxgk/xinxifenlei/fdzdgknr/fgzc/bmgz/202306/t20230601_186416.html
Ministry of Science and Technology (2023b) Notice on Updating Service Guidelines for Administrative Licensing Matters and Record-filing and Prior Report Scope and Procedures, https://www.most.gov.cn/tztg/202307/t20230714_187075.html
Ministry of Science and Technology (2023c) Notice on the Release of Frequently Asked Questions about Human Genetic Resource Management, https://www.most.gov.cn/tztg/202309/t20230912_187931.html
Ministry of Science and Technology (2023d) Announcement on Adjustment to Matters Related to the Management of Human Genetic Resources, https://www.most.gov.cn/tztg/202306/t20230627_186788.html
National People's Congress (2023) State Council Institutional Reform Plan. http://www.gov.cn/xinwen/2023-03/11/content_5745977.htm
Patchen T (March 6, 2022) Frank Zhang is in the clear after Chinese authorities close investigation without charges, Company Says. Endpoints News. https://endpts.com/frank-zhang-is-in-the-clear-after-chinese-authorities-conclude-investigation-without-charges-company-says/
Standing Committee of the National People's Congress (2021) Biosecurity Law
State Council of the People's Republic of China (2024) Order 777 Decision on Amending Administrative Regulations, https://www.gov.cn/zhengce/content/202403/content_6939590.htm.
Wu J, Wang J, Nicholas S, Maitland E, Fan Q (2020) Application of big data technology for COVID-19 prevention and control in China: lessons and recommendations. J Med Internet Res 22(10):e21980. https://doi.org/10.2196/21980. PMID: 33001836; PMCID: PMC7561444. https://www.ncbi.nlm.nih.gov/pmc/articles/PMC7561444/
Xin N, Wang X (eds) (2023) Institutional reform of the State Council, approval of drug companies has been "accelerated," Caixin https://m.huxiu.com/article/815419.html

Food Regulation in China 8

8.1 Introduction

Food regulation in China shares many similarities with the other spaces discussed above, but it is also different in certain key ways. Food regulation has been a trend setter in the China FDD area, because efforts to improve and modernize the food area started roughly around 2004, after the reforms prompted by World Trade Organization (WTO) accession, before plans to reform the health care system and improve innovation in the drug and device spaces. As this chapter will discuss, food reform was prompted not by a desire for innovation but by serious safety concerns about the food supply. The food regulatory space was the first area to experiment with new types of penalties for non-compliance, including at the individual level (i.e., responsible managers), to undertake a comprehensive re-examination of the standards governing the area, and to use consumer complaint and rights of action to supervise the industry.

The food regulatory space is also different in other ways. First, although certain foods are pre-approved, general foods are not subject to a licensing or other pre-approval process, whereas all drugs, devices, and cosmetics go through some pre-market approval process. Second, although authority over devices and cosmetics has been moved around between agencies, food has been a particularly fragmented area of regulation, having fallen under the jurisdiction of various agencies, from the Ministry of Health (MOH), to Administration of Quality Supervision Inspection and Quarantine (AQSIQ), to the State Food and Drug Administration (SFDA), to MOH, to the China Food and Drug Administration (CFDA) and now shared regulation between the State Administration for Market Regulation (SAMR) and the National Health Commission (NHC).

Third, food safety is a significant social issue in China. Drug safety and health care resources have provoked public debate, even anger, but level of scrutiny not been the same as foods. The infant formula, pet food, contaminated pork, and gutter oil scandals have hit deep nerves in China, raising suspicions both inside and outside of China and deeply embarrassing the country. For that reason, there are unique

J. C. Balzano, *China Food and Drug Law*, Law for Professionals,
https://doi.org/10.1007/978-3-031-61901-4_8

features of food regulation and challenges for stakeholders in this area, including features of regulation and decisions about labeling and ingredients driven by consumer complaints and individual lawsuits (to a greater extent than in other areas).

8.2 History of Food Law and Regulation

Food safety in China became a visible and significant issue in China when the first of several infant formula scandals occurred in 2004, leaving a number of infants suffering from malnutrition. Food safety became a global issue when authorities discovered melamine in pet food, and other food-safety incidents that resulted in injuries raised fears China's exports. Since that time, stories have proliferated about food scandals in China, such as trash in candy, used gutter oil to produce food, and diseased meat. The Chinese media has portrayed these scandals as due to a series of bad actors (companies and the responsible individuals within them) in the food production and distribution system, who took short cuts or ran outright criminal enterprises.

China dealt with these scandals by enacting a series of laws and regulations to enhance penalties, including criminal penalties, to punish bad actors. Food safety is (as drugs and devices also are) subject to the "four most stringents" policy (the strictest standards, the strictest supervision, the strictest penalties, and the strictest accountability in the areas of food and drug safety) to "guarantee," as the government professed, the "safety of that on the tip of the tongue of the people."[1] The government prosecuted executives at the heart of milk scandals and corrupt FDA officials, implemented legal provisions that provided for a stronger range of individual liability, and enacted rules to increase cooperation between food regulatory authorities and the police force to identify and then impose such punishments.

These incidents also revealed other shortcomings in China's food regulatory system, including a system of confusing food standards governing products, ingredients, and testing. And, perhaps more importantly, they illuminated structural issues in the system, including that the system is fed by a number of small farms of around a bit more than an acre, which over utilize pesticides and fertilizers.[2] These farms' overuse of animal drugs, including clenbuterol to increase muscle mass and antibiotics, contributes to public health concerns.[3] The Covid-19 pandemic, 2003 SARS outbreak, and Avian Flu have also heightened concerns about small-scale operations and their handling of animal and plant ingredients.[4]

These small animal and agricultural farmers feed products of varied quality into a supply chain that still lacks traceability, which makes it challenging to locate the source of problems and eliminate them. This supply chain also still lack sufficient

[1] Xinhua News Agency (January 28, 2016).
[2] Minter (March 20, 2021).
[3] Jing et al. (September 2020).
[4] Reuters (March 19, 2021).

cold storage to transport and store various food products. China's system of open markets for small-scale vendors contributes to these issues.

Recognizing that even increasingly serious penalties were not enough to deal with these structural deficiencies, China embarked on an overhaul of its food-safety standards. This aimed to organize and update the standards and to establish a stronger system to license and inspect food manufacturers and distributors of prepacked foods, including restaurants, and to upgrade "agricultural product quality."[5] Certain restaurants, including in Shanghai, installed surveillance cameras in their kitchens to allow consumers to have a view of cleanliness practices.[6]

Alongside these issues, a debate has emerged about what is good quality and healthy food. Articles and commentary have described China's rising trend of obesity and related diseases, such as diabetes, which has become a focus of drug development.[7] The World Health Organization found that obesity rates in China have reached 20% in certain cities.[8] In 2011, China enacted some of its first nutritional labeling standards and issued and revised Dietary Guidelines for its citizens, encouraging them to eat a balanced and diverse diet. More recently in 2017, China has also established standards concerning dietary intakes (e.g., Chinese Dietary Reference Intakes, Part 1: Macronutrients, WS/T 578.1—2017),[9] which inform the content of standards governing the formulation and production of food products, such as infant formula. China increasingly embraced other trends, such as plant-based meat, and considered (although they have not approved) cultivated meat.

This brief section scarcely does justice to what are much larger trends and debates in China. The following sections explore how China's food laws and regulations have reacted to and shaped the approaches to these issues and debates.

8.3 General Principles of Food Regulation

To understand China's food-safety system, a few general principles are useful. First, there is a regulatory divide between prepackaged food regulated by SAMR, with ingredients and standard making handled by the NHC, and domestic raw agricultural and aquacultural products regulated by Ministry of Agriculture and Rural Affairs (MARA) and other agencies. MARA administers the Agricultural Product Quality and Safety Law, which has a similar purpose to the Food Safety Law, which is discussed throughout this Chapter.[10] This Law provides that standards enacted under the Food Safety Law must be followed if applicable during the raw agricultural food production process, but otherwise creates a separate system of production, treatment

[5] National People's Congress Standing Committee (2022) Agricultural Product Quality Law.

[6] Hu (April 17, 2019).

[7] *See, e.g.,* Wang et al. (2021).

[8] World Health Organization (2003); Hemmingsson (2021).

[9] National Health Commission (2017a) Chinese Dietary Reference Intakes: Part 1 Macronutrients.

[10] National People's Congressional Standing Committee (2022).

and distribution of these products. The system of licensure for food distributors does not apply to edible agricultural products.[11]

The China Customs Administration and its quarantine-related bureaus initially regulate imported processed ingredients, raw materials, and finished products; SAMR takes over to regulate that food as it circulates within China. So there are different regulatory spheres, under the command of more agencies, than in the drug, device, and cosmetics spaces. Although customs certainly plays a role, the medical products administrations still play a substantial role in the import process; drafts of standards, such as the China Pharmacopeia, are handled by the China Pharmacopeia Commission or the National Institutions of Food and Drug Control, which are both public institutions affiliated with NMPA.

Second, the Food Safety Law governs three types of products: food, food ingredients (e.g., food additives), and food-related materials. Food consists of "finished products" intended for eating and drinking and includes items that have been traditionally used for both food and medicine, but which are not intended for the purpose of treatment of disease.

Ingredients include food additives, which are both synthetic and natural substances that are added to food for purposes of improving its quality, color, smell and/or taste, and those substances that are added as necessary for preservation (i.e., prevention of decay) and processing (i.e., processing aids). The term food additives also includes nutritional fortifiers (e.g., vitamins and minerals). Ingredients can also be "new" food raw materials (see discussion below), but the Food Safety Law focuses on additives.

Food-related materials are those materials, such as packaging, containers, detergents, disinfectants, and tools and equipment that are used for food production and directly contact food or food additives. These materials are governed by special ingredient and labeling standards, and their manufacturing is controlled *in part* by food law and regulation; the other relevant regime is that on general industrial and consumer products. Various other standards contain other, similar, at times overlapping, definitions of these terms. Another standard sets forth a more specific definition of "food-contact substances," which is discussed below.

The category of food includes both general food and special foods. General foods require no approval, while special foods all require some form of pre-market approval, whether registrations or record-filings. Special foods include health foods (dietary supplements), infant formula products, and foods formulated for special medical purposes. The pre-approval processes for special foods emphasize safety and functionality in a way similar to the processes that China designed for drugs, medical devices, and cosmetics, although requirements are tailored for the different risk levels of these products. The divide between general and special foods also has significance post market. Special foods are subject to different types of inspections and testing and restrictions on advertising and marketing.

[11] State Administration for Market Regulation (2023) Food Distribution License and Filing Measures.

Third, the various technical aspects of special foods, food ingredients, and food-related materials are governed by mandatory and, in some cases, voluntary food-safety standards. The Food Safety Law provides that food-safety standards may cover the following areas:

- Limits on pathogens, pollutants, heavy metals, and pesticide and veterinary medicine residues;
- Food additives, including their permitted usage level and scope of application;
- Labels and marks, including information as to health and nutrition;
- Hygiene requirements for food production and distribution;
- Food quality control;
- Testing methods and procedures.[12]

Food safety standards have existed for many years, and many have or had not been updated for decades. The NHC has been engaged in an effort to reorganize the considerable body of food-safety standards by canceling outdated and/or redundant standards and amending and establishing new standards. Following State Council orders in 2004 to restructure the food-safety system, the health authorities began a plan to "clean up" and modernize the standard system.[13] The NHC and its Center for Food Safety Risk Assessment and other relevant NHC divisions have a full agenda of standards to consider each year. The 2022 Food Safety Standard Plan contained 41 separate items for enactment or amendment (almost double from the 22 in 2021).[14]

Considerable work has been done. In 2022, the NHC (which coordinates the drafting of these food-safety standards) released an inventory of 1419 such standards, divided into various sections such as "common criteria" (e.g., bacteria levels, pesticide residues, general and nutritional labeling), "product standards" (e.g., soybean products, nut and seed products), and "food additive quality specifications" (some 646 standards governing the specification requirements for additives, such as colors and preservatives).[15] The standard GB/T 1.1 (which is based on ISO and International Electrotechnical Commission (IEC) guidelines) provides guidelines and rules for writing standards.[16]

There are also product standards. A general food's label must indicate the product standard on which it is based. Importation of a product must be based on standard compliance, and it must use ingredients pursuant to standards (e.g., that are on positive lists of food nutrients or food additives) and be tested and produced in

[12]National People's Congressional Standing Committee (2021), Article 26.

[13]Ministry of Commerce (2005).

[14]National Health Commission (2022b) Food Safety Standard Plan; National Health Commission (2021b) Food Safety Standard Plan.

[15]National Health Commission (2022a) Catalogue of National Food Safety Standards (1419 as of Feb. 2022).

[16]National Health Commission (2020).

accordance with standards. Manufacturers are required to ensure that foods comply with these standards from the earliest stages of development and production to the time when the food is placed on the table.

Although food-safety standards can be prescriptive, they are not necessarily a recipe for the product and may be more flexible. Product standards usually define the product, include specific microbial and impurity limits and sensory or physio-chemical indicators, guidelines on use of additives or nutrients (e.g., that their use must be according to China's established positive lists of ingredients), and perhaps also labeling and packaging instructions or guidelines (e.g., references to general standards on packaging requirements or measurement). Some labeling sections may mandate specific warnings. The problem with the product standards is often that there is not one to fit innovative products or that there is no freedom to vary from certain aspects, even with a showing that such a variance will not affect safety or quality. The answer in some cases is the is the enterprise standard (see below).

For example, the standard for Jelly GB 19299-2015 defines "jelly" generally as "jell or gelatin food, with water, sugar, or other such ingredients as the main raw materials, supplemented with food additives such as thickeners, with or without adding fruit and vegetable products, milk and dairy products and other raw materials, that are processed through sol, blending, filling, sterilization, cooling and other such processes." It then goes on to define other specific types of jelly, such as "dairy jelly," and then to set forth sensory indicators, impurity and microbial limits, ingredient guidelines (i.e., to reference positive lists), and some other general guidelines on the size of the inner diameter of packaging, among other aspects.[17]

Although China's food-safety standard system is more modern and complete than at the beginning of the twenty-first century, there is still substantial work to be done. The enactment of a food-safety standard can take years and is largely government-led.

Domestically controlled food industry organizations and laboratories, among other quasi-governmental organizations, drive this system. While local companies can become involved in these organizations, overseas companies that do not manufacture products in China may face challenges as they may not be able to join the domestic trade associations or may be limited to observer status.

Another type of standard is an enterprise standard. Enterprise standards may be formulated by domestic manufacturers in accordance with national standards and record-filed with local governments in China where the products are produced (perhaps after a local expert review process). Older regulations that were not specific to food from the early 1990s provided that enterprise standards could be used under a variety of circumstances.

- Formulated if there are no national standards, industry [agency level] standards or local standards for the products produced by the enterprise;

[17] National Health and Family Planning Commission (2015).

- Formulated in order to improve product quality and technological progress, or to create standards that are stricter than national standards, industry standards or local standards;
- Formulated to cover the selection of national or industry standards, or to supplement national or industry standards;
- Formulated to cover processes, equipment, semi-finished products and methods;
- Formulated as management or work standards covering parts of production and distribution activities.[18]

However, the Food Safety Law has narrowed the official purpose of enterprise standards to set standards/levels of certain product aspects (e.g., permitted microbial levels) at points that are more stringents than that permitted under mandatory standards.[19] As long as an enterprise standard meets this criterion it can also be used to create a different type of food product based on compliance with existing standards. Imported food products are not eligible for enterprise standards, and if there is not a suitable product standard for them their manufacturers or brand owners must go through other processes to seek the enactment of a new standard based on international experience (discussed below).

8.4 Food and Food Ingredients

Food is defined under Chinese law as "various finished products and raw materials for human consumption or drinking, as well as articles that are traditionally both food and Chinese medicinal materials, but does not include articles for the purpose of medical treatment."[20] There is a regulation and a positive list of ingredients that help to determine whether ingredients may be used in both Chinese medicine and food. When used in food, these ingredients must be used in appropriate amounts and the food may not make therapeutic claims of curing or preventing disease.[21]

General foods may also not intrude upon the scope of special food claims. For example, certain health foods may make claims of regulating intestinal flora. While general foods can use certain probiotic ingredients, they may not make claims about the use of these ingredients in regulating the bacteria in the gut.

China's regulations concerning what ingredients can be used in what products and in what amounts still require some reform. The system relies heavily on positive lists. In other words, manufacturers cannot use ingredients in food unless there is some express basis for doing so in a certain amount(s) and specific types of products.

[18] State Technology and Quality Bureau (1991), Administrative Provisions on Enterprise Standards, Article 6.

[19] Shanghai Health Commission (2020).

[20] National People's Congressional Standing Committee (2021), Article 50.

[21] National Health Commission (2021a) Administrative Regulations on the Catalogue of Substances that are Both Foods and Chinese Medicinal Materials in Accordance with Tradition.

There is no equivalent or arguably close analogue to the system in the United States for ingredients that are Generally Recognized as Safe.

Article 37 of the Food Safety Law makes it clear that food manufacturers may not use ingredients that are not approved for food in China. Therefore, use of an ingredient that is not prohibited may still provoke enforcement action.

In China, there are essentially four ways to obtain permission to use an ingredient. The first and likely most popular, as discussed, is to rely on an existing standard. The second is to obtain approval for the ingredient as a new food additive. "New" can include an entirely new additive or an expansion of the approved scope (i.e., amount of use) of an existing food additive.[22]

Food additives are defined under the Food Safety Law, as "synthetic or natural substances, including nutritional fortifiers, that are added to food for the purpose of improving the quality, color, aroma and taste of food, as well as for the needs of preservation, preservation and processing."[23] As indicated, this category also includes "nutritional fortifiers," which are vitamins, minerals, and other nutrients that stakeholders are permitted to add. A list of permitted food additives is contained in the food-safety standard GB 2760, which underwent its last full scale update in 2014,[24] although new additives have been added by individual notices that can be found on the NHC's website.[25] Nutritional fortifiers have their own separate standard, GB 4880-2012, which states the name of the nutrient and the type of foods in which it is permitted to be used.

Food additives must be produced in facilities that have food additive manufacturing licenses.[26] The additives themselves must be labeled in accordance with a separate food-safety standard on the labeling of food additives. From 2018 to 2022, NHC approved between 20 and 32 food additives per year.[27]

The third method is to obtain approval for a "new" food raw material. A new food raw material is "new" because it is not a material that is part of traditional eating habits in China. Traditional eating habits means that the food has been in production and distribution in fixed or non-standard prepacked form in a given province and it is not listed in the China Pharmacopeia. New food raw materials are animals, plants, and microorganisms, components of them, materials that have changed the original structure of the raw material, and other newly developed food raw materials.[28] New food raw materials do not include genetically modified foods.

[22] National Health and Family Planning Commission (2017a) Administrative Measures on New Varieties of Food Additives, Article 2.

[23] National People's Congressional Standing Committee (2021), Article 150.

[24] National Health and Family Planning Commission (2014).

[25] National Health Commission, Notices of new food additives.

[26] State Administration for Market Regulation (2020b) Food Manufacturing Measures, Article 15.

[27] CIRS (2022) 2018-2022, Summary of Acceptance and Approval of New Food Additives Over Five Years.

[28] National Health and Family Planning Commission (2017b), Administrative Measures on the Safety Review of New Food Raw Materials Article 2.

The Food Safety Law describes no clear line between a food additive and a new food raw material but, generally speaking, food additives serve auxiliary functions, such as preservatives, flavors, or even processing aids that may largely dissipate before the final product, whereas new food raw materials are more likely to be a main functional component of the food.

Finally, no approval is needed if the raw material is not "new." This means that the food manufacturer can show that the food, either prepackaged retail items of a fixed amount or prepackaged food of a non-fixed amount, is part of traditional eating habits and has been produced and/or distributed in a specific province for 30 years or more.

Only certain varieties of genetically modified foods are permitted to be produced or imported. The different varieties fall into five different categories: soybeans, rapeseeds, corn, cotton, and tomatoes; only cotton and papaya are approved for planting, whereas soybeans, corn, cotton, rapeseed, sugar beets, and papaya are permitted to be imported for further processing.[29] The types of food permitted and the specific companies that are permitted to produce or import genetically foods are approved by the Ministry of Agriculture and Rural Affairs. Genetically modified foods must also be labeled as such, e.g., raw materials are genetically modified soybeans.[30] If a particular variety of GMO foods is not approved, then the non-GMO variety may not state as much on the label as that could be considered confusing or misleading.[31]

8.4.1 Mitigating Risk of Using Unapproved Ingredients

Use of not only a prohibited but also an unpermitted ingredient can bring significant penalties, including recalls, fines, seizures of product, and even penalties against individuals, such as criminal terms. Therefore, companies need to be careful to find some legal basis for both using an ingredient and using that ingredient in the specific type of food that the business is producing. For example, as an initial step they would check the general food additive standard (GB 2760) and related notices, a nutritional fortifier standard (GB 14880) and notices, and food raw material documents and notices. Both GB 2760 and GB 14880 act like positive lists and are supplemented with notices that update them.

In addition, other food standards related to ingredients may provide some basis to use an ingredient, including local standards, collective standards, and enterprise standards. As we saw above, the NHC has published lists of these standards that can be queried, although many of the lower-level standards may require additional

[29] Ministry of Agriculture and Rural Affairs (2023).

[30] State Council (2017) Administrative Regulations on the Safety of Agricultural Genetically Modified Organisms.

[31] Ministry of Agriculture and Rural Affairs (2023).

research. The site Foodmate.com provides access to a number of these standards, or as do the State Standards Administration and industry association websites.

Another important inquiry is whether an ingredient is expressly prohibited or restricted from use in food, or in a certain type of food. In this respect, certain ingredients may be used in both foods and health foods, whereas others are only for health foods, and some are prohibited in health foods. NMPA maintains a list of those ingredients,[32] and in 2010 the National Health Commission issued an explanation that health food ingredients could be approved for use in general foods by going through the approval process for new food raw materials.[33] If they have not gone through that process—and do not have a raw material notice associated with them—then those ingredients are only for health foods and may not be used in general foods.

Furthermore, certain other ingredients may be reserved only for use in drugs, including western drugs and Chinese medicine. In that case, the Drug Code, a/k/a the China Pharmacopeia, may provide insight. Further, companies may refer to lists of ingredients that are prohibited in foods,[34] and there may be news stories or published administrative penalties or court cases that speak to whether others have been penalized for the use of certain ingredients.

These principles are illustrated in a case in 2013 that ultimately became a "Guiding Case," in which individuals at a local company were convicted of adding toxic ingredients to food in violation of China's Criminal Code.

The company obtained ingredients that contained buformin, which is an antidiabetic drug, and added them to health food. The case is interesting because, at that time, buformin was not specifically on any of the negative lists of ingredients prohibited in food products and, as discussed above, the Food Safety Law and other rules acknowledge that use of an ingredient in drugs does not necessarily disqualify it from use in food under proper conditions. However, in finding the defendants guilty, the court relied on a provision in the judicial interpretation of the Supreme People's Court and the Supreme People's Procuratorate on Several Issues Concerning the Application of Law in Handling Criminal Cases of Endangering Food Safety. That Interpretation stated that "toxic food," as used in the Criminal Code, includes those substances that are specifically prohibited on negative lists and "other substances that may harm human health." Relying on an expert report from a Judicial Institute—one of the expert institutes used to inform courts on technical aspects of FDD cases—articulating the potentially harmful effects of consuming buformin, the courts involved concluded that the individuals and entity could be charged and the individuals would serve criminal sentences on that basis.[35] The

[32] Ministry of Health (2002).

[33] Ministry of Health (2010).

[34] National Health and Family Planning Commission (2011).

[35] Beijing Sunshine Yibai Biotechnology Development Co., Ltd., et al., Production and Sales of Toxic and Harmful Foods (2018).

individual defendants received personal fines, two of over a million dollars (15 million and 9 million RMB), and between 5 and 15 years in prison.

In some cases, however, the concerns about use of dual use ingredients can be misplaced or even opportunistic. For example, in 2012, a consumer alleged that the addition of caffeine to health food products that also contained the commonly used and permitted preservative, sodium benzoate, would turn the product into a drug because it would cause the formation of "sodium caffeine," which can be treated as a controlled substance. In this case, both the use of caffeine and sodium benzoate were expressly permitted by a health food approval and food-safety standards, respectively. After investigations by provincial and national food and drug authorities, the allegations were proven baseless. The investigation showed that the amounts used were in permitted and safe ranges and manufacturing process itself was insufficient to actually produce the drug, including the specific conditions called for by the China Pharmacopeia to create sodium caffeine.[36]

8.4.2 Food Related Materials and Food Contact Substances

As discussed above, the Food Safety Law also covers a class of materials, including tools, equipment, and containers, that are known as "food-related materials." Food-contact substances, including containers and packaging and wrapping materials, are covered by a separate set of food-safety standards governing additives, labeling, and testing. These standards were revised or established in 2016, providing a much more comprehensive and unified regime over these products.

Food contact materials are defined as: materials that are in, or may come into, contact with food or food additives, or those materials whose ingredients have the potential to migrate to foods or food additives. They include packaging, containers, and kitchen utensils, as well as paints, coatings, and adhesives that may come into contact with food or food additives, but they do not include disinfectants, detergents, or public water facilities.[37] It should be noted that the National Health Commission administers a separate regulatory regime for the approval, safety assessment, and labeling of disinfectants, including hygiene wipes and other cleaners sold to consumers.[38]

The 2016 standards made clear the general principles as to food-contact materials, including assessment of permitted migration of these substances into food and establishment of positive lists of permitted adhesives[39] and other ingredients,[40] as

[36] State Food and Drug Administration (2012).

[37] National Health and Family Planning Commission (2016a).

[38] National Health Commission (2018).

[39] National Health and Family Planning Commission (2016b) GB9685-2016, Positive List of Food Contact Material Additives.

[40] National Health and Family Planning Commission (2016c) GB 4806.6-2016 Permitted Resins in Plastic Food Contact Materials.

well as principles of labeling.[41] For example, among other principles, these materials should be designed to minimize the potential that migration of food-contact material ingredients could harm human health and avoid changes to the composition, structure, and properties, such as taste or smell, of the food that they contact.

As discussed below, these food-related materials are not subject to the same manufacturing and distribution rules as food and food ingredients. For manufacturing, they are treated as industrial products and the facilities do not require food manufacturing licenses; instead, food-contact material production facilities receive industrial-product manufacturing licenses. However, this regime and its standards continue to develop and may undergo changes in the years to come.

In 2022, the State Administration for Market Regulation released a rule governing quality practice of food contact manufacturers and distributors. Among other things, the rule requires that these facilities have procedures for receiving and inspecting ingredients and raw materials to ensure their safety and conducting monitoring of conditions in the facility that could lead to contamination.[42]

8.5 General Foods Versus Special Foods

Food regulatory enforcement often focuses on the line between general foods and special foods. Because the time and costs of approval of a health food can be considerable, authorities are concerned companies may market what is essentially a health food under a general food category in order to save time and money and avoid compliance with more onerous post-market rules. China has engaged in campaigns for several years to surveil general foods that fraudulently make health food claims and health foods that exceed the scope of the permitted claims that they were registered with.[43] These campaigns consist of surveillance of marketing of potentially violative foods on the Internet, through phone calls, and in print news.

Despite this focus, the line between general food claims and health food claims is not always clear. As discussed above, general foods must rely on product standards to come to market, but these food product standards do not articulate a clear range of permitted claims for each category. When product standards do not provide clear guidance, companies have often used local standards or compiled their own "enterprise standards." Enterprise standards exceed the requirements of national standards and were record-filed with various localities after review by an expert committee. To be accepted, an enterprise standard must had to have one aspect that is more stringents than a national standard (e.g., lower tolerance for microbial content or higher physio-chemical indicators), but the rest of the standard may use existing

[41] Standard Administration Commission (2014) GB/T 30643-2014, Principles of Food Contact Substance Labeling.

[42] State Administration for Market Regulation (2023a) Interim Measures for the Supervision and Administration of Food-Related Product Quality and Safety.

[43] State Council (2018).

elements of other standards to create categories specific to novel foods or beverages. An amendment to the National Standardization Law abolished the record-filing system, and companies are required to publish their standards online. These standards are published on the Enterprise Standard Public Service Platform. In general, campaigns to root out improper marketing of general foods as health foods may not second-guess those foods that are covered by enterprise standards.

As noted, imported foods cannot be covered by an enterprise standard. Previously, any imported general foods not covered by an existing standard was required to submit a license application, supported by safety data, to the National Health Commission. Currently, under the Food Safety Law, imported foods without existing standards in China may receive permission to enter the country if the applicant can present evidence of compliance with international or foreign standards. The National Health Commission's Center for Food Safety Risk Assessment has developed and refined this process, including the solicitation of public opinion and review of technical materials; if there is basis to do so, it can issue a temporary standard while more long-term solutions are being considered.[44]

8.5.1 Health Foods

China has regulated dietary supplements according to strict registration criteria since 2005. In China, the products that we think of as dietary supplements elsewhere are referred to as "health foods." Various other names are thrown around for this category of foods, such as functional foods and dietary supplements. However, the most accurate name and that used by the government and stakeholders alike is health foods.

What is a health food? The fullest definition of health food appears in the 2005 registration regulations that the former State Food and Drug Administration released and the National Food Safety Standard on Health Food GB 16740-2014. According to those sources, a health food is a food that "claims to have specific health functions [set forth in a list of accepted claims] or can supplement vitamins and/or minerals. Health foods are suitable for certain groups to improve bodily functions without serving a therapeutic function and do not cause any acute, sub-acute, or chronic harms to the human body."

As this definition indicates, health foods really consist of two varieties in China. First are vitamin and mineral supplements, which make claims of supplementing or fortifying certain key vitamins. They differ from general foods that are permitted to make claims that a food contains or is a source of vitamins, both in terms of the permitted claim and the permitted amount of vitamins (which is pre-approved). Vitamin and mineral supplement health foods may be subject to a less stringents approval process than the second type of health foods discussed below. These

[44]National Health and Family Planning Commission (2017b) Notice on Review Work of Importation of Foods that Do Not Have a Food Safety Standard. No. 14.

vitamin and mineral supplements may be permitted to come on the market after they submit a filing, if they use existing ingredients (i.e., ingredients that are approved for use or have been used in an approved health food before).

The second type of health foods are the functional claim health foods, which make one or more of 27 permitted structure function claims, e.g., enhance immunity or improve digestion. These do not include therapeutic claims that a food prevents or cures a specific disease (e.g., prevents heart disease is prohibited, but reduces blood lipids is not). Functional claim health foods are subject to more rigorous application requirements and, in some cases, a licensing process that requires a technical review.

Tables 8.1 and 8.2 shows original approved 27 health food functional claims.

In September 2023, SAMR issued a notice containing an adjusted set of permitted health food claims and guidelines providing information on the methods for evaluating those claims, including guidelines on ethics committee approval for human studies. This Notice adjusted the wording of the claims (making some less absolute) and removed three of the claims, for which there was "low consensus and

Table 8.1 Original 27 health food functional claims

Strengthen immune function, immunomodulary	Weight loss
Adjusts blood lipids (reduces overall cholesterol, reduces triglycerides), Aides blood lipid reduction	Improves growth and development
Adjusts blood sugar, Aides blood sugar reduction	Increases bone density
Anti-aging, Antioxidation	Improves nutritional anemia, Improves iron deficiency anemia
Improves memory, Helps improve memory	Assists in protecting against chemical liver damage, Protects against chemical liver damage
Improves vision, relieves eye fatigue	Cosmetic (removes acne), removes acne
Promotes lead excretion	Cosmetic (removes chloasma), removes chloasma
Clears and moisturizes the throat, Clears Throat	Cosmetic (improves skin moisture/oil), Improves skin moisture
Adjusts blood pressure, Aides blood pressure reduction	Improves skin oil content
Improves sleep	Improves gastrointestinal function (regulates intestinal flora), Regulates intestinal flora
Promotes lactation	Improves gastrointestinal function (promotes digestion), Promotes digestion
Relieves physical fatigue, Antifatigue	Improves gastrointestinal function (soothes and laxative), Laxative
Improves hypoxia tolerance, Hypoxia resistant	Improves gastrointestinal function (Protects against gastric mucous membrane injury), Protects injury to gastric mucosa
Anti-irradiation, Aides protection against irradiation hazards	

Table 8.2 Revised health food claims as of September 2023[a]

Helps strengthen immunity	Helps improve chloasma
Helps with anti-oxidants	Helps improve skin moisture
Helps improve memory	Helps regulate intestinal flora
Relieves eye fatigue	Helps with digestion
Clears and moisturizes the throat	Helps relieve constipation
Helps improve sleep	Helps protect gastric mucosa
Relieves physical fatigue	Helps maintain healthy blood lipid (cholesterol/triglyceride) levels
Hypoxia resistant	Helps maintain healthy blood sugar levels
Helps control body fat	Helps maintain healthy blood pressure levels
Helps improve bone density	Assists in protecting against chemical liver damage
Improves iron deficiency anemia	Assists with protecting against ionizing radiation hazards
Improves acne	Aides lead excretion

[a] State Administration for Market Regulation, National Health Commission, et al. (2023) Announcement on the release of the "Catalogue of Health Functions Allowed for Health Food Claims of Non-nutrient Supplements (2023 Edition)" and Supporting Documents

an unclear health need:" "improves growth and development," "promotes lactation," and "improves skin oil content."[45]

Although not part of the formal definition, in addition to these types of claims health foods may also be defined by their dosage forms. For example, SAMR issued a set of rules on ingredients and types of forms that health foods that go through the record-filing pre-approval process may take, as well as their manufacturing methods. When general foods take these forms, they can be questioned in stores or during other inspections, such as at distribution centers or during the importation process:

- Tablets: [main production processes, include] crushing, sieving, mixing, granulating, drying, tableting, coating, packaging
- Hard capsules: [main production processes, include] crushing, sieving, mixing, granulating, drying, encapsulation, packaging.
- Soft capsules: mixing, homogenizing, filtering, pilling, drying, packaging, etc.
- Oral solution: [main production processes, include] mixing, dissolving, preparing, filtering, filling, packaging (if the process involves sterilization, the specific sterilization method and process parameters should be reported, such as moist heat sterilization, hot pressing sterilization, circulating steam sterilization).
- Granules: [main production processes, include] crushing, sieving, mixing, granulating, drying, packaging.

[45] State Administration for Market Regulation (2023b) "Permitted Claims for Health Foods in Non-Nutrient Supplement Health Function Catalogue (2023 Version)" Released.

- Gel candy: [main production processes, include] sol, sugar, boiling, mixing, deployment, filtration, inflation, molding, drying, sand mixing, coating, polishing, coating, packaging.
- Powder: [main production processes, include] crushing, sieving, mixing, subpackaging, packaging.[46]

Finally, as the last part of the health food definition states, these products can make structure function claims about the human body, but not claims that they prevent, treat or cure diseases. This can be a difficult line to implement in practice. Generally, health food labels or advertisements that mention specific diseases, treatment, or compensation for injuries can be at risk of enforcement.

8.5.2 Health Food Applications

The data required for functional claim health foods can be extensive and, depending on the claim, requires some combination of research articles, animal data on the specific product, and human testing. SAMR or, more specifically, the Center for Health Food Evaluation, sets specific population, subject headcount, and endpoint requirements depending on the claim. Despite advances with drugs and devices, no well-established system for submitting foreign data to meet this requirement has been developed; the expectation is that these data are generated in China.

In addition to these safety and efficacy data, applicants must order a series of laboratory testing at an accredited laboratory in China. These laboratory tests include toxicology testing, stability testing to determine shelf life, and anti-doping testing to determine whether certain illegal drug substances are present.

The 2015 Food Safety Law (FSL) drew a stark distinction between imported and domestically made health foods in terms of application requirements, when it introduced a filing (i.e., notification system) for certain types of health foods. Under the filing system, an applicant needs to complete the same application requirements as described but could be granted a marketing certificate (i.e., a filing certificate) once the application requirements are complete.

For imported health foods, under the new system all functional claim health foods would continue go through the technical review and receive a license. Imported vitamin and mineral supplement health foods that use certain accepted ingredients in a certain health-food catalogue can go through a filing process. Domestic products that use existing ingredients in one of the promised health food ingredient catalogues are permitted to use the filing system.

[46] State Administration for Market Regulation (2021b) Record-Filed Health Food Dosage Form Technical Requirements.

8.5.3 Health Food Ingredient and Claim Reform

Health foods are permitted to use the same ingredients that all general foods are, plus additional ingredients that are only permitted for health foods. Prior to the 2015 FSL, these health-food ingredients were encapsulated in a variety of haphazard notices and regulations (e.g., the regulation on probiotic health foods from 2005, which list certain accepted strains of probiotics permitted in health foods).

The new amendment promised catalogues of both ingredients and claims that would coordinate the ingredients permitted and for what claim and what population (e.g., adults versus children). Moreover, a claims catalogue would address what data and materials are required to support functional claims. To be approved, as a health food ingredient the ingredient must have a history consumption in China and/or abroad, have proven safety, and have been used in health food that has been approved for registration (i.e., that went through a review). The function of the ingredient, i.e., its efficacy must be related to a claim that is included in the health food claims catalogue. Pursuant to a rule on the formulation of the health food ingredient catalogue, any individual or entity can apply to adjust the ingredient catalogue, which is jointly managed by SAMR, the NHC, and the State Administration for Traditional Chinese Medicine.[47]

To be a permitted health food claim, the following criteria must be met: The purpose must be to supplement nutrients, maintain and improve bodily health, or reduce risk factors of disease. But the claims may not involve the preservation, treatment or diagnosis of disease. Vulgar or superstitious claims are also prohibited. Otherwise, the claim must include clear and easily understood consumption instructions, be supported by scientific evaluation, be guided by health care (or Chinese medicinal) theory, and have clearly defined intended populations, e.g., children, adults.

In the early years after the 2015 FSL, CFDA issued several proposals for both catalogues and a regulation governing the making of the ingredient and claims catalogues. These regulations have been gradually finalized over a period of years and the catalogues have slowly emerged. For example, in 2021 SAMR released list of excipients for use in record-filed health foods, which included the applicable food-safety standard and maximum usage amounts.[48] The catalogue for nutritional supplement ingredients and one for claims were also released have been updated regularly (seemingly biannually or yearly since 2020). Most recently, drafts of these catalogues were released for public comment in 2022, including the nutrient supplement ingredient catalogue, the claims catalogue, and the protein ingredient catalogues for soy and whey, and these drafts were finalized in mid-2023, further

[47] State Administration for Market Regulation (2019a) Administrative Measures on the Health Food Raw Material Catalogue and Health Food Claims Catalogue.

[48] State Administration for Market Regulation (2021a) Provisions on Record Filed Health Food Product Excipients.

completing what turned out to be a rather protracted implementation process for the reforms commenced in 2015.[49]

The data requirements for health foods remain relatively significant (some combination of literature, testing, animal and/or human studies depending on the claim). Domestically made products, which must also comply with specialized health food manufacturing regulations. For imported functional claim products or domestic products that contain new ingredients, the wait for a technical review can be three to four years. Further, because there is no well-established process for approving a new health-food ingredient, it may take two to three years to obtain approval through the new food raw material process. It remains to be seen whether the guidelines that were released in 2023 (see above) will improve these circumstances change this picture.[50] A in 2021 study found that only 4.68% of all health food registration applications (including amendment and renewal) were for "new health foods."[51]

8.5.4 Health Food Manufacturing

As will be discussed below, health-food makers in China must have a food manufacturing license, like any other food. However, the scope of that license requires that the manufacturer meet special standards, some of which can be more stringents than general food standards.[52] Like all manufacturers, these manufacturers are subject to regular and for-cause inspections (as discussed below). Importers of finished health foods do not require a manufacturing license or registration. SAMR reserves the right to inspect overseas establishments but does not currently have a robust program similar to the one that NMPA has for drugs and devices.

8.5.5 Health Food Distribution

Similarly, health-food makers must have food distribution licenses and the scope must include health foods (see discussion below). China also places limitations on the retail of these products. For example, there are restrictions on the display of health foods that prohibit their display along with general foods. Retailers may only

[49] State Administration for Market Regulation (2021c) Nutrient Supplements in the Catalog of Health Food Raw Materials (2022 Edition) (Draft for Comment), Catalogue of Nutrient Supplements with Health Functions Allowed for Health Food Claims (2022 Edition) (Draft for Comment) and Protein in the Catalog of Health Food Raw Materials (Draft for comments); State Administration for Market Regulation (2023c) Four Health Food Catalogs Including "Health Food Raw Materials Catalog Nutrient Supplements (2023 Edition)" Have Been Released.

[50] CIRS Group (2023).

[51] CIRS Group (2022).

[52] Ministry of Health (1998) Health Food Good Manufacturing Standard, GB 17405-1998; See also China Food and Drug Administration (2016c) Health Food Manufacturing License Review Implementing Rules.

be permitted to display them behind special counters and need to separate them from general foods. Placement of health foods and general foods side-by-side may be flagged during routine store inspections.[53]

8.5.6 Health Food Labeling and Advertisement Restrictions

Health foods are subject to advertising restrictions that are similar to those for drugs and devices. There are strict content restrictions and a pre-approval process. Labels are subject to the general labeling restrictions that all foods must meet and a set of specific labeling and naming restrictions that prohibit certain words and phrases. Health food labels must also contain additional information. For example, those labels must have a list of "functional ingredients," a section that indicates the "registered claim" or claims, and a statement as to the groups for which these foods are intended, e.g., children, adults (and likewise any for which they are not).[54] In 2018, NMPA began requiring that claims also include a reference to the animal data supporting them, and there have been other proposals to have labels include a more detailed description of the basis for the registered claim, e.g., human data, literature, or animal data.

8.6 Infant Formula

Infant formula has been at the center of high-profile food incidents in China, including two widely publicized incidents in 2004 and 2008. In 2004, substandard infant formula was responsible for "big head" disease among malnourished infants in Anhui province. In 2008, the use of melamine formula led to widespread kidney damage in infants throughout the country and led to a range of criminal prosecutions and an administrative compensation fund that established important precedents for food safety in China.

What of the actual regulation of infant formula then? Early regulations treated formula as a type of general food subject to various standards. After 2008, the government revised and added to these standards, including revisions in 2021. Infant formula is also subject to a variety of manufacturing, formulation, and ingredient standards.

In 2015, the Food Safety Law was amended to establish pre-market registration requirements for powered formula and a new series of regulations on registration, manufacturing, ingredient listing, and labeling were gradually introduced thereafter. The Food Safety Law and these regulations included a unique prohibition on more than one formula being associated with a specific brand name and on sub-packing of these products. Regulations more specifically required that there be significant,

[53] Yan (November 25, 2022).

[54] Ministry of Health (1996).

scientifically proven differences between two or more formulas for the same age group, and each enterprise may have no more than three formula series with nine formulas, separated out by age group: 0–6 months, 6–12 months, and 12–36 months.[55] In 2024, a draft to amend the Food Safety Law to cover liquid infant formula was proposed.[56]

There are different product standards governing different types of infant formula, including infant formula, older infant formula, and infant formula for special medical purposes. The standards detail the required nutrition and features formulation of the products, as well as special requirements on labeling.[57] Most infant formula in China is cow's milk powder, with sheep's milk in second place, and goat's milk gaining in popularity.[58]

Applications for formula registration include product and ingredient standards, a research and development report, manufacturing information, and a product test report to ensure compliance with quality and safety standards. The approval process can include onsite verification for domestic facilities (overseas facilities are registered) and product sampling by the government or government authorized entities.

As part of the registration process, the applicant must submit (and SAMR, Center for Food Evaluation will review) labeling for the product. There are various special labeling requirements for infant formula, set forth in the regulations, and specific standards, including requirements on statements about animal-derived formulas and prohibitions on therapeutic claims. Like other areas of food labeling, the labels may not emphasize that certain ingredients are not added, or the product contains "zero,"[59] as those claims may be considered misleading.

China's health policy generally revolves around the principle that breast milk is the ideal source of nutrition for infants. The National Health Commission has stated:

> [b]reast milk is the ideal food for infants, but when a small number of nursing mothers cannot breastfeed due to illness, insufficient or no milk secretion, they need to provide nutrients for infant growth and development through infant formula. Infant formula is an important, or perhaps the only, source of nutrients for infants who cannot be breastfed.[60]

---

[55] China Food and Drug Administration (2016) Infant Formula Registration Administrative Measures.

[56] State Administration for Market Regulation (2024a) Notice on Publicly Soliciting Opinions on the "Food Safety Law of the People's Republic of China" (Draft for Comments on Amendments).

[57] National Health Commission, Department of Food Safety Standards, Monitoring and Evaluation (2021) Q&A on Infant Formula Food Series.

[58] Wang and Jin (2021).

[59] China Food and Drug Administration (2016) Infant Formula Registration Administrative Measures, Article 34.

[60] National Health Commission, Department of Food Safety Standards, Monitoring and Evaluation (2021) Q&A on Infant Formula Food Series.

Perhaps for this reason, the Advertisement Law prohibits publication of advertisements for infant formula or other beverages and foods in mass media or other public places that claim that the infant formula or other such products can replace breast milk in whole or in part.[61] That Law subjects such conduct to removal of the advertisement from circulation and potentially substantial fines.[62]

As with other products, these registrations must be maintained through amendments, if certain major or minor items change. They must also be renewed every five years, like drugs and devices.

As these restrictions grew on infant formula imported through traditional channels and made domestically, and due in part to the scandals surrounding domestic brands, infant formula increasingly came into the country through other means, including personal importation and cross-border e-commerce (CBEC). Personal importation (the original basis for CBEC in many respects) became a significant source of foreign infant formula into China. This led to legislative restrictions, such as the 2012 Hong Kong regulation, which limits the amount of closed infant formula that individuals may bring on planes between HK and destinations in the mainland.

Later, as CBEC grew, infant formula arrived via the direct mail and bonded warehouse channels described above. In 2016, the government briefly ended CBEC.[63] Specifically, it stated that all infant formula imported through CBEC would have to be registered. However, the government retracted this restriction shortly thereafter. Today, foreign-labeled infant formula may be ordered online by individual consumers and shipped into China legally, subject to limits on individual shipments and amount ordered per person annually.[64]

Regarding traditional (non-CBEC) channels of importation and distribution, statistics also show that registrations of domestically produced infant formula products are outpacing those for imported products. In 2017, there were 209 imported product registrations and 695 domestically produced registrations; in 2021, these numbers were 6 and 35, respectively. Also apparent is the significant decline in applications overall, as each manufacturer is only allowed to have three series and 9 formulas overall.[65]

[61] National People's Congress Standing Committee (2018) Advertisement Law. Article 20.

[62] National People's Congress Standing Committee (2018) Advertisement Law, Article 57.

[63] Du (2016).

[64] Ministry of Commerce (2019) Cross-Border E-Commerce Positive List.

[65] Wang and Jin (2021).

8.7 Foods for Special Medical Purposes

The category of foods formulated for special medical purposes ("FSMP") was originally a category did not require pre-approval. The National Health Commission administered a standard for these foods, and they could be made per compliance with that standard.

FSMPs are currently defined under regulations established in 2016 (after the 2015 amendment of the Food Safety Law that added the category) as:

> FSMPs refer to foods formulated for the special needs in the form of nutrients or meals or special processing for those with dietary restrictions, digestion and absorption disorders, metabolic disorders, or other specific disease states. It includes infant formula for special medical purposes intended for infants who are 0 to 12 months of age and FSMPs suitable for individuals over 1 year of age.[66]

This category also includes "total nutrition foods," which are foods that can be used as the sole nutritional source of a designated population, in a specific disease state. Special development requirements apply to these total nutrition foods, including a clinical trial that is conducted in accordance with specialized FSMP Good Clinical Practice rules.[67]

Under the 2015 Food Safety Law, FSMPs became a special food category subject to pre-approval. In some ways, FSMPs are treated more like drugs than foods. Like other products that need pre-approval, the applicant must submit research and development materials, information about manufacturing, a product standard with relevant specifications, a laboratory testing report, samples of labels according to FSMP labeling standards, and proof of the applicant's design and production capacity. The approvals are valid for five years and must be renewed prior to license expiry. Amendments to the registrations are required if items change in the interim.

The labels must be in accordance with relevant standards, such as National Standard for Labeling of Foods for Special Dietary Use and the Sample Requirements for FSMP Labels and Instructions,[68] and the FSMP registration rules require that the labels contain warnings, including that the product should be used under the guidance of a doctor or clinical nutritionist, used only on the designated suitable populations, and not used for parenteral nutrition support.[69] FSMPs must be made in accordance with special FSMP Good Manufacturing Practice Rules, and advertisements for FSMPs must follow the drug rules on publication in medical journals and be individually pre-approved.

---

[66] China Food and Drug Administration (2016b), Article 48.

[67] China Food and Drug Administration (2016d) Administrative Measures for the Registration of FSMPs. Article 50.

[68] China Food and Drug Administration (2016b) Foods Formulated for Special Medical Purposes.

[69] China Food and Drug Administration (2016d) Administrative Measures for the Registration of FSMPs, Article 38.

As discussed, total nutrition FSMPs require clinical trials. According to specialized Good Clinical Practice (GCP) rules, FSMP clinical trials do not require NMPA approval, but must take place at hospitals that are permitted to conduct drug clinical trials and have the appropriate nutrition department and personnel with experience related to FSMPs. These institutions can be subject to inspection by FSMP regulators. Unless there is an exemption, these should be randomized controlled trial, and sponsors should observe safety (tolerance), nutritional adequacy, and clinical effect. The trial design should be approved by the ethics committee(s) of the relevant institutions. Subjects must give informed consent prior to receiving the test product and receive appropriate treatment for the injuries they have incurred during the trial; serious events should be reported to the ethics committee.[70]

The first approvals of FSMPs under the new rules occurred in 2017 (3 approved), and increased year over year, with 18 approved in 2018 and 22 approved in 2019. During the pandemic in 2020, the number approved dropped to 12, but rebounded to 24 in 2021. In addition, as with infant formula, the number of domestically produced FSMPs approved per year was 92% of the total as of 2020, indicating that, likely for a variety of reasons, China-based applicants are dominating the field.[71] In 2021, 100% were domestically produced products.[72]

8.8 Food Labeling

Food labeling has become a critical aspect of regulation both because labels are a critical way to ensure consumers are made aware of the nature and content of the food they are buying (aspects that have not always been clear and have led to safety issues) and because consumers have seized on labels as a basis for both justified and frivolous litigation.

Regardless of whether the food is a general or special food, there are certain basic principles of labeling that are set forth primarily in a national standard GB 7718 on Prepackaged Food. This standard sets forth requirements on general principles of clarity (e.g., clear and non-misleading labels), food naming (trademarks and non-proprietary descriptive names), ingredient naming and listing, and net weight, among others. General foods must also comply with nutrition labeling requirements that are set forth in a separate standard, GB 28050 Nutritional Labeling for Prepackaged Foods, which sets the daily nutrient reference values and standards for nutrient content claims about fat, fiber, protein, and vitamin content, among others. It also contains a list of predetermined claims that may be made about the effect of certain nutrients on the body, e.g., Vitamin D Facilitates the Absorption of

[70]China Food and Drug Administration (2016a) Explanation of Good Clinical Practice Rule for Food Formulated for Special Medical Purposes (Trial Implementation).

[71]Foodmate Network News (2021).

[72]Ma (2022).

Calcium or Dietary Fiber Helps Maintain Normal Intestine Function.[73] A health food registration is required for a claim of supplementing vitamins or minerals and, as discussed above, special foods may have their own nutrition labeling standards.

These basic standards are one layer of labeling, but other standards and new food raw material announcements contain additional requirements. An important example is the new food raw material announcement for cultivated ginseng. Ginseng is a dual-use ingredient in food and Chinese medicine. Its use in food is permitted, but because an overdose of ginseng can be dangerous to pregnant women and cause side effects, such as nausea, vomiting, fever, and increased blood pressure, among others, an individual must be careful not to consume more than a certain amount per day. When "cultivated panax ginseng (five years or less)" was approved as a new food raw material in 2012, the notice announced certain labeling requirements, including that the labels should state that the product is not suitable for consumption by pregnant or lactating women or individuals under the age of 14. It also states that the label should indicate a consumption limit. The recommendation in the notice is that individuals should not consume more than three grans per day.[74] Ginseng labeling has produced a number of cases through the years, due to these concerns about its safety and related restrictions.

The label can also be an advertisement and a form of publicity. This means that the claims on the label are subject to various other laws, including the Advertisement Law (see Chap. 10) and the Anti-Unfair Competition Law. The content requirements in those laws are similar to those in the food-labeling standards, including prohibitions on superstitious, pornographic, and violent material. And it includes broad prohibitions on misleading content. This can lead to overzealous officials or consumers alleging that content on a label that amounts to routine puffery is misleading because it is not literally true.

8.9 Food Manufacturing

Unlike in the drug, device, and cosmetic spaces, there is not a marketing authorization holder system for food. The primary responsible entities under the law are licensed manufacturers and distributors. In some cases, brand owners without these licenses may entrust functions to license entities, but the law is not well-established on this point so practice may differ from province to province. For imported products, China focuses on the manufacturers.

Using the Food Safety Law as a foundation, China has gradually built a body of rules for standardizing practices for food production, including food preparation in the restaurant retail environments. The food-safety law requires that all entities engaged in the production of food (an undefined term, but which can be inferred to include testing and storage to some degree), must meet various general criteria.

[73] Ministry of Health (2011).

[74] Ministry of Health (2012).

These include having sufficient facilities and a layout to review and process raw materials, process food, and package and store the final product, while ensuring protection against cross contamination and pollution by hazardous substances. The Food Safety Law expressly requires that the water used in the production of food meet China's standards for drinking water. These facilities must also have compliant equipment, tools, and containers, including those for disinfection, and appropriately trained manufacturing and quality personnel.

A new rule issued in 2022 defines the various roles and responsibilities of those personnel responsible for the entity and the food safety managers and related personnel and their potential liability for regulatory compliance. The later type of personnel must identify and propose resolutions for food safety issues that may arise. As discussed in Chap. 3, failure to fulfill individual responsibilities can lead to personal liability for these personnel, particularly when that failure leads to unsafe food that causes or could cause human harm.[75]

While the Food Safety Law sets forth high-level rules, the various food-safety standards articulate more specific requirements. A set of General Principles on Food Manufacturing Hygiene GB 14881-2013, released in 1994 and updated in 2013, delineates basic requirements for facility layout and equipment, separation of processes and hazardous substances, and measures and procedures for cleanliness. Other standards and good manufacturing practices, such as those for health foods and infant formula, apply to those product areas. Additional requirements are also articulated by the general food manufacturing rules and inspection rules.

Like other regulated products, a food manufacturing license that attaches to the facility is required for all foods. Arguably, any facility that does anything from processing raw materials (including additives) to producing finished foods, including labeling and packaging them, should have such a license. If the entity is not producing food, but is instead handling, transporting, serving, or selling it, a food distribution license is required (see below). For both manufacturing and distribution licenses, there is an exception for small-scale enterprises.

The application for this manufacturing license is a local process. For special foods, the licenses are issued by the provincial-level local governments, but for general foods it can be done by lower levels of local governments.[76] Under applicable rules, there are 28 types of food manufacturing licenses, e.g., grains, edible oils, dairy products, beverages, alcohol, and vegetable products. Each type of license may open the door to yet another set of rules. For example, the predecessor of NMPA, the China Food and Drug Administration (CFDA), released a set of detailed rules for beverage products. Each rule articulates different levels of requirements for facilities (e.g., different clean room requirements), equipment, and personnel for types of

[75] State Administration for Market Regulation (2022).

[76] State Administration for Market Regulation (2020a) Food Manufacturing Licensing Administrative Measures, Article 7.

beverages, bottled water, carbonated beverages, tea beverages, fruit and other beverages, protein drinks, and solid beverages.[77]

Obtaining a food manufacturing license requires an inspection to confirm that the facility meets the relevant standards to support its scope. For example, a facility that makes beverages must have a scope that includes beverages and, more specifically, fruit beverages or colas (as categorized in the relevant beverage standards). The standards will dictate the content of the facility, including air handling, space, and environmental requirements.

As described above, food-related products require industrial-product licenses.

In 2024, the SAMR released a draft rule to govern entrustment of food manufacturing; the comment period was closed in February 2024. Like in the drug, device, and cosmetics areas, this draft set obligations for the entrustor and entrusted entity. The credentials for the entrustor of general food manufacturing for distribution in China must be an entity in China holding its own food manufacturing or distribution license (or in the case of distribution a record-filing). The entrustor for food manufacturing for export may be an overseas entity. The entrustor for production of a health food for distribution in China must be a health food registration holder and hold a manufacturing or distribution permission. Entrusted manufacturing is prohibited for infant formula and foods formulated for special medical purposes.

The entrustor must execute a contract with the entrusted entity, which must be appropriately licensed to produce the food, and that contract must have certain content specified by the draft rule, and the arrangement must be reported to the local food safety regulators for each entity. And the entrustor must supervise the entrusted contract manufacturer, either through dispatching its own inspectors or a third party to observe conditions in the facility. Both entrustor and entrusted are jointly liable for non-compliance with the Food Safety Law. Either can be sued and then seek compensation from the other, depending on responsibility.[78]

8.10 Food Transportation and Distribution

Safe transportation of food remains a problem in China, due to the absence of certain infrastructure for distribution, including appropriate cold storage to transport food throughout the larger country and distribution establishments that do not have appropriate capacity in terms of personnel and procedures.

The Food Safety Law (FSL) established national provisions on food distribution, requiring entities involved in transport, sale, and catering (e.g., restaurants) of food (but not food ingredients) to obtain food-distribution licenses. Under those national-level regulations, food distributors are required to have food distribution licenses

[77] China Food and Drug Administration (2017) Beverage Manufacturing License Review Rules.

[78] State Administration for Market Regulation (2024b) Administrative Measures on Supervision of Entrusted Food Manufacturing (Draft for Comment).

issued by the local level of government. Small street vendors are subject to a separate regulatory regime, as are those who sell edible agricultural products and prepackaged foods only, which were clearly exempted from licenses under a 2023 amendment to the rules on food distribution licenses and record-filings. The FSL and the national regulations also set requirements for the basic systems that distributors of food must have, including systems to receive and check food and the credentials of manufacturers or other distributors who sell it to them for compliance with law.

Distributors can be subject to civil and administrative liability (without indemnification) if they fail to check the credentials of the preceding entity in the distribution chain and end up selling food products that do not meet food-safety standards.

Despite this national-level system, food-distribution licenses are issued by the provinces or even municipalities in some cases. In 2018, pursuant to the administrative policies described above, China began to implement an electronic system for applying for and renewing food distribution licenses. These must be renewed at least once every five years, although if the lease on the premises is shorter, a renewal may be required more frequently. In some places, distribution licenses have been replaced for simplified record-filings for pre-packaged foods.

Restaurants hold food-distribution licenses, but there is a detailed Food Catering Standard that may be applicable to these establishments. This Catering Standard sets forth a number of detailed standard operating procedures needed to establish a food safety management system.[79] The Standard encourages food service providers to adopt this system according to its guidelines.

The provinces define more intricate requirements for food-distribution establishments, depending on the category that they fall into. Despite there being no national-level definition of food distribution, provinces have indicated that a broad range of restaurants, cafeterias, wholesalers, and even holding companies that merely book sales are subject to these regulations.

8.11 Food Importation and Export

Importing food into China, whether processed food or raw agricultural produce, can be a difficult process. A few structural points may be helpful. First, under the government reorganization that occurred in 2018 (See Chap. 2) regulation of imported and exported food rests generally under the China General Customs Administration. Customs has various powers to approve and inspect both the companies that import and export food and their products. SAMR does not have jurisdiction.

In 2021, Customs cancelled several specific rules on the import and export of products like honey, dairy products, meat products, and aquatic products and issued general rules on food import and export. In terms of import, under those rules

---

[79] State Administration for Market Regulation (2018) Code of Practice for Food Safety in Catering Services.

Customs evaluates both the country and the enterprise from which the goods are coming. Specifically, it evaluates the food safety system of those countries prior to permitting them to import food as a general matter. These conformity assessments of individual countries can be revisited if there is a material change in the law or an outbreak of disease or other serious food safety issue. Customs can restrict different types of food coming from a country for quarantine reasons.[80] A list of countries that are eligible to export meant to China and the types of meats they can export (including quarantines due to illness) can be found on the customs website.[81] For example, beef from the United States and countries in Europe have been banned from time to time, due to concerns about mad cow disease.[82]

Customs then evaluates the enterprise from which the products are coming. They do this through a facility registration process on which Customs also issued new rules in 2021. There are two types of registered facilities: Those that must be recommended by the authority in a given country or region that is responsible for the health and safety of food producers. Foods that require this type of home country recommendation for facility registration in China are meat and meat products, sausage casing, aquatic products, dairy products, edible bird's nests, bee products, eggs and egg products, edible fats and oils, flour products, grains, fresh and dehydrated vegetables, dried beans, seasonings, nuts and seeds, dried fruits, unroasted coffee and cocoa beans, prepackaged foods for special dietary uses, and health foods.[83] Other types of food facilities can apply for registration directly through an agent. In registering a facility, Customs can review documentation and inspection them onsite or virtually using cameras.

The facility registration system has been criticized by the United States as creating too substantial an administrative burden without an adequate connection to food safety. Despite those concerns, approximately 79,000 entities had been registered by the fall of 2022.[84]

Customs also reviews documentation, conducts sampling and inspections and approves importation at the product level. There are no pre-approvals for general foods, as there are for health foods or other special foods, but customs will review the information about the importer, the facility, the documentation about the food itself, including any quarantine approval and its certificate of conformity, and the results of any sampling inspection.

A quarantine approval is required for certain types of foods, for example, those that may contain foreign animal or plant material. This includes meat, dairy products

[80] General Customs Administration, Measures on the Safety of Imported and Exported Foods (2021).

[81] General Customs Administration, List of meat eligible for export to China from countries or regions that meet the requirements of assessment and review (2023).

[82] BBC News (June 27, 2018).

[83] General Customs Administration, Provisions on Registration of Overseas Food Production Enterprises (2021).

[84] Whitworth (2023).

(e.g. milk), fresh eggs, and aquatic products, as well as plant derived food, such as grains and beans. As all foods go through a review process at the border, failure to comply with applicable requirements above (many of which may change frequently or be subjectively applied) can result in rejection of a food product.

Similarly, the government can tests pursuant to testing standards for the products and ingredients—e.g., zero-tolerance standards for substances for which a zero reading may not be scientifically possible. More recently, concerns about cold-chain foods that China argues allow COVID-19 to survive have created new requirements for companies to certify that their products and supply chain are COVID-free.[85] All of that said, surveys since show that China rejects only about 1 percent of the foods imported, and often times the rejections are based on labeling or other documentation errors.[86]

Although China is fairly reliant on imports for its own food security,[87] China's exports of food have grown considerably. It is currently the 7th largest exporter of foods to the United States with substantial shares of the markets for garlic, apple juice and frozen shrimp, among other products.[88] Concerns have been raised about it as a source of exports, because of past problems in food safety, particularly around the time of the milk and pet food stories discussed earlier in this chapter.

The regime on export of foods is similar to that for import. Companies are expected to follow the requirements of the destination country and any applicable contracts or international agreements. Both farms and food production enterprises that produce food for export must submit record-filings with local customs authorities. Lists of the entities that are approved for filing can be found online, and China's quasi-governmental industry associations may release preferred lists. For example, the China Chamber of Commerce for Import and Export of Food released a "Preferred Chinese Export Brands List."[89] Customs can exercise supervision over these operations, including through inspections. This would be in addition to any production license that the entity holds, which would be issued by local administrations for market regulation (related to SAMR) where the facility is (see above).

Customs' regulations also require general health and safety measures, such as supplier and material evaluation systems and accurate production and inspection records, as well as methods to trace the products. These producers must apply for export and subject to inspection and quarantine, including onsite, and then again at port. Under the regulations, these food production enterprises can also be investigated if international organizations or overseas health authorities report safety issues with their products.

---

[85] Xinhua News Agency (May 23, 2022).

[86] Gale (2021).

[87] Liu (2023).

[88] USTR, People's Republic of China (2020).

[89] Ministry of Commerce (2021).

8.12 Food Regulation Enforcement

The government surveils and enforces against food manufacturers, distributors, and other parties through a variety of testing, inspection, complaint handling, and investigation mechanisms that can culminate in either administrative sanctions, such as fines, seizures of product, debarment (as in the other spaces), or criminal enforcement, which the government has increasingly deployed in the food space, pursuant to the four most stringents policy.

8.12.1 How Does Enforcement Start?

The government gets information about alleged food-safety violations in several ways. Some of the more common ways are consumer complaints and whistleblower tips, food-testing programs (i.e., a program that tests products in stores or at ports of entry), and facility inspections.

Food facility inspections are an important part of enforcement. Local market regulation officials inspect stores or warehouses and potentially test products as specified by the government testing program, which includes provisions for re-testing and sets forth an enforcement process.

The national and local governments formulate food plans, including, among other things, the types of food to be tested, the sampling environment and methods, and testing methods, as well as methods for assessing the risk to determine the focus of the plans.[90] However, in some cases the inspectors may simply review products on the shelves and ask questions about the labeling or presentation (e.g., general foods next to health foods). The stores/distributors are the immediate targets of these questions, but they may contact the manufactures for more information. Based on their distribution agreements, they will likely be indemnified if the issues relate to labeling or quality defects produced in the manufacturing environment. Retail stores or wholesale distributors have their own obligation to inspect products for compliance and, if they fail to do so properly, can be subject to penalties.

Consumer complaints are another common trigger for investigations that culminate in penalties is. Consumers can bring complaints about the products that they purchase from stores and lodge them with local AMRs. The AMRs are then obligated to investigate the issue by issuing questions and document requests to the distributor and/or manufacturer. Many of these complaints are frivolous—and indeed, authorities may disregard complaints if the food was purchased simply for lodging a complaint—but if there are grains of truth to the complaints it can result in a larger investigation and ultimately penalties, perhaps in addition to compensation to the consumer.

[90] State Administration for Market Regulation (2019c) Food Sampling and Testing Administrative Measures.

While individual complaints are permitted and even encouraged (if valid), what the government views as "activism" or group action against food issues is discouraged. Such was the case with the group of complaints that were brought in response to the 2008 milk scandals, when individuals were detained for filing court cases or forced to seek compensation through administrative means. Some went so far as to sue a parent company in the United States of one of the milk companies implicated, but the United States Court of Appeals for the Fourth Circuit dismissed the suit on forum non conveniens grounds, concluding that the plaintiffs had adequate remedies in China because they could participate in a compensation fund administered by the government or suing in courts on an individual basis.[91] Technically judicial remedies were available under the law, but the government retaliated against the most vocal parents,[92] even jailing one of them for allegedly extorting one of the milk companies for five years.[93]

8.12.2 Food Litigation

Despite prior issues, in addition to consumer complaints to administrative agencies, the litigation of food-safety issues in the courts, often small claims, has become very popular. The critical provision is Article 148 of the Food Safety Law, which provides for nominal damages for proof of any food-safety standard violation and punitive damages if a consumer can prove that a manufacturer or distributor willfully breached food-safety standards.

> Article 148: Any consumer that is damaged by food that does not comply with food-safety standards may claim compensation against either the distributor or Manufacturer. Any producer or distributor that receives claims for compensation from consumers shall be liable for compensation first and may not refuse to compensate. If it is attributable to the producer, the distributor may recover such compensation from the producer; if attributable to the distributor, the producer may recover such compensation from the distributor.
>
> In the event that any manufacturer produces food that does not conform to food-safety standards or distributes food while being aware of its nonconformity with food-safety standards, the customer can demand the producer or distributor to pay a penalty of 10 times the paid amount or 3 times of the loss, in addition to the compensation for the loss thereof. If the additional compensation is less than RMB 1,000, such additional compensation shall be increased to RMB 1,000; unless the defects are contained in labels and instructions of food that will neither affect food safety nor mislead consumers.[94]

[91] Jiali Tang v. Synutra International, Co., Ltd. (2011).

[92] Jacobs (November 10, 2010).

[93] Qiao and Mudie (April 7, 2017).

[94] National People's Congressional Standing Committee (2021). Translation from: USDA Foreign Agricultural Service (2015) Food Safety Law of the People's Republic of China.

This provision has been combined with plaintiff-friendly judicial interpretations to permit a variety of consumer suits to proceed. Unfortunately, while it has allowed for consumers to be heard in various channels, it has also encouraged a class of "professional consumers" who engage in a combination of manipulation and harassment to obtain compensation for nominal or non-existent violations. Manipulations include actions like using cleaner to wipe the expiration dates off products or buying products and holding them past their expiration date before bringing a claim. Harassment can include a deluge of complaints filed by individuals in league with each other (sometimes incorporated as a company of complainers), overwhelming the legal resources of manufacturers and distributors.

While China's food-safety system needs strong consumer supervision, professional consumers have annoyed officials and created cynicism in companies, diminishing the effect that a complaint might have on stakeholders and pushing parties to resort to contorted settlements with little benefit for product safety. To paraphrase an opinion of the Supreme People's Court to the National People's Congress in 2017:

> Current judicial practice related to consumer rights protection shows that the behavior of knowingly defrauding and buying fakes has become a trend in commerce, and an increasing number of professional anti-counterfeiters and anti-counterfeiting companies (i.e., groups) have emerged. Their motivation is not to purify the market but to use available punitive damages for their own profit or to extort retailers. Moreover, even when these professionals have won their lawsuits and received compensation, they buy another such product to make more profit. This behavior seriously violates principles of good faith, ignores judicial authority, and wastes judicial resources. We do not support this method of punishing evil with evil and drinking poison to quench thirst.[95]

The Supreme People's Court proposed to curb this opportunist behavior in areas *other than* food and drug consumption. For food, the relatively friendly rules of court remained. For example, the following judicial interpretation, which was further supported by a judicial interpretation in 2021, demonstrates a plaintiff-friendly stance.[96] Although this judicial interpretation covers both food and drugs (and references cosmetics), it focuses heavily on food.

> The Provisions of the Supreme People's Court on Several Issues Concerning the Application of Laws in the Trial of Food and Drug Disputes Cases was adopted by the Supreme People's Court Judicial Committee at its 1599th meeting on December 9, 2013. It is hereby promulgated and will be implemented as of March 15, 2014.

[95] Supreme People's Court (2017).

[96] General Office of the Supreme People's Court (2017).

Provisions of the Supreme People's Court on Several Issues Concerning the Application of Law in the Trial of Food and Drug Disputes

(Adopted at the 1599th meeting of the Judicial Committee of the Supreme People's Court on December 9, 2013)

These provisions are formulated for the purpose of properly handling food and drug disputes, in accordance with the Law of the People's Republic of China on Tort Liability, the Contract Law of the People's Republic of China [both the Tort and Contract Laws are now part of the Civil Code], the Law of the People's Republic of China on the Protection of the Rights and Interests of Consumers, the Food Safety Law of the People's Republic of China and judicial practice.

Article 1: If a consumer files a civil lawsuit on the basis of a food and drug dispute and the case meets the conditions for acceptance under the Civil Procedure Law, a people's court shall accept that case.

Article 2: Consumers who suffer harm due to quality problems with food and drugs may sue the seller or the manufacturer separately or at the same time.

If a consumer only sues the seller or the producer, a people's court may add relevant parties to participate in the lawsuit if it deems necessary.

Article 3: In disputes concerning the quality of food and drugs, in which the purchaser seeks to assert their rights against producers and/or sellers, and the producers and sellers defend the case on the grounds that purchaser knowingly purchased the food with a quality problem, the court will not support such a defense.

Article 4: In the case that consumers assert their rights against food and drug manufacturers and sellers based on the quality problems of gifts that caused the consumer damage, the People's Court shall not a support the defense of exemption from liability on the grounds that the consumer did not pay a price for the food or drug gift.

Article 5: If the consumer proves that the facts of the purchased foods and medicines and the purchased foods and medicines do not conform to the contract, and the producers or sellers of the food and medicines that are the object of the claim will assume liability for breach of contract, then people's court shall support them.

If the consumer proves that they have been damaged by the consumption of the food or use of the drug, and initially proves that the damage is causally linked with the food or drug, and requests the producer or seller of the food or drug to bear the tort liability, the people's court shall support it, unless the food or drug producer or seller can prove that the damage is not caused by the product's failure to meet quality standards.

Article 6: Producers and sellers of food shall bear the burden of proving that the food meets quality standards. Determining whether the food meets quality standards shall be based on national standards; if there is no national standard, it shall be based on local standards; if there is no national or local standard, it shall be based on enterprise standards. If the standard adopted by the producer of food is higher than the national standard or local standard, the determination shall be based on the enterprise standard [embodying the higher standard]. Where there is no such standard, the determination shall be based on the relevant provisions of the Food Safety Law.

Article 7: Although food and medicine obtain a testing qualification certificate prior to sale, and at the time of consumption or use the product is within the warranty period, if the product is determined through testing to be non-compliant, and the producer or seller defense the case using the testing qualification certificate demonstrating compliance, the people's court shall not support that defense.

Article 8: If the founders of centralized trading markets, lessors, of counters in those markets, and organizers of trade fairs fail to perform their obligations, such as review of credentials [of those who participate in the market], inspection, management, as provided in the Food Safety Law, and food-safety accidents occur that result in personal injury to consumers, and consumers request that the founder, the counter lessor and/or the organizer of the trade fair of the centralized trading market bear joint liability, the court shall support that claim.

Article 9: If consumers who purchase food and drugs through an internet trading platform suffer damage, and the network trading platform provider cannot provide the real name, address and effective contact information of the producer or seller of food or medicine, and the consumer requests the internet trading platform provider to bear liability, then the court shall support the claim.

After the provider of the internet trading platform assumes the liability for compensation, if it exercises the right of recourse against the producer or seller, the people's court shall support that claim.

If the provider of the online trading platform knows or should have known that the producers and sellers of food and drugs use their platforms to infringe on the legitimate rights and interests of consumers and fail to take necessary preventive measures, thereby causing damage to consumers, and consumers demand that those producers and sellers assume joint liability, then the court shall support that claim.

Article 10: If individuals, enterprises, or other organizations that have not obtained the qualifications for food production and sales and that rely on the qualifications of other producers and sellers produce and sell food that causes damage to consumers, and consumers request that those relying on the qualifications and those possessing the qualifications bear joint liability, the court shall support that claim.

If the consumer only sues [one of the two entities described above], the people's court may add relevant parties to participate in the lawsuit if necessary.

Article 11: If due to false advertisements of foods and medicines with quality problems, consumers are subject to harm, and consumers request that the advertisement distributor or publisher assume joint and several liability according to the relevant requirements of the Consumer Protection Law and other such laws, the people's court shall support that request.

If social groups, other organizations, or individuals recommend foods and medicines to consumers in false advertisements, and consumers are subject to harm, and consumers request food and drug producers and sellers to assume joint and several liability in accordance with the relevant provisions of the Consumer Protection Law and other such laws and regulations, the court will support that request.

Article 12: If a food and drug testing institution deliberately issues a false inspection report, causing damage to consumers, the court will support a request by a consumer that it bear joint and several liability [with other parties].

If the food and drug testing institution negligently issues a false inspection report, and such action causes damage to consumers, the court shall support a request by the consumer that the testing agency bears corresponding liability.

Article 13: If a food institution agency deliberately issues a false certification, causing damage to consumers, the court shall support a request by the consumer that it bear joint and several liability [with other parties].

If the food accreditation institution negligently issues a false certification, causing damage to consumers, the court shall support a request by the consumer that the institution bear corresponding liability.

Article 14: If there are quality problems in the production and sale of food and medicine, producers and sellers are required to bear civil, administrative, and criminal liability the same time. If their property is insufficient to pay, in accordance with the Tort Liability law, and the parties request that the food producer or seller first bear civil liability, then the court shall support that request.

Article 15: For production of food that is non-compliant with food-safety standards or knowing sale of food that is non-compliant with food-safety standards, in addition to claiming compensation for losses, if consumers may a claim for 10 times the price of the food or other compensation standard in accordance with the law, the court will support such a claim.

Article 16: Producers and sellers of food and drugs shall, in the form of format contracts, notices, declarations, notices, exclude or restrict consumer rights, reduce or exempt distributor's responsibilities, or increase consumer responsibilities, among other such measures, and are unfair or unreasonable to consumers, the court will support a request to determine that such content is invalid.

Article 17: Disputes between consumers and producers, sellers, advertisement distributors, advertisement publishers, endorsers, and testing institutions of other products, such as cosmetics and health care products, shall be governed by these Provisions.

Where a consumer association initiates a public interest litigation in accordance with the law, these Provisions shall apply.

Article 18: These Provisions shall apply to the first-instance and second-instance cases that are being tried by the people's court after the implementation of these Provisions.

These Provisions shall not be applicable when the final review has been carried out before the implementation of these Provisions and the parties have applied for retrial after the implementation of these Provisions or to retrial in accordance with the trial supervision procedures.

In contrast to the rules of court noted above, in 2019 the State Administration for Industry and Commerce released new complaint rules that allowed consumer complaints to be rejected if the consumer purchased the product with the intention of suing.[97] This was the culmination of years during which some of the more creative courts and administrative agencies took similar approaches to dismiss suits; as China has no mechanism to dismiss inadequately pleaded or vexatious complaints there were few avenues to remove these cases, absent a full, time-consuming, and costly hearing. The SAMR complaint rules provided a basis to dismiss frivolous administrative complaints immediately.

8.13 Administrative and Criminal Sanctions

Like other spaces, there are various administrative and criminal sanctions for food-safety violations. These range from fines to debarment and even administrative detention on the administrative side, and production and sale of unsafe food can lead to criminal penalties, particularly when individuals are harmed.

[97] State Administration for Market Regulation (2019d).

8.13.1 Criminal Law and Food Safety

As with drugs and devices, the Criminal Code contains provisions on unsafe food. Production or sale of food that does not meet food-safety standards and causes harm due to food-borne illness or contamination can be subject to prison terms of three to seven years, depending on the severity. Production or knowing sale of toxic or harmful food, e.g., using unlawful ingredients, that causes death can lead to life in prison or the death penalty.[98]

The case of the Hebei Province Shijiazhuang Sanlu Group Co., Ltd. executives illustrates the role that the criminal law plays in punishing food-safety law violations. As discussed above, many companies were implicated in this scandal. In the Sanlu case, four executives were held criminally liable for failing to stop the production and sales of substandard and contaminated infant formula, despite knowing of problematic testing data. This case relates to the high-profile scandals involving melamine and other contaminants in milk in 2008, which affected hundreds of thousands of infants around the world and caused a number of injuries and deaths in China, due to kidney damage caused by these contaminants.

The evidence in the case showed that since December 2007 the Company had received complaints from consumers that infants consuming their product had red precipitate in their urine. Two of the defendants and others on the leadership team were informed of these complaints and in May 2008, organized testing to determine the cause. The testing showed melamine in 15 of the 16 batches of product tested. With this knowledge in hand, the company temporarily sealed its product in the warehouses and tested for melamine content. The product testing at 10 mg/kg or below was allowed to be distributed to the market.

The company decided to gradually withdraw other products on the market through offering exchanges to consumers. However, it continued to produce and sell infant formula contaminated with melamine, until they were ordered by the government to cease production on September 12, 2008. During this time, infants continued to experience symptoms, some were hospitalized, and some died. Seventy-two batches were produced between the time of the testing and the government's order to cease production. As the court stated, "The state has invested a huge amount of money in the examination and medical treatment of sick infants and young children, and the normal production and operation of many dairy companies and dairy farmers have been greatly affected, resulting in huge economic losses."[99]

Defendants who knew that the milk contained illegal and harmful additives (i.e., melamine) and continued to order its production and distribution were criminally charged, convicted, and sentenced to between life in prison and 5 years in prison. This included both leadership at the company that made the decision to continue to distribute the milk and those who carried out the illegal activities. The chairman and

[98] China Criminal Code (2020) Articles 141–144.

[99] Hebei Higher People's Court (2013).

legal representative of the company played an "organizing and directing" role, while others on the leadership team and in the company were directly responsible because they implemented the arrangement by which the contaminated milk was continuously distributed. Another individual who accepted instructions to distribute the milk was charged and convicted as an accomplice with a lesser sentence.

The next sections illustrate some of the procedures that the defendants in the Sanlu case failed to follow upon discovery of the contaminated milk.

8.14 Food Safety Accidents

Food safety accidents are incidents in which food causes or could cause harm to human health, such as incidents resulting from food-borne illness or contamination with harmful substances.[100] Food-borne illness is defined to include infection, poisoning, or other disease resulting from pathogens in food. The source of food poisoning can include bacteria, toxic chemicals, or animals or plants containing toxic materials (e.g., puffer fish, fish with histamines, or poisonous mushrooms), and causes can include consumption of uncooked food, improper storage, contamination by food-preparation personnel, expired food, failure to property heat or cook food, and naturally occurring toxic substances that are not properly removed during the preparation process.[101]

Pursuant to the Food Safety Law, both the government and individual enterprises have obligations to report and promptly handle these types of incidents.

8.15 Food Recalls

Like other areas, food and food-related material manufacturers are required to recall their products if they discover that they are unsafe, and food distributors and retailers are required to report issues that they find to the manufacturers and assist in implementing those recalls. Recalls of food (including special food and food additives) are governed by specialized regulations,[102] whereas food-related material recalls are governed by recall rules on general consumer products.[103]

Under the specialized recall rules, unsafe food is food, the manufacture and/or distribution, of which are prohibited or that *may* endanger human health.[104] Food Safety Law itself sets forth a broader standard that food manufacturers must recall food if it does not comply with food-safety standards; arguably, this could include

[100] National People's Congressional Standing Committee (2021), Article 150.

[101] Pingrang Market Supervision Bureau (2020).

[102] China Food and Drug Administration (2015).

[103] State Administration for Market Regulation (2019b) Administrative Provisions on Consumer Product Recalls.

[104] China Food and Drug Administration (2015), Article 2.

issues that are not safety related.[105] There may be some room for certain lesser defects that would not cause damage to health (it is unclear how minor the potential defect would have to be, but it would likely include formatting issues or typographical errors on the label) to be corrected by stickering product in the market.[106] China's regulators can support stickering if the sticker is sufficient to cover the error and is firmly affixed to the underlying product.

When a food manufacturer learns of potentially unsafe food, either through its own investigation or outside complaints, China's laws require that it stop producing that food, inform distributors and the authorities, and initiate procedures to notify and recall the food from consumers. Food that is not in the hands of consumers can be withdrawn from the distribution chain and contained.[107] The authorities may also order food manufacturers to recall food.

Once reported, food recalls occur under the supervision of the local authorities, which must review and approve the various forms of documentation, including the recall plan and the recall notice to consumers. Like drugs and devices, there are classes of recalls, with the first class being the highest risk, for foods that have already caused or could cause serious illness or death.

- Class one recalls require recall initiation and a recall plan within 24 h of becoming aware of the food-safety risk.
- Class two recalls are for actual or possible cause of "general health damage," and the window for initiation and recall plan is 48 h.
- Class three recalls are for foods with defective or false labels that may cause health damage, e.g., because the consumer would be misled or not have the proper warnings of certain ingredients. The window for initiation of Class three recalls is 72 h.

Although the timelines in the regulations are tight, food recalls can take some time to commence. The government considers itself the approver of the recall plan and notice, and it may take 2–3 rounds of revisions before the discussion with the manufacturer is complete and the recall can begin. The plan generally contains a description of the safety issue and the range and volume of product affected, its distribution in China, and a plan for notifying consumers and obtaining and replacing or refunding the affected product, as well as plans for its destruction or quarantine. Manufacturers implementing a recall must also update the authorities on its progress and submit a summary report at its conclusion.

If a recall is evidence of a violation of a prohibited act under the Food Safety Law (e.g., distribution of food that is not compliant with standards), the authorities can pursue a penalty against the company. Although the food-recall regulations include a clause, consistent with the general penalty provisions under Chinese law

[105] National People's Congressional Standing Committee (2021), Article 63.

[106] China Food and Drug Administration (2015), Article 13.

[107] China Food and Drug Administration (2015), Article 7.

(as described in Chap. 3, on administrative penalties), that calls for reduction or elimination of a penalty if the company takes steps to contain harm and eliminate harm, in practice, officials may still impose a penalty. Other recalls proceed in a routine fashion and without penalty, unless the manufacturer or distributor fails to notify the recall or comply with procedures for its implementation. Although notification of the issue can lead to penalties and/or consumer lawsuits, recalls are increasingly routine (various companies large and small have gone through them), and they are not a death knell for business, although some manufacturers continue to view them in that light.

Still, concern about whether the aforementioned mitigation principle will apply in practice to reduce or eliminate penalties can discourage companies from recalling food. Companies are also faced with difficulties tracing the issue with distributors and/or convincing distributors to comply and remove affected product from the market. If distributors discover the issue, they may be reluctant to inform authorities for fear that they could be implicated in the issue or damage business relationships.

References

BBC News (June 27, 2018) China Lifts Ban on Imports of British Beef. https://www.bbc.com/news/business-44629275.

China Food and Drug Administration (2015) Administrative Measures on Food Recalls. http://gkml.samr.gov.cn/nsjg/fgs/202011/t20201103_322883.html#:~:text=%EF%BC%88%E4%B8%80%EF%BC%89%E4%B8%80%E7%BA%A7%E5%8F%AC%E5%9B%9E%EF%BC%9A,%E7%AE%A1%E7%90%86%E9%83%A8%E9%97%A8%E6%8A%A5%E5%91%8A%E5%8F%AC%E5%9B%9E%E8%AE%A1%E5%88%92%E3%80%82.

China Food and Drug Administration (2016a) Explanation of Good Clinical Practice Rule for Food Formulated for Special Medical Purposes (Trial Implementation). http://www.gov.cn/xinwen/2016-11/01/content_5127060.htm.

China Food and Drug Administration (2016b) Administrative Measures for the Registration of Foods Formulated for Special Medical Purposes. NMPA No. 24.

China Food and Drug Administration (2016c) Health Food Manufacturing License Review Implementing Rules. https://www.samr.gov.cn/tssps/zcwj/202108/t20210819_333759.html.

China Food and Drug Administration (2016d) Administrative Measures for the Registration of FSMPs

China Food and Drug Administration (2017) Beverage Manufacturing License Review Rules. http://www.cfe-samr.org.cn/zcfg/spjc_152/gfxwj_156/202104/t20210416_3575.html.

CIRS (2022) 2018-2022, Summary of Acceptance and Approval of New Food Additives Over Five Years. https://www.cirs-group.com/cn/food/2018-2022-jin-5-nian-shi-pin-tian-jia-ji-xin-pin-zhong-shou-li-he-shen-pi-qing-kuang-hui-zong.

CIRS Group (2022) 2021 Annual Health Food Registration Circumstance Analysis. https://www.cirs-group.com/cn/food/2021nbjspzcqkfx.

CIRS Group (2023) 2022 Annual Analysis of Registration Circumstance of Health Foods. https://www.cirs-group.com/cn/food/2022-nian-bao-jian-shi-pin-zhu-ce-qing-kuang-fen-xi.

Du H. People's Daily (2016) Cross Border E-Commerce Can Still Sell Powdered Infant Formula. http://www.gov.cn/xinwen/2016-04/14/content_5063847.htm.

Foodmate Network News (2021) 2020 Registration and Approval Information of Formulas for Special Medical Purposes (Annual Inventory). https://www.sohu.com/a/442611273_172731.

Gale F (2021) China Safety Requirements Pose Challenges for Food Exports, Economic Research Service. https://www.ers.usda.gov/amber-waves/2021/may/china-s-safety-requirements-pose-challenge-for-food-exporters/.

General Customs Administration, List of meat eligible for export to China from countries or regions that meet the requirements of assessment and review (2023). http://english.customs.gov.cn/statics/1a11198b-def3-4bd2-9157-6fc147a30717.html.

General Customs Administration, Measures on the Safety of Imported and Exported Foods (2021). http://www.qdh.gov.cn/art/2023/1/31/art_1229715711_59014912.html

General Customs Administration, Provisions on Registration of Overseas Food Production Enterprises (2021). http://www.gov.cn/zhengce/2021-03/12/content_5723374.htm.

General Office of the Supreme People's Court (2017) Reply of the General Office of the Supreme People's Court to Recommendation No. 5990 of the Fifth Session of the Twelfth National People's Congress (Regulation of Professional Anti-counterfeiting). No. 181. https://www.sohu.com/a/151299563_169411.

Hebei Higher People's Court (2013) The case of Shijiazhuang Sanlu Group Co., Ltd. and relevant responsible personnel. http://wap.lsbar.com/phone/context/2/5387.

Hemmingsson E (2021) The unparalleled rise of obesity in China: a call to action. Int J Obes 45: 921–922. https://doi.org/10.1038/s41366-021-00774-w.

Hu M (April 17, 2019) Cameras Keep Watch Over Restaurant Kitchens. Shine. https://www.shine.cn/news/metro/1904173159/.

Jacobs A (November 10, 2010) China Sentences Activist in Milk Case to Prison. The New York Times. https://www.nytimes.com/2010/11/11/world/asia/11beijing.html.

Jiali Tang v. Synutra International, Co., Ltd. (2011) 656 F.3d 242, 246-247 (4th Cir.)

Jing X et al (September 2020) Antibiotic Use on Chicken Farms in Northwest China. https://aricjournal.biomedcentral.com/counter/pdf/10.1186/s13756-019-0672-6.pdf.

Liaoning Provincial Court Beijing Sunshine Yibai Biotechnology Development Co., Ltd., et al., Production and Sales of Toxic and Harmful Foods (2018). http://ln.lncourt.gov.cn/article/detail/2018/11/id/4458520.shtml.

Liu Z, China Increasingly Relies on Imported Food. That's a Problem (2023). https://www.cfr.org/article/china-increasingly-relies-imported-food-thats-problem

Ma D (2022) 2021 Registration and Approval Information of FSMPs (Annual Inventory). Food Partner Network News. http://news.foodmate.net/2022/01/618033.html.

Ministry of Agriculture and Rural Affairs (2023) How Does my Country Regulate the Labeling of Genetically Modified Products. http://www.moa.gov.cn/ztzl/zjyqwgz/kpxc/202302/t20230217_6420890.htm.

Ministry of Commerce (2005) Opinion on the Construction of the Food Safety Standard System. http://www.mofcom.gov.cn/aarticle/b/d/200501/20050100329049.html.

Ministry of Commerce (2019) Cross-Border E-Commerce Positive List. http://images.mofcom.gov.cn/cws/202001/20200110143527533.pdf.

Ministry of Commerce (2021) China Chamber of Commerce of Import and Export of Foodstufs, Native Produce and Animal By-Products Issues China Food Export Preferred Brand List. http://www.mofcom.gov.cn/article/zwjg/zwsq/zwsqyz/202112/20211203233573.shtml.

Ministry of Health (1996) Health Food Labeling Rules. https://www.cfe-samr.org.cn/zcfg/bjsp_134/gfxwj/202104/t20210409_684.html.

Ministry of Health (1998) Health Food Good Manufacturing Standard, GB 17405-1998. Available at http://down.foodmate.net/wap/index.php?moduleid=23&itemid=1270.

Ministry of Health (2002) Notice on Further Standardizing the Management of Health Food Raw Materials. http://www.nhc.gov.cn/sps/s3593/200810/bc239ea3d226449b86379f645dfd881d.shtml.

Ministry of Health (2010) Explanation Regarding General Food, Health Food and New Food Raw Materials. http://www.nhc.gov.cn/bgt/s3578/201306/068ee16797814e8dbd3675c37a8cb256.shtml.

Ministry of Health (2011) National Food Safety Standard for Nutrition Labeling of Prepackaged Foods (GB-28050-2011). https://www.fsis.usda.gov/sites/default/files/media_file/2021-03/gb-28050-2011.pdf.

Ministry of Health (2012) Announcement on Approving Ginseng (Cultivated) as New Food Raw Material (No. 17). http://www.nhc.gov.cn/sps/s7891/201209/e94e15f2d9384b6795597ff2b101b2f1.shtml.

Minter A (March 20, 2021) China Can No Longer Ignore Food Safety Issues. Bloomberg. https://www.bloomberg.com/opinion/articles/2021-03-30/china-can-no-longer-ignore-its-food-safety-issues?leadSource=uverify%20wall.

National Health and Family Planning Commission (2011) List of Non-Food Substances and Illegal Food Additives that Can be Easily Abused. http://www.nhc.gov.cn/zwgkzt/spaqxx/201406/38e5c8a53615486888d93ed05ac9731a.shtml.

National Health and Family Planning Commission (2014) Standard on the Use of Food Additives GB 2760-2014. http://down.foodmate.net/standard/sort/3/42543.html.

National Health and Family Planning Commission (2015) GB 19299-2015. http://down.foodmate.net/standard/sort/3/48423.html.

National Health and Family Planning Commission (2016a) General Safety Requirements of Food-Contact Materials and Articles (GB 4806.1-2016). Translated by USDA Foreign Agricultural Service. https://apps.fas.usda.gov/newgainapi/api/report/downloadreportbyfilename?filename=General%20Safety%20Requirements%20of%20Food-Contact%20Materials%20and%20Articles_Beijing_China%20-%20Peoples%20Republic%20of_3-31-2017.pdf.

National Health and Family Planning Commission (2016b) GB9685, Positive List of Food Contact Material Additives. http://down.foodmate.net/standard/sort/3/49855.html.

National Health and Family Planning Commission (2016c) GB 4806.6-2016 Permitted Resins in Plastic Food Contact Materials, available at http://down.foodmate.net/standard/down.php?auth=49848.

National Health and Family Planning Commission (2017a) Administrative Measures on New Varieties of Food Additives. http://www.nhc.gov.cn/fzs/s3576/201808/94cd77df782a418c9ff96ad5ad5789ee.shtml.

National Health and Family Planning Commission (2017b) Administrative Measures on the Safety Review of New Food Raw Materials. http://www.nhc.gov.cn/fzs/s3576/201808/b6894ec987b64af3ae75f5140130c869.shtml.

National Health Commission (2017a) Chinese Dietary Reference Intakes: Part 1 Macronutrients. http://www.nhc.gov.cn/wjw/yingyang/201710/fdade20feb8144ba921b412944ffb779/files/0fa10dfb812a48b483d931972df1ccb8.pdf.

National Health Commission (2017b) Notice on Review Work of Importation of Foods that Do Not Have a Food Safety Standard. No. 14. http://www.nhc.gov.cn/sps/s3593/201704/15031a65db34492eab8fb5819e9cb466.shtml.

National Health Commission (2018) Disinfectant Measures (NHC No. 18 2018). http://www.nhc.gov.cn/zhjcj/gongwen1/201806/047c54980196495ab95856cc4839f3cc.shtml.

National Health Commission (2020) GB/T 1.1-2020. http://www.zjsm.org/file/news/637231474274567101.pdf.

National Health Commission (2021a) Administrative Regulations on the Catalogue of Substances that are Both Foods and Chinese Medicinal Materials in Accordance with Tradition. http://www.nhc.gov.cn/sps/s7892/202111/1b3e18ba75f142f99a4a15ce0d1660f3.shtml.

National Health Commission (2021b) Food Safety Standard Plan. http://www.nhc.gov.cn/sps/s7891/202108/1f6bf27d4c6c498ea4094893b4510aaa.shtml.

National Health Commission (2022a) Catalogue of National Food Safety Standards (1419 as of Feb. 2022). http://www.nhc.gov.cn/sps/s7891/202202/abb7090ad744405fba8244893839206d.shtml.

National Health Commission (2022b) Food Safety Standard Plan. http://www.gov.cn/zhengce/zhengceku/2022-08/06/content_5704446.htm.

National Health Commission, Department of Food Safety Standards, Monitoring and Evaluation (2021) Q&A on Infant Formula Food Series. http://www.nhc.gov.cn/sps/s7886t/202103/16cf6e65740e42b5a3451fe044e19c70.shtml.
National Health Commission, Notices of new food additives. https://zwfw.nhc.gov.cn/kzx/tzgg/sptjjxpzsp_224/.
National People's Congress (2018) Advertisement Law
National People's Congress China Criminal Code (2020) Articles 141-144
National People's Congress Standing Committee (2022) Agricultural Product Quality Law. https://gkml.samr.gov.cn/nsjg/bgt/202211/t20221103_351312.html.
National People's Congressional Standing Committee (2021) Food Safety Law. http://www.linan.gov.cn/art/2022/5/17/art_1367594_59068207.html.
National People's Congressional Standing Committee (2022) Agricultural Product Quality and Safety Law, amended. http://www.gov.cn/xinwen/2022-09/03/content_5708127.htm.
National Technical and Quality Supervision Bureau (1991) Administrative Provisions on Enterprise Standards. https://gkml.samr.gov.cn/nsjg/bzcxs/201903/t20190313_291966.html.
Pingrang Market Supervision Administration (2020) Foodborne Illness, Food Safety Incidents, and Food Poisoning. https://www.zjpy.gov.cn/art/2020/11/25/art_1418068_58994304.html#:~:text=%E9%A3%9F%E5%93%81%E5%AE%89%E5%85%A8%E4%BA%8B%E6%95%85%EF%BC%8C%E6%8C%87%E9%A3%9F,%E5%8F%AF%E8%83%BD%E6%9C%89%E5%8D%B1%E5%AE%B3%E7%9A%84%E4%BA%8B%E6%95%85%E3%80%82.
Qiao L, Mudie L (April 7, 2017) Court in China's Guangdong Acquits Milk-Scandal Father Jailed For Five Years. Radio Free Asia. https://www.rfa.org/english/news/china/court-in-chinas-guangdong-acquits-milk-scandal-father-jailed-for-five-years-04072017094814.html#:~:text=Guo%20Li%2C%20a%20simultaneous%20interpreter,became%20ill%20with%20kidney%20stones.
Reuters (March 19, 2021) China to Crack Down on the Use of Leanness Enhancers in Cattle and Sheep. https://www.reuters.com/article/us-china-livestock-clenbuterol/china-to-crack-down-on-use-of-leanness-enhancers-in-cattle-and-sheep-idUSKBN2BB1Q2.
Shanghai Health Commission, How to File Food Safety Enterprise Standards (2020). https://www.pudong.gov.cn/019023004/20220102/410866.html.
Standards Administration of China (2014) GB/T 30643-2014, Principles of Food Contact Substance Labeling. http://www.designartj.com/uploadfile/bzgcysb/20161209/GBT%2030643-2014%20%E9%A3%9F%E5%93%81%E6%8E%A5%E8%A7%A6%E6%9D%90%E6%96%99%E5%8F%8A%E5%88%B6%E5%93%81%E6%A0%87%E7%AD%BE%E9%80%9A%E5%88%99.pdf
State Administration for Market Regulation (2016) Infant Formula Registration Administrative Measures. https://gkml.samr.gov.cn/nsjg/tssps/201903/t20190329_292466.html.
State Administration for Market Regulation (2018) Code of Practice for Food Safety in Catering Services. https://www.samr.gov.cn/spjys/tzgg/201902/t20190226_291361.html.
State Administration for Market Regulation (2019a) Administrative Measures on the Health Food Raw Material Catalogue and Health Food Claims Catalogue. https://gkml.samr.gov.cn/nsjg/fgs/201908/t20190820_306117.html.
State Administration for Market Regulation (2019b) Administrative Provisions on Consumer Product Recalls. https://gkml.samr.gov.cn/nsjg/fgs/201911/t20191126_308824.html.
State Administration for Market Regulation (2019c) Food Sampling and Testing Administrative Measures. https://www.samr.gov.cn/zw/gz/zjhzd/202111/P020211126531169612412.pdf.
State Administration for Market Regulation (2019d) Tentative Measures on Complaint and Report Administration. http://www.gov.cn/gongbao/content/2020/content_5483878.htm.
State Administration for Market Regulation (2020a) Food Manufacturing Licensing Administrative Measures. https://www.samr.gov.cn/zw/gz/zjhzd/202111/P020211127347394500267.pdf.
State Administration for Market Regulation (2020b) Food Manufacturing Measures. http://www.gov.cn/gongbao/content/2020/content_5509732.htm?ivk_sa=1024320u.

State Administration for Market Regulation (2021a) Provisions on Record Filed Health Food Product Excipients. https://gkml.samr.gov.cn/nsjg/tssps/202102/W020210220533475984445.docx.
State Administration for Market Regulation (2021b) Record-Filed Health Food Dosage Form Technical Requirements. https://gkml.samr.gov.cn/nsjg/tssps/202102/t20210220_326169.html.
State Administration for Market Regulation (2021c) Nutrient Supplements in the Catalog of Health Food Raw Materials (2022 Edition) (Draft for Comment), Catalogue of Nutrient Supplements with Health Functions Allowed for Health Food Claims (2022 Edition) (Draft for Comment) and Protein in the Catalog of Health Food Raw Materials (Draft for comments). https://www.samr.gov.cn/hd/zjjg/202209/t20220927_350342.html.
State Administration for Market Regulation (2022) Administrative Measures on the Supervision of Implementation of Primary Food Safety Responsibilities. https://www.samr.gov.cn/zw/gz/zjhzd/202209/P020220926434327688550.pdf.
State Administration for Market Regulation (2023a) Interim Measures for the Supervision and Administration of Food-Related Product Quality and Safety. https://www.gov.cn/zhengce/2022-10/28/content_5726886.htm.
State Administration for Market Regulation (2023b) "Permitted Claims for Health Foods in Non-Nutrient Supplement Health Function Catalogue (2023 Version)" Released. https://www.gov.cn/lianbo/bumen/202309/content_6901834.htm.
State Administration for Market Regulation (2023c) Four Health Food Catalogs Including "Health Food Raw Materials Catalog Nutrient Supplements (2023 Edition)" Have Been Released. https://www.gov.cn/lianbo/bumen/202306/content_6888155.htm.
State Administration for Market Regulation (2024a) Notice on Publicly Soliciting Opinions on the "Food Safety Law of the People's Republic of China" (Draft for Comments on Amendments). https://www.samr.gov.cn/hd/zjdc/art/2024/art_b143a2ba4ca74088bc8471833bce5063.html.
State Administration for Market Regulation (2024b) Administrative Measures on Supervision of Entrusted Food Manufacturing (Draft for Comment). https://www.moj.gov.cn/pub/sfbgw/lfyjzj/lflfyjzj/202401/t20240109_493113.html
State Administration for Market Regulation Foods Formulated for Special Medical Purposes. http://www.cfe-samr.org.cn/xzzq/tsyxytpfsp_123/.
State Administration for Market Regulation, National Health Commission, et al (2023) Announcement on the release of the "Catalogue of Health Functions Allowed for Health Food Claims of Non-nutrient Supplements (2023 Edition)" and Supporting Documents. https://www.cifst.org.cn/a/news/fagui/20230831/2852.html.
State Council (2017) Administrative Regulations on the Safety of Agricultural Genetically Modified Organisms. http://www.gov.cn/zhengce/2020-12/26/content_5574537.htm.
State Council (2018) The Food Safety Office of the State Council Notice on Continuing Good Work to Rectify Food and Health Food Fraud and False Propaganda. http://health.people.com.cn/n1/2018/0807/c420159-30214799.html.
State Food and Drug Administration (2012) Test Results for Beverage Case. http://www.gov.cn/gzdt/2012-02/15/content_2067769.htm.
Supreme People's Court (2017) The Reply of the General Office of the Supreme People's Court to Proposal No. 5990 of the Fifth Session of the Twelfth National People's Congress (Regulating Professional Counterfeiters). https://www.sohu.com/a/151299563_169411.
USDA Foreign Agricultural Service (2015) Food Safety Law of the People's Republic of China. https://apps.fas.usda.gov/newgainapi/api/report/downloadreportbyfilename?filename=Amended%20Food%20Safety%20Law%20of%20China_Beijing_China%20-%20Peoples%20Republic%20of_5-18-2015.pdf
USTR, People's Republic of China (2020). https://ustr.gov/countries-regions/china-mongolia-taiwan/peoples-republic-china.
Wang Y, Jin Z (2021) Annual Inventory of Infant Formula Information. Food Partner Network. http://news.foodmate.net/2022/01/617507.html.

Wang Y, Zhao L, Gao L et al (2021) China's Obesity Problem: Time to Act. Lancelet Diabetes Endocrinol. https://www.thelancet.com/journals/landia/article/PIIS2213-8587%2821%2900150-9/fulltext

Whitworth J (2023) U.S. Repeats Concerns about Chinese Import Rules at WTO Meeting. Food Safety News. https://www.foodsafetynews.com/2023/04/u-s-repeats-concerns-about-chinese-import-rules-at-wto-meeting/#:~:text=The%20United%20States%20has%20said,food%20safety%20and%20public%20health.

World Health Organization (2003) Obesity and Overweight Factsheet. https://web.archive.org/web/20170713190601/http://www.who.int/dietphysicalactivity/media/en/gsfs_obesity.pdf.

Xinhua News Agency (January 28, 2016) Xi Jinping Issues Important Instructions on Food Safety Work. http://www.xinhuanet.com//politics/2016-01/28/c_1117928490.htm.

Xinhua News Agency (May 23, 2022) China Strengths COVID-19 Control Over Cold Chain Food Imports. https://english.www.gov.cn/statecouncil/ministries/202205/23/content_WS628b8349c6d02e533532b2de.html.

Yan L (November 25, 2022) Ordinary Food Mixed into Health Food Shelves. Sina News. https://k.sina.cn/article_2381872931_8df87f230190108qa.html?from=news&subch=onews.

Cosmetics Regulation in China

9

9.1 Introduction to the Cosmetics Space

Unlike drugs and medical devices, cosmetics have not been subject to policies with special incentives to encourage innovation. To be sure, cosmetics are a vibrant and flourishing industry in China with products not only sold in stores, but also increasingly online through domestic and cross-border e-commerce websites. The value of sales has grown 150% between 2012 and 2020 to be the second largest market in the world for these products after the United States.[1] Influencers, both men and women, demonstrate lipstick and blush online to millions of followers. However, unlike drugs and medical devices, cosmetics contend with anti-luxury trends that have flowed through politics and society in China, including during the pandemic.

In China, cosmetics are subject to a more stringents regulatory filing and approval process, for both products and ingredients, than in many other countries. These pre-market approval processes require in-depth manufacturing, ingredient origin, and quality and formulation information, as well as standard product testing and specific literature, animal, and/or human studies to support specific functional claims. Furthermore, certain products that are regulated in China under more rigorous rules for "special" cosmetics are in other countries considered over-the-counter drugs, cosmeceuticals, or quasi-drugs. In this respect, China takes somewhat less stringents approach to these products.

Discussions in policy-making forums in this area have focused on whether and to what extent companies can engage in self-regulation and be subject to less stringents supervision from the government (as under the cosmetics regimes of other jurisdictions, such as the United States). Although the current framework regulation, the Cosmetics Supervision and Administration Regulation (CSAR), contains a general principle of reliance on industry regulation, the implementing rules remain stringents and prescriptive. Because the process for developing and obtaining

[1] U.S. Department of Commerce, International Trade Administration (2020).

J. C. Balzano, *China Food and Drug Law*, Law for Professionals,
https://doi.org/10.1007/978-3-031-61901-4_9

approval of cosmetics can be quite expensive, it can be difficult for small- and medium-sized companies to enter the market. The cross-border e-commerce pathway, which does not require pre-market approval provided that the products comply with regulation in the country of origin, has helped smaller companies access the China market, albeit on a limited basis. As discussed in other chapters, cross-border e-commerce rules set limits on the amount per shipment and the amount per purchaser, per year that can enjoy reduced customs duties.

For more than twenty years, until a major reform in 2020, cosmetics was a neglected area of regulation. The previous framework regulation, the Cosmetics Hygiene Management Regulation (CHMR), was enacted in 1989 and did not undergo significant amendment for decades. Although the National Medical Products Administration (NMPA) increasingly issued implementing rules to shape cosmetics regulation in the 2010s, many of the true innovations in regulation have only recently occurred.

Cosmetics regulation was originally under the jurisdiction of the Ministry of Health and the General Administration for Quality Supervision, Inspection, and Quarantine (AQSIQ) but, in 2013, was transferred to the China Food and Drug Administration (CFDA) and handled, in part, under the drug registration division.[2] There it has remained for several years while CFDA and now NMPA, which now has a sub-division dedicated to cosmetics supervision, work on reform plans for cosmetics regulation, testing various reform measures; many drafts (but few final rules) emerged between 2013 and 2020.

One of the most significant regulatory issues during this time was CFDA's unwillingness or inability (it is not clear which) to approve new ingredients. As with other areas, use of cosmetic ingredients was subject to positive lists of existing ingredients that do not require pre-approval, and new ingredients required pre-approval. However, until 2015, CFDA did not frequently update the positive list to reflect new existing ingredients that had been approved as part of product applications over the years, nor did it accept independent applications to approve new ingredients. These issues created a substantial problem for some manufacturers, which were forced to use older, outdated ingredients in products for China without a pathway for novel ingredients or updated products.

Another problem was that imported products had to demonstrate both approval and marketing history in other countries. In 2013, CFDA issued a brief circular indicating that all products must present a certificate of free sale, indicating that the product they were importing into China was *actually* sold in foreign countries. Some companies were essentially forced to use different formulations abroad because of the limitations on ingredients. This prompted a new jostle for creative solutions to permit products to be imported to China even though they might contain different ingredients from those sold elsewhere.

[2]State Food and Drug Administration (2013c) Notice on Further Improving Work Related to Cosmetics Production Licenses.

Slowly things began to change as CFDA grew into its role as a cosmetics regulator. As CFDA was issuing new drafts of the framework regulation—retitled the Cosmetics Supervision and Administration Regulation (CSAR)—in 2014, 2015, 2018, and 2019, it gradually began to enact new regulations and policies. For example, it issued new rules on cosmetics manufacturing, a new comprehensive set of standards, and new regulations on product and ingredient testing and laboratories. It also issued a list of existing cosmetic ingredients in 2015, the Inventory of Existing Cosmetic Ingredients (IECI). The IECI included various ingredients that had either been approved individually, approved as part of product applications, or otherwise permitted for use in cosmetics.

NMPA also revised the Hygiene Standard (2007)—a standard containing testing monographs for products and ingredients and lists of prohibited and restricted ingredients—for the first time in eight years. The newly named Cosmetics Safety and Technical Standard was issued in 2015.[3] Like its predecessor, the newly revised Cosmetics Safety and Technical Standard included required tests on animals for many products. Alternatives to animal testing were only permitted for low-risk, domestically produced products as of 2013; certain imported products could not receive an exemption until 2021.

Another series of reforms allowed record-filings in place of approvals for certain imported and domestic cosmetics. The domestic reform occurred in 2013 and permitted products to come onto the market immediately after filing, acting like a notification. Four years later, the program expanded to products imported into Shanghai's Pudong New Area and ultimately spread to all provinces and directly administered municipalities.

NMPA proposed various other regulations. It issued a draft (never finalized) of controversial labeling regulation in 2014, which would have required pure Chinese labels for cosmetics. It proposed regulations on cosmetic claims, intended to provide guidance on what data, including clinical data and real-world studies, would be required to support claims on cosmetic labels and in advertising.

These efforts culminated in a final version of the CSAR being passed by the State Council in late 2019 and issued publicly in 2020, and they influenced the implementing rules later issued pursuant to CSAR. CSAR was a step forward in regulation because it created a risk-based system that removed constraints on new ingredient approvals, leaving lower risk ingredients to be approved via a record-filing/notification. The CSAR allowed ingredient makers to file information through a kind of masterfile system (like that of drugs and devices) that could be referenced by product makers, thereby potentially protecting the proprietary information. It also expanded the notification system for normal cosmetics nationwide, allowing those products to come onto the market more quickly after completing all required materials. To balance this ostensible reduction in pre-approvals, CSAR established a cosmetovigilance system with requirements for post-market safety reporting, risk assessment, and recall of defective products. It also established a requirement for

[3]China Food and Drug Administration (2015c). Cosmetics Safety and Technical Standard.

compliance with uniform quality standards: the cosmetics good manufacturing practice rules.

The implementing rules for CSAR did not, however, reflect CSAR's seemingly more flexible structure. They increased application requirements, calling for substantial amounts of information about product and ingredient manufacturing processes, quality standards and systems, personnel, and suppliers. The CSAR implementing rules established a system for effectiveness claims that in some cases called for verification in the form of substantial human, animal, and laboratory studies, limited claim language to that on positive lists, introduced a complex new system of product categorization codes, and generated stricter label requirements, including limits on the use of foreign language on the package. In short, CSAR implementation arguably took the system away from risk-based performance standards and self-regulation and decreased flexibility by subjecting products and ingredients to a more stringents system of regulation.

9.2 Definition of Cosmetics and Product Classification

Although, as we have seen in prior Chapters, the Chinese system often relies on positive lists (referred to as catalogues) to define product categories, the definition of cosmetics in the CSAR is still significant. It makes clear that cosmetics are topical products that can only make beautification, skin protection (not tied to any disease), and limited cleansing claims:

> The term "cosmetics" means daily used chemical products applied on the surface of any parts of the human body, such as skin, hair, nails and lips, by spreading, spraying or similar methods for the purposes of cleaning, protection, beautification and modification.[4]

Despite the topical nature of cosmetics, CSAR also asserts that toothpastes are subject to cosmetic regulation. Other oral-care products are not. Despite some proposals to put the entire oral-care category under the cosmetics regime, these products are regulated by a set of standards issued by the National Health Commission (NHC). As will be discussed below, although cosmetics are not for ingestion, any products used on the lips or mouth only should have additives that are approved for use in foods.

As is made clear in other parts of CSAR and its claims, categorization, and labeling regulations, cosmetics are not permitted to make therapeutic claims—although, as we shall see, some products are permitted to make what are considered drug claims in other countries (e.g., sunscreen). Indeed, permitted toothpaste claims include anticaries, anti-plaque, prevention of dentin sensitivity, and gingivitis treatments,[5] and special cosmetics are permitted to make claims of hair coloring,

[4] State Council of the People's Republic of China (2020), Article 3.

[5] State Council of the People's Republic of China (2020), Article 77.

hair perm, whitening and spot removal, sun protection, and hair loss prevention. Further, a product with a function claim not on this list or in the cosmetic claims evaluation guidelines (see below) would be approved as a special cosmetic.

Whitening products deserve special mention. In 2013, the NMPA issued a notice that spot-removal products would include popular skin whitening and brightening products, including products that produce a whitening effect via a coating. For kits of products containing a whitening product, at least that product must be approved as a special-use cosmetic, even if the other products are non-special use.[6]

Notably, cosmetics with the original special function claims of hair growth, hair removal, breast health, body building, and deodorizing that received approval prior to January 1, 2021 may be manufactured, imported, and sold as cosmetics until December 31, 2025. After that point, these licenses will not be honored, and during the transition period expiring licenses may not be supplemented or renewed. At the time of this writing, it is not clear under existing guidelines how NMPA plans to regulate products that make these claims,[7] but those products that come within the scope of cosmetic regulation, as articulated under CSAR, must be approved as special cosmetics (e.g., new efficacy) or record-filed as general cosmetics, if they qualify.

Another category of potential overlap is the category of disinfectants. Cosmetics can make claims of cleansing and washing but cannot make claims of disinfecting or killing germs. These are reserved for separate classes of disinfectant and hygiene products regulated and approved by the NHC.[8] Disinfectant products include "disinfectants, disinfection instruments (including biological indicators, chemical indicators and sterilization article packaging), sanitary products and disposable medical products."[9] Soap occupies a unique position. CSAR states that soap that only perfumes the body is excluded from cosmetic regulation (presumably being regulated as a consumer product), but if it makes any special cosmetic/functional claims it must be pre-approved as a cosmetic.

The status of a product as a cosmetic is judged by its *intended* use. Although there is not substantial guidance on what this means, if cosmetics are treated like other product categories, intended use will be judged by claims made on the physical label of the product and those made in advertisements and promotional materials.

---

[6]State Food and Drug Administration (2013a) Circular on Adjusting Cosmetics Registration and Filing Management.

[7]NMPA (2021a) Announcement on Further Clarifying the Transitional Management of Original Special Purpose Cosmetics and Other Related Matters.

[8]National Health and Family Planning Commission (2009) Catalogue of Disinfectants.

[9]General Office of the National Health Commission (2018).

9.3 Non-Special Use ("General Cosmetics") Versus "Special Use"

Once classified as a cosmetic, as alluded to above, these products are then further subdivided into two broad categories, non-special use and special use. This distinction was fundamental to the 1989 CHMR and continues in the CSAR. Non-special use or general cosmetics are defined by not being the in the "special" category. In other words, everything that does not fall into the special-use category is considered a general cosmetic.

Special cosmetics include products that make specific claims. In the CHMR, the original nine special use categories were:

- Hair growth
- Hair permanent
- Hair dye
- Hair Removal
- Whitening and spot removal
- Deodorant (which has been interpreted to include antiperspirant)
- Slimming
- Breast health
- Sunscreen[10]

As discussed above, in the CSAR, the following claims require special cosmetic registration:

- Hair coloring
- Hair permanent
- Whitening and spot removal
- Sunscreen
- Hair loss prevention (distinct from hair growth and hair nourishment, which are general cosmetic claims)

The special-use designation means that applicants must put forth specific data to support these claims and undergo an expert review before receiving a license/registration. In contrast, general cosmetics go through a simpler process discussed below, after the applicant submits the same materials. There is not an expert review prior to marketing general cosmetics, but local authorities will continue to review the documentation for three months after the product launches in the market. If there are problems, authorities may ask the applicant further questions and could order the product off the market. Indeed, anytime local authorities see a fundamental issue with safety and/or quality they have the authority to ask the applicant to remove the product from the market and be re-evaluated.

[10] State Council (1989), Article 10.

Special-use cosmetics must also produce a testing report from an accredited state laboratory. Like other products, such as drugs, devices, and health foods, China requires a report produced by government labs that follow a standardized protocol. These reports include various types of testing depending on the product and ingredient and include anti-microbial testing and toxicology testing, including skin and eye irritation tests. These tests are contained in the Cosmetics Safety and Technical Standard.

As discussed below, what efficacy data are needed to support certain claims is subject to an ongoing discussion; in 2021, NMPA issued a claims evaluation guideline that set forth the different types of data that are accepted for a list of permitted claims, as well as other requirements for study design and data collection. The specifics of this claims process are discussed in greater detail below.

Both imported and domestic special cosmetics require a license and go through a similar licensing procedure. Although imported products require additional materials, such as proof of marketing and sale abroad, the differences do not relate to the special-use claim. The line between imported and domestic cosmetics is set by the location of the last step in manufacturing that touches the content of the cosmetic. If fill-and-finish steps occur inside China, for example, the product would be considered domestically produced.

Imported and domestic special-use cosmetics are both approved by NMPA, whereas general cosmetics are record-filed with the local medical products administrations where the filer is established or at the port of entry for imported products.

9.3.1 Non-Special Use ("General") Cosmetics

CSAR refers to non-special use cosmetics as "general cosmetics." Since 2013, general cosmetic approval has changed more than special use. General cosmetics also have more significant differences in requirements for imported versus domestic products.

In 2013, NMPA issued a circular that permitted domestically produced general cosmetics to be marketed immediately after completing a simple filing with the provincial level cosmetics regulatory authority.[11] If the documents were complete, the product could be marketed. Once the products are on the market, the authorities may continue their review of the materials and raise questions or remove the product from the market if they discover issues.

The same notice that offered the record-filing procedure for domestic general cosmetics also introduced another significant change: It stated that, in lieu of a toxicology test report, provinces would accept a self-completed safety risk assessment by the applicant according to NMPA's guidelines. The significance of this

[11] State Food and Drug Administration (2013a) Circular on Adjusting Cosmetics Registration and Filing Management.

change was that domestic non-special products could be placed on the market without toxicology tests conducted on animals. The animal testing exemption was significant for companies that took pride in their cruelty-free methods. Except for cosmetics sold via cross border e-commerce, which require no testing, this 2013 exemption was for several years the only way to sell cosmetics in the China without animal testing.

In 2016, the State Council, in accordance with broader administrative law policies on the reduction of unnecessary pre-approvals, called for a pilot program that would permit notifications of imported non-special use products. The pilot program was to be run out of the Pudong New Area Zone in Shanghai. Despite the issuance of the policy document in 2016 and calling for this program to begin shortly thereafter—sun setting in 2018—CFDA did not issue a notice initiating the program until early 2017.[12] The Pudong New Area Zone issued its more detailed implementing guidance shortly after that, allowing the program to proceed.

Among other requirements, foreign companies using the pilot needed to have a distributor or their own distribution company in the zone. The program did not give the same flexibility with respect to animal testing. Applicants had to submit their products for full testing (including toxicology) to approved laboratories.

Otherwise, the program functioned in a similar manner as the domestic program. Application materials were submitted electronically through a portal run by the Pudong New Area Zone. Once complete, the local cosmetics regulatory authority issued a filing certificate, permitting the products to enter the country.

The Pudong New Area Pilot for non-special use products was expanded to several other cities and in 2018 it was announced it would apply throughout China, utilizing the guidance developed by the local cosmetics regulatory authorities in Shanghai.[13] Finally, CSAR made local filings the method of pre-market review for all non-special use cosmetics, both imported and domestically produced.

Although there is a difference in the level of scrutiny of general versus special-use cosmetic applications, the materials supporting them are generally the same, other than to support specific claims associated with a given product. Like special cosmetics, general cosmetic applicants must present test reports and safety assessments, as well as manufacturing and formulation information, among other materials.

9.4 Children's Cosmetics

There are special rules governing children's cosmetics. These are any cosmetics that target children ages 12 and under and are intended for cleaning, moisturizing, and refreshing the skin or sunscreen. Products that state they are for the whole family or

[12]China Food and Drug Administration (2017b) Shanghai Pudong New Area Imported Non-Special Use Cosmetics Filing Management.

[13]NMPA (2018).

for use by all groups are regulated as children's cosmetics. In 2021, NMPA enacted rules aimed at regulating the formulation, claims, labeling, and safety reporting of these products.[14] These rules were in response to (among others) concerns about the vulnerability of this group of users and also practices in which toys that are not for use on human skin could be mispresented as cosmetics.

9.5 Marketing-Authorization Holder System

Like other product areas, under pre-CSAR rules there was not a clear distinction in the rules between a manufacturer and a brand or product owner. The earlier pre-CSAR rules did not distinguish between a manufacturer and a product permission holder. Imported products obtained "hygiene permits" for products that were finished abroad and imported into China, and domestic companies held both product and manufacturing permits. When registering products, there were spaces on the forms for "manufacturers" and, in some cases, "actual manufacturers." But the rules did not provide guidance on the roles and requirements of each.

Prior to CSAR, there were not clear requirements for the capacity of a permission holder, such as having a quality team. For example, in 2013, when companies were using fill-and-finish procedures in China to qualify their products as domestic for purposes of animal testing (see below), NMPA allowed for local distribution companies (which required no extra license) to hold the permission for those products, instead of the actual manufacturing facilities.[15] In addition, a company in China was allowed to hold the permission for a cosmetic but to use contract manufacturers outside of China to produce it.

Under the revised CSAR guidelines, product permission holders are referred to as "registrants and filers" (R/Fs). There is flexibility regarding the location of the R/F, relative to the manufacturing operation. A company outside of China can be the R/F but manufacture the product in China, and a company in China can be the R/F for products made outside of it. While there may be other reasons to structure these arrangements in a different way, such as other regulatory, tax, or corporate governance issues, there is more freedom than there is with other regulated products.

If the R/F is ex-China, then it must appoint an agent (a "responsible person"). The agent must be an entity in China to assist with the fulfillment of regulatory obligations on product registration and filing, quality, and cosmetovigilance.[16] The agent must assist with these obligations, and if it fails to provide such assistance it can be held liable. However, unlike for drugs, there is no provision in the CSAR that

[14] National Medical Products Administration (2021k) Administrative Provisions for the Supervision of Children's Cosmetics.

[15] State Food and Drug Administration (2013a) Circular on Adjusting Cosmetics Registration and Filing Management.

[16] State Administration for Market Regulation (2021b) Cosmetics Registration and Record-Filing Measures. Article 8.

makes the R/F and the agent jointly and severally liable for the same violations. These responsible persons are liable for their own failure to assist.[17]

CSAR and other rules, such as cosmetics good manufacturing practices (GMP), set capacity requirements for the R/F. It must be a company and not an individual, it must have the appropriate quality management system for the products that it is marketing (even if it entrusts manufacturing), and it must have the ability to monitor adverse reactions of cosmetics. While practice is still unfolding, these requirements narrow the field and, like in other spaces, seek to prevent the use of entities that are shells or are not associated with established quality systems. The responsible person, as defined in implementing rules, include handling registration and record-filings, assisting with the reporting of adverse reactions and recalls, monitoring and handling safety and quality issues, and cooperating with NMPA inspections.

The same R/F-responsible person system also applies for registrations and record-filings of new cosmetic ingredients. The same responsible person requirements apply in the ingredient context, too.

9.6 Application Requirements

While the CSAR sets general application requirements, NMPA releases detailed rules as to the specific materials and format. Those requirements include information about the R/F and its quality personnel and the manufacturer (if different), a safety assessment of the product conducted according to safety assessment guidelines that NMPA released pursuant to CSAR, a testing report from an accredited China Inspection Body and Laboratory Mandatory Approval "CMA" laboratory, the product's formulation, including the percentage of each ingredient, the standard to which the product will comply (i.e., a set of specifications as dictated by mandatory standards in China), and samples of product labels for China.[18]

Although NMPA has long required that the test report come from a third-party lab that the government accredits, in January 2023 NMPA issued a notice that allows domestic producers of general cosmetics to issue self-test reports if they meet laboratory requirements and other quality system requirements.[19] If this self-testing were to become the norm, it would take a substantial burden off of these CMA testing labs that serve many other purposes.

Tables 9.1, 9.2, 9.3, 9.4, 9.5, and 9.6 are example tables that are used to determine the testing items for various types of products, pursuant to a rule adopted prior to the CSAR in 2019. Although the rule is still effective, some of these types of products no longer belong in the special or even the cosmetics category. A circle denotes that it is a required tested item. Certain notes to these items are not presented here.

[17] State Council (2020), Article 70.

[18] State Council (2020), Article 19.

[19] National Medical Products Administration (2023a, c).

Table 9.1 Microbial testing items[a]

Testing items	Ordinary cosmetics ①②	Special use cosmetics								
		Hair growth	Hair dye	Hair perm	Hair removal	Breast health	Slimming	Deodorant	Spot removal	Sunscreen
Total colonies	○	○				○	○	○	○	○
Number of molds and yeasts	○	○				○	○	○	○	○
Thermotolerant Choliforms	○	○				○	○	○	○	○
Staphylococcus aureus	○	○				○	○	○	○	○
Pseudomonas aeruginosa	○	○				○	○	○	○	○

[a]National Medical Products Administration (2019a). Cosmetics Registration and Record-Filing Testing Work Standard

Table 9.2 Physio-chemical testing items[a]

Testing items	Non-special use	Special use cosmetics								
		Hair growth	Hair dye	Hair perm	Hair removal	Breast health	Slimming	Deodorant	Whitening	Sunscreen
Mercury	○	○	○	○	○	○	○	○	○	○
Lead	○	○	○	○	○	○	○	○	○	○
Arsenic	○	○	○	○	○	○	○	○	○	○
Cadmium	○	○	○	○	○	○	○	○	○	○
Methanol										
Dioxane										
Asbestos										
Formaldehyde								○		
Thioglycolic acid				○	○					
Sunscreen										○
Hair dye			○							
pH Value				○	○				○	
Alpha hydroxy acids										
Anti-dandruff agents										
Acrylamide										
Preservatives	○	○	○	○	○	○	○	○	○	○
Anti-UVA ability parameters-critical wavelength										

[a]National Medical Products Administration (2019a). Cosmetics Registration and Record-Filing Testing Work Standard

Table 9.3 Non-special use cosmetics toxicology testing items[a]

Testing item	Hair use	Skin care		Makeup			Finger and toe nails	Aromatic products
	Hair products that contact eyes	Ordinary skin care	Skin care products that touch eyes	Ordinary makeup	Eye makeup	Lip care and lip		
Acute skin irritation test	○						○	○
Acute eye irritation test	○		○		○			
Multiple skin irritation		○	○	○	○	○		

[a]National Medical Products Administration (2019a). Cosmetics Registration and Record-Filing Testing Work Standard

Table 9.4 Special use cosmetics toxicology testing items[a]

Testing items	Hair growth	Hair dye	Hair perm	Hair removal	Breast health	Slimming	Deodorant	Whitening	Sunscreen
Acute eye irritation test	○	○	○						
Acute skin irritation test			○						
Multiple skin irritation test	○				○	○	○	○	○
Skin allergy test	○	○	○	○	○	○	○	○	○
Skin toxicity test	○							○	○
Salmonella Typhimurium/Reverse mutation test	○	○			○	○			
Chromosomal aberration assay in mammalian cells in vitro	○	○			○	○			

[a]National Medical Products Administration (2019a). Cosmetics Registration and Record-Filing Testing Work Standard

Table 9.5 Human safety testing items[a]

Testing items	Non-special use	Special use cosmetics						
		Hair growth	Hair removal	Breast health	Sliming	Deodorant	Whitening	Sunscreen
Human skin patch test						○	○	○
Human trial safety evaluation		○	○	○	○			

[a]National Medical Products Administration (2019a). Cosmetics Registration and Record-Filing Testing Work Standard

Table 9.6 Human functionality evaluation testing items[a]

Sunscreen cosmetics	Testing items
	Sun Protection Factor(SPF value)Determination
	(PFA) Determination of Long-wave Ultraviolet Protection Index
	Determination of Waterproof Performance

[a]National Medical Products Administration (2019a). Cosmetics Registration and Record-Filing Testing Work Standard

9.7 Animal Testing

Pursuant to the Cosmetics Safety and Technical Standard, the toxicology tests above are conducted on animals. As noted above, starting in 2013, general cosmetics produced domestically could submit risk assessments in lieu of toxicology testing. Under CSAR, NMPA expanded the exemption to all general cosmetics. In addition to the risk assessment, a certification of the manufacturing quality system by a competent cosmetics regulator is required. For domestic products, this simply means that the facility has a cosmetics manufacturing license, but the requirement is more complex for imported products. For these, the applicant must find a regulator to certify that its facility meets a set of GMP manufacturing requirements. As different countries have varying levels of scrutiny of cosmetics manufacturing (particularly when compared to drugs), this requirement can present problems. Attention to the quality system is also puzzling because toxicology testing generally relates to the safety of the formulation, so it is not clear how assessment of the manufacturing quality system is relevant, other than to ensure that production is also safe. Still, this seems redundant as the quality system is more heavily assessed in other parts of the applications.

There are also a set of exceptions for R/Fs and products that cannot qualify for the animal testing exemption. Products with claims for infants and children (children's cosmetics) and those that contain new ingredients that are still under a monitoring period (i.e., within the first three years after approval of the ingredient) are excluded from the exemption. R/Fs under special scrutiny in terms of quality, as determined by the government, are also excluded from applying.[20] This might occur for high-risk products or those companies with products that have been the subject of serious enforcement.

[20]National Medical Products Administration (2021h) Provisions on Cosmetics Registration and Record-Filing Materials.

9.8 Certificates of Free Sale

Like other products, China has traditionally required proof that imported cosmetic products are legal in the country of origin, adding a burden for offshore manufacturers to obtain documentation. Not all countries issue proof that cosmetics comply with the law or that manufacturing facilities are compliant (another document that NMPA requires), so NMPA has had to accept documentation from other third parties such as industry associations.

Until 2013, NMPA accepted certificates indicating that a product could be legally marketed in the country of origin, even if they were not actually sold there. This flexibility was helpful to companies because of China's failure to approve new ingredients, manufacturers perhaps need to use older ingredients for products in China that they no longer used in, for example, the United States, Europe, and Japan. In 2014, NMPA changed course to require that the certificates indicate that the product is actually marketed and sold in the country of origin—a certification much more difficult to make.[21] There was no apparent reason for this change other than, perhaps, a change in NMPA's interpretation of the existing CHMR. This left manufacturers with a choice: to start selling a product abroad, so they could truthfully assert that the product was on the market overseas, or using other options, such as rearranging manufacturing so that the product could be considered domestically produced obviating the need for a certificate of free sale.

Since 2014, there has been some progress on this issue. The CSAR provides that a foreign manufacturer may either provide a certificate of free sale or data indicating that the product marketed abroad is safe. Still, with China's ingredient list remaining behind the rest of the world, there may be circumstances in which these options are an awkward fit.

9.9 Cosmetic Ingredients

Under CSAR, NMPA regulates cosmetic natural and synthetic ingredients in two ways. First, it approves "new" ingredients that have not yet been used in cosmetics marketed in China, either through a registration or a record-filing process. An ingredient is a "new" ingredient if it (1) is not on the Inventory of Existing Cosmetic Ingredients (IECI), a list of existing ingredients that may be used in all cosmetics in China according to certain conditions (e.g., leave on/rinse off values); (2) is not prohibited as a cosmetic ingredient altogether by the prohibited list in the Cosmetics Safety and Technical Standard; and (3) meets the definition of a cosmetic ingredient,

[21] State Food and Drug Administration (2013b) Letter on Strictly Circumscribing Documentation of Proof of Approval and Sale for Imported Cosmetics.

i.e., meets the intended use and application (i.e., topical) requirements and is not, for example, used for the treatment of disease or for activating or regenerating cells.[22]

This process for approving many new ingredients via record-filing, and implementation guidelines to match, is a CSAR innovation and perhaps one of the most important contributions of the 2020 revision. Ingredients that are intended to be used as preservatives (anti-corrosion), sun protection, coloring, hair coloring, or whitening (i.e., freckle removal and skin whitening) require pre-market registration (i.e., materials are submitted and technically evaluated prior to approval). All other new ingredients require a record-filing prior to being used.

Once a new ingredient is approved it is automatically placed under a three-year "monitoring period" during which the ingredient R/F must provide regular reports of safety information, including safety information from the product R/Fs, to which they have granted permission to use the new ingredient. At the end of the monitoring period, if there are no serious safety issues the ingredient may be added to the IECI. As noted above, a more comprehensive IECI was established in 2015 (previously, there were scattered lists) and has been updated periodically since then. The IECI has evolved to become more specific regarding ingredient names (use of internationally accepted and Chinese standard names) and recently, in the 2021 revision, added conditions of use.[23]

NMPA released extensive guidelines outlining the testing and other requirements for new ingredient applications shortly after CSAR.[24] These details helped make the approval of new ingredients possible for the first time in a long time. However, the guidelines were not clear on the avoidance of animal testing—they required testing under the Cosmetics Safety and Technical Standard and, while offering possible substitutes/alternatives to animal tests, included no blanket exemption on the same grounds as existed for other products. As such, if an applicant cannot chart a course to approval using non-animal methods, animal testing may apply for new ingredients. Furthermore, even if the ingredient application does not include animal testing, as discussed above, products with newly approved ingredients under a monitoring period may not be exempt. This type of one-step-forward-one-step-back approach is characteristic of CSAR implementation.

The second way that NMPA regulates ingredients is to review their safety and quality in product applications. Each product application requires a listing of the ingredients in its formulation, including their percentage, and the suppliers from which the product R/F obtains these ingredients. The way in which an ingredient is listed must comply with restrictions under the IECI, as well as separate lists that are contained in the Cosmetics Safety and Technical Standard on prohibited and restricted ingredients. Certain ingredients can be restricted to certain amounts, for

[22]National Medical Products Administration (2021j) Q&A on the Registration and Filing Policy for New Cosmetic Ingredients.

[23]National Medical Products Administration (2021f) Inventory of Existing Cosmetics Ingredients.

[24]National Medical Products Administration (2021i) Provisions on the Registration and Filing Materials for New Cosmetic Ingredients.

example. While beyond the scope of this discussion, there are also rules administered by environmental agencies that require notifications for new chemicals entering China.

Since CSAR, NMPA also requires a set of safety and quality information about each ingredient. This information is substantially more than required before—arguably an overreach by the agency—including inquiring deeper into the composition and manufacturing process and quality control techniques of individual ingredients and requiring amendments should a raw material supplier associated with a product application change. This information can be presented in one of two ways, either by direct submission with the product application or through the ingredient manufacturer or its authorized representative submitting to the Cosmetics Safety Information Registration Platform (CRSIP).[25] Once registered with CRISP, the system will generate a code that a product R/F can reference in their application. This is still an evolving online platform at the time of this writing.

9.10 Cosmetic Manufacturing

CSAR introduced a much more complete system for the regulation of manufacturing and quality control for cosmetic products. Cosmetics GMP cover both finished and semi-finished products, and cosmetics manufacturers in China must qualify as GMP compliant and receive a manufacturing license prior to commencing production for the market.[26]

As with other regulated products, whether the product is considered domestically produced (versus imported) depends on where the product is finished and, more specifically, where the last step that touches the content of the cosmetic takes place. That determines certain requirements. Although many of the requirements for imported and domestic products have been merged under CSAR, databases still divide products into imported and domestic.

Under the CSAR, the R/F and the manufacturer can be separate and can have separate roles as defined the regulations and by the parties' own quality agreement. The R/F is responsible for the safety and overall quality of the product that is released to the market, whereas an entrusted manufacturer (if any) must produce the product in accordance with the GMP and the R/F's instructions. The name of the R/F and the entrusted manufacturer need to be on the product's label. Both are also listed in NMPA's public database, as is the case for other products.

In 2021, NMPA issued a separate set of rules governing the obligations of distributors and manufacturers. For manufacturers, it also includes the criteria and procedure to apply for cosmetic manufacturing facility licenses. Having these

[25]National Medical Products Administration (2021e) Cosmetics Ingredient Safety Information Registration Platform Online.

[26]National Medical Products Administration (2022c) Cosmetics Good Manufacturing Practice.

separate rules is significant because it adds substance to the roles, particularly of distributors, that did not exist before.[27]

As will be discussed further below, cosmetics GMP are also divided into responsibilities demanded of the non-manufacturer R/F and the entrusted manufacturers. If an R/F itself is a manufacturer, then it must observe both sets of obligations. As noted above, the GMP and related implementing rules require a quality agreement between the R/F and its manufacturer that sets forth their respective roles and responsibilities.

9.11 Manufacturer Facility Licenses

Cosmetics produced in China must be manufactured at licensed facilities. However, until recently, China had not developed a clear system of good manufacturing practice certification for cosmetics, like that for drugs or medical devices. There were many small and medium-sized cosmetics manufacturers that performed simple steps of filling small containers from bulk stock, packaging, and labeling. These enterprises held hygiene production permits for products, which may have come from the Ministry of Health, and industrial product manufacturing licenses, which were issued by the former agency of the General Administration for Quality Supervision Inspection and Quarantine.

The proliferation of smaller cosmetics makers may have been due, in part, to the demand from overseas manufacturers for small fill-and-finish jobs, in order to have their products qualify as domestically produced.

In 2014 and 2015, however, NMPA issued additional cosmetics manufacturing regulations. It required that all cosmetics makers apply for cosmetics manufacturing licenses, and those who held industrial product licenses would receive the cosmetics manufacturing licenses when they applied for renewal. All labels needed to reflect cosmetics manufacturing license numbers by January 1, 2017.[28] A the same time, NMPA issued a Standard for Cosmetic Manufacturing Licenses,[29] which sets forth substantive requirements for personnel, systems, and facilities—until CSAR, the closest to a cosmetics GMP that NMPA had come.

[27] State Administration for Market Regulation (2021a) Cosmetics Manufacturing and Distribution Supervision Administrative Measures.

[28] China Food and Drug Administration (2015a) Announcement on Matters Concerning Cosmetics Production License.

[29] China Food and Drug Administration (2015b) Cosmetics Manufacturing Licensing Work Guideline (Trial Implementation).

9.12 Cosmetics GMP

The issuance of cosmetics GMP marked an important step in establishing specialized rules and design considerations to produce cosmetics. In this respect, the GMP articulates detailed requirements for facilities, personnel, systems, and testing for manufacturers. The cosmetics GMP—effective July 2022—sets forth not only requirements for those companies that have facilities for production, i.e., the actual manufacturers (whether it is an R/F doing the manufacturing or a contract manufacturer), but also for the R/Fs that supervise manufacturing. It covers specific personnel requirements and those for establishing and verifying core manufacturing procedures. Shortly after the GMP was issued, NMPA released an inspection scoring system and an extensive checklist of items for both R/Fs and actual manufacturers to meet.[30]

The requirement for R/Fs to meet these standards means that smaller R/Fs must consider whether to add personnel and systems to meet these capacity requirements. The R/F must establish a quality system (with a responsible person) for choosing a competent actual manufacturer and supervising it. The R/F is also responsible for conducting a review of quality information during the product's release to market, preserving market samples for later testing, and establishing a general recordkeeping system.[31] One of the primary responsibilities of the R/F is to assess the capacity of the entrusted manufacturer and supervise their compliance with production specifications; important tools for supervision are a quality agreement between the R/F and the manufacturer setting forth roles and responsibilities and training and audits of the manufacturer.[32]

Although implementation is still ongoing at the time of this writing, stakeholders may find the GMP prescriptive compared to other jurisdictions; this approach is not uncommon in Chinese standards, which set specific limits and requirements rather than focusing on performance to a general standard, allowing the use of methods more tailored to products and their risk level. For example, the GMP have specific requirements for clean rooms,[33] sanitary facilities (e.g., changing rooms and shoe

[30] National Medical Products Administration (2022d) Cosmetics GMP Inspection Main Points and Determination Principles.

[31] NMPA distilled the R/F responsibilities into a brief document. *See* National Medical Products Administration (2022g) Policy Interpretation—Cosmetics GMP Freshly Released—Main Responsibilities of Entrusting Parties (Cosmetics Registrants and Filers).

[32] Ibid.

[33] National Medical Products Administration (2022c) Cosmetics Good Manufacturing Practice, Appendix 2.

cabinets),[34] water systems and purification,[35] temperature,[36] and pressure[37] that may not be tailored to the risk level of cosmetics products.

The cosmetics GMP does not limit its applicability to domestic manufacturers, nor does it expressly state that it applies to overseas production facilities. However, if cosmetics follows the trend in other spaces (i.e., drug and device) overseas R/Fs and manufacturers will likely be subject to its requirements. Furthermore, other rules make it clear that overseas products makers will need to observe these requirements in addition to certifying that they meet the requirements of the home jurisdiction.

Earlier drafts of the CSAR contained a provision on foreign inspections, and in 2019 NMPA released a draft regulation on foreign inspections that mirrored the provisions of overseas inspection regulations for drugs and devices described in prior chapters.[38] That draft made it clear that the basis for evaluating an overseas cosmetics manufacturing facility would be the China cosmetics GMP. It is not yet clear where this program will lead, particularly when the Center for Food and Drug Inspection (CFDI) is arguably already strapped for resources. Like all health authorities around the world, it is challenging, if not impossible, to inspect every facility that imports regulated FDD products into a country. As of 2022, there were some signs that overseas cosmetics GMP inspections could occur, such as the potential for a new inspection rule.

9.13 Cosmetics Distribution

Unlike all other regulated products discussed herein (foods, drugs, devices), cosmetics do not require distribution licenses or record-filings. Cosmetics trading companies are permitted to distribute these products provided they have an ordinary business license with a scope of business that includes cosmetics distribution. They are required to distribute their products in accordance with storage conditions. Distributors, including retailers, cannot engage in mixing or processing activities, because CSAR provides that entities that have these facilities must have a manufacturing license.[39]

An important channel for cosmetics distribution is online. This includes both domestic e-commerce (i.e., websites in China that distribute products approved in China) and cross-border e-commerce, or CBEC (i.e., websites outside of China that

[34] Ibid., Article 20.

[35] Ibid., Article 26; State Food and Drug Administration (2022) Cosmetics GMP Inspection Main Points and Determination Principles, Appendix 1, No. 41.

[36] National Medical Products Administration (2022d) Cosmetics GMP Inspection Main Points and Determination Principles, Appendix 1, No. 31.

[37] Ibid., Appendix 1, No. 32.

[38] National Medical Products Administration (2019b) Rules on Cosmetics Overseas Inspections (Draft for Comment).

[39] State Council of the People's Republic of China (2020), Articles 38–39.

distribute foreign labeled products). As discussed in Chap. 8, CBEC has grown from a make-shift model that used personal-use exemptions for mailing product to China and bonded-warehouse zones that exempted bulk product from restrictions to an institutionalized pathway for bringing foreign-labeled products into the country via various pathways, including special CBEC zones, which have expanded from 5 to 132.[40] E-commerce generally is a huge industry for China; in 2020, cross-border e-commerce had a nearly 1.7-trillion-yuan import and export volume, growing 18–19% per year since 2015.[41] The growth of CBEC brands is overtaking those in the traditional channel environment.[42]

For cosmetics, CBEC allows products with new ingredients made by small and medium sized enterprises to enter China without going through the expensive registration and filing process. Previously, the expense of registration and filing and the infrastructure needed to maintain those approvals put the China market out of reach for those companies.

CBEC cosmetics do not require registration or filing, are not subject to clear ingredient restrictions (absent use of clearly prohibited or quarantined ingredients), and do not require the battery of tests set forth above. Thus, they do not require product animal testing, including animal testing that would be required of products with new ingredients, and are viewed as perhaps the most cruelty-free pathway to bring cosmetics into the country. Instead, CBEC products must observe special restrictions, which may include being subject to a positive list of permitted CBEC products, filings and reporting to customs, compliance with country-of-origin standards and label requirements, limits on individual shipments and annual purchase limits for each consumer, and special rules on platforms (i.e., websites), including surveillance by the platform operator.[43]

9.14 Cosmetics Claims and Advertising

Under CSAR, there is now a more structured regulatory regime for supporting and approving specific claims as part of the registration process. Each product receives a code indicating its features and claims, further defining approved scope of the registration or filing, according to which the product may be marketed.[44] To assemble a product code, the R/F applicant must enter numbers on the product application form that correspond to the applicable efficacy claims, application area of the body, product dosage form, applicable group of users, and application method. Those numbers combine to form the classification code, although it is still unclear

[40] Ibid.

[41] Xinhua News Agency (April 26, 2022); Ernst and Young (2021).

[42] Ibid.

[43] Ministry of Commerce (2018).

[44] National Medical Products Administration (2021c) Cosmetics Categorization Principles and Categorization Catalogue.

exactly what significance this code has in practice, other than as an exercise in defining certain aspects of the product.[45]

The NMPA also released a claims guideline that provides specific requirements to support various types of permitted claims, e.g., human data, laboratory testing, and/or literature.[46] These claims guidelines also dictate where certain testing must take place (i.e., laboratories with a certain accreditation in China) and require that R/Fs complete a form that summarizes the basis for the claim; this will be published online for consumers to evaluate. Unfortunately, the coding guideline and the claims guideline do not always match in terms of language. The following are the permitted claims according to the claims guideline:[47]

	Claim name	Human efficacy evaluation tests	Consumer use test	Laboratory test	Literature materials or research data
1	Whitening and freckle removal	√			
2	Sunscreen	√			
3	Hair removal	√			
4	Acne	√			
5	Nourishing	√			
6	Repair	√			
7	Anti-Wrinkle	*	*	*	△
8	Tightening	*	*	*	△
9	Soothing	*	*	*	△
10	Oil control	*	*	*	△
11	Exfoliation	*	*	*	△
12	Anti-breakage hair products	*	*	*	△
13	Dandruff removal	*	*	*	△
14	Moisturizing	*	*	*	*
15	Hair care	*	*	*	*
16	Specific claims (sensitive skin or tear free)	*	*		
17	Specific claims (ingredient efficacy)	*	*	*	*
18	Claims of mildness (non-irritating)	*	*	*	△
19	Claims with quantitative indicators (e.g., time, statistical data)	*	*	*	△
20	New efficacy claims	Select the appropriate basis according to a given efficacy claim			

[45] National Medical Products Administration (2021l) Cosmetics Registration and Filing Dossier Provisions, Appendix 10.

[46] National Medical Products Administration (2021d) Cosmetics Claims Evaluation Standard.

[47] Ibid.

Notes:

1. Items marked with a √ are mandatory items.
2. Items marked with an * are optional items, but at least one item must be selected [for a given claim].
3. The items marked with a △ are optional items, but they must be used together with human efficacy evaluation tests, consumer use tests, or laboratory tests [for a given claim].

Specific labeling regulations guide the content and wording of claims on cosmetics' labels and other promotional materials. While new labeling rules were released in 2021, under CSAR, this regime had already evolved a fair amount prior to this time. A domestic labeling rule, a labeling standard, a naming guideline, and draft labeling regulations collectively governed. They produced answers—sometimes conflicting—on various questions.

For cosmetics claims regulation the devil was and continues to be is in the details. For example, the naming guideline (pre-CSAR) contained a list of forbidden words for cosmetics names and therefore labels.[48] Similarly, NMPA issued a draft labeling regulation for cosmetics, which included an appendix of prohibited words and phrases. Although never finalized—in part because it included a controversial prohibition on over-labeling of foreign products with Chinese stickers—its prohibited word list is still used and remains one of the most detailed sources of guidance on objectionable language for cosmetics advertisements and labeling.[49] Also, NMPA has its own archive of prior decisions on specific wording, some of which companies may have partially gathered from their own rejections from the agency. Starting from sources like these, companies have developed substantial lexicons of prohibited or restricted words or phrases that trigger scrutiny either because they are considered therapeutic claims reserved for drugs and medical devices or because they are questionable under the other prohibitions in Article 9 of the Advertisement Law (Chap. 10).

The labeling rules pursuant to CSAR did not include a lengthy prohibited word list but did not necessarily move away from this approach of focusing on the details of words and phrases rather than the message. The labeling rules under CSAR include a provision that requires that all terms on a label in foreign language that are related to safety and functionality or effectiveness of the product must be consistent with the Chinese. Without further explanation or narrowing, this provision can have broad application (various arguments can be made as to whether a term is related to safety), perhaps pushing companies to adopt a Chinese-only label, cover all foreign language and design, or edit their English labels to match the Chinese. A

[48] State Food and Drug Administration (2010).

[49] China Food and Drug Administration (2014).

requirement that Chinese be on the primary package (e.g., the bottle inside the box) compound the pressure for China-specific packaging.[50]

The new cosmetic rules also restrict different types of words and phrases through broader prohibitions. For example:

- No express or implied medical terms or claims, including names of doctors or drugs to imply medical function.
- No false, exaggerated, or absolute claims.
- No use of trademarks, patterns, font colors, sizes, color contrasts, homophonic texts, or symbols that imply medical functions.
- No use of false information to degrade other legal products and deceive consumers.
- No use of false or unsupported scientific data or research achievements.
- No guarantees of safety or effectiveness.
- No superstitious, vulgar, or other language that violates public order and good morals or customs.[51]

This is not an exhaustive list. Some of these restrictions come from the Advertisement Law and drug and device regulations on advertisements (e.g., guarantees of safety and effectiveness). The prohibitions on medical claims are certainly fair, but any mention of a doctor (even if not related to prevention or treatment of a disease) can be prohibited.

In addition to claims on the label, general content restrictions under the Advertisement Law apply to cosmetics advertisements. This means the content should be clear and non-misleading, avoid references that are against the social order, and follow the rules on endorsements (see Chap. 10).

Unlike drugs, medical devices, and special foods, there is no pre-approval process for cosmetic advertisements. In 1993, the predecessor agency of SAMR, the State Administration for Industry and Commerce, issued a rule specifically regulating the content and qualifications for publishing cosmetics advertisements, but that rule is not regularly referenced in practice.[52]

A central theme with respect to claims is that, even under the new rules, China takes a granular and prescriptive approach to regulating the language on labels and in promotional materials. It tends to try to dictate the specific content of the claims, i.e., prohibited or permitted language, rather than allowing brand-owners to select language that conveys a message consistent with the permitted claim.

[50] National Medical Products Administration (2021g) Measures on Cosmetics Labeling, Articles 5–6.

[51] Ibid., Article 19.

[52] State Administration for Industry and Commerce (1993).

9.15 Cosmetovigilance (Post-Market Surveillance)

Until CSAR, the post-market surveillance and enforcement regime for cosmetics was under-developed compared to drugs and devices. CSAR and its implementing rules have crystallized a set of basic requirements, some of which arguably already existed in various vaguely worded regulatory documents. There are four basic aspects of cosmetovigilance: adverse reaction reporting, product re-evaluation, defect evaluation and recall, and inspections (both in-market product testing and facility inspections).

9.15.1 Adverse Reactions

NMPA has taken a broad approach in CSAR and in subsequent implementing regulations regarding adverse reactions: all adverse reactions (whether serious or not) must be reported. An adverse reaction is defined as "lesions to the skin and its accessory organs [or] local or systemic damage to the human body caused by normal use of cosmetics." There are three aspects to this definition: (1) the harm, (2) the link to the product, and (3) the conditions of use.

The harm element is broad and could apply to any harm that meets those conditions, whether serious or not. In other words, a small skin lesion that is short term and does not require medical intervention could qualify. The causation element is a higher standard, requiring that the product have a link with the injury, which could be difficult to establish. A proposed guidance document indicates that a relation in time between use of the cosmetic and the injury could be a factor in determining cause, but fails to state whether time alone suffices.[53] Such a vague approach to causation—any number of injuries could coincide in time with the use of a daily-use product—will likely lead to confusion that may need to be sorted in practice or through provincial rules, as has occurred in other areas. The last element, of normal use, can be fulfilled by showing that the individual was using the cosmetic according to the instructions on the label.

The adverse-reaction implementing regulations require that R/Fs monitor for adverse events, meaning that they collect them, report them, analyze and evaluate them, investigate their sources, and handle any necessary control measures, such as a recall and/or changes to the product. R/Fs must report all suspected non-serious adverse reactions within 30 days. A suspected adverse reaction generally means that the R/F has received enough information to make a reasonable determination that the elements of an adverse reaction have been met. Reporting generally occur via a national website, with which all R/Fs must register.[54]

[53] National Medical Products Administration (2022f) R/F Cosmetic Adverse Reaction Collection and Reporting Guiding Principles (Trial Implementation) (Draft for Comment).

[54] National Medical Products Administration (2022f) Cosmetic Adverse Reaction Administrative Measures.

R/Fs are also required to report "serious" reactions within 15 days. Those reactions are those that, to paraphrase, cause temporary or permanent loss of functions and affect normal human and social functions, cause systemic damage to the human body, result in hospitalization or hospitalization is deemed necessary by medical institutions, or cause other serious damage to the human body that endangers life or causes death.

R/Fs must also report adverse reactions that "have a significant social impact." These are loosely defined in the regulations as events that result in some type of significant social impact by causing serious injury to a number of people in a certain area as a result of normal use of the same cosmetic.

In addition to R/Fs, entrusted manufacturers have an obligation to report adverse reactions related to the products that they produce.

Reports of adverse reactions are handled by a network of monitoring institutions in different provincial-level localities. Shanghai has 17 such institutions at the district level, and it has 26 medical institutions with reporting sentinels, i.e., hubs for reporting.[55] The monitoring institutions receive and analyze the reported reactions, including from smaller hospitals and distributors that may not have a registration with the national platform and must submit a paper form.[56] The monitoring institutions work with local government and pass information to the national level, including NMPA and its affiliated centers, for analysis of national trends.

9.15.2 Recalls

Adverse reactions and other non-compliance, e.g., with standards or product registration, can be cause for a recall. Prior to CSAR there was not a clear recall standard or reporting procedure for cosmetics, although authorities could order a recall if they determined that there was a safety issue. Some local MPAs had forms and reporting procedures for cosmetics.

Under CSAR, it appears that a system for recalls is emerging. R/Fs have an obligation to recall when a cosmetic could cause harm to human health due to quality defects or other problems,[57] and manufacturers and distributors have an obligation to assist with such recalls. Unlike drugs and medical devices, the term "defect" is not defined in CSAR or its implementing rules. This standard is broad and not currently refined in a meaningful way in CSAR implementing rules, so opening the door to judgment calls by R/Fs and regulators. R/Fs could try to use other existing definitions as a guide, e.g., the general definition of "defect" in the Product Quality Law: "unreasonable dangers existing in products that threaten safety of persons or properties or products that do not conform to national or industrial

[55] Sina (May 25, 2021).

[56] National Medical Products Administration (2022b) Announcement of the National Cosmetics Reaction Monitoring System.

[57] State Council of the People's Republic of China (2020), Article 44.

standards if there are any...."[58] Consumer product recall rules have a similar definition.

Based on the Product Quality Law definition, there is an argument that a cosmetic, like a drug or a device, could be recallable if it does not meet applicable national or industrial standards. It is less clear whether a recall is required if the product does not meet its own registration or record-filing, which are technically a form of product standard. In the past, when authorities have discovered through testing that products did not meet their registrations, they asked for recalls, either because it was considered a safety issue or non-compliance with a required standard.

There is no clear penalty tied to the failure of R/Fs to voluntarily implement this standard. However, if they do not implement a recall the government can order one, and there are strong penalties tied to the failure of a business to implement such an order. In practice, if a local government requests that a business recall a cosmetic product (or even an ingredient) it would likely be fairly damaging to that business's relationship with local officials to fail to act on such a request. Those local officials have many levers from which to choose to make operations difficult for such a business, including criminal referrals if they consider there to be a safety issue. Therefore, if the government requests a recall, businesses will likely comply.

9.15.3 Cosmetic Safety Re-Evaluation

When safety issues emerge regarding the fitness of a product for the market, the government can order the R/F to re-evaluate the basis for the product, or the government can conduct that re-evaluation independently. Such a re-evaluation may result in the registration or filing being cancelled.[59] There are not detailed rules governing the scope and content of a re-evaluation, but if other spaces are a guide, this measure is not frequently applied in practice.

9.15.4 Testing and Inspections

NMPA and the Provincial-Level Medical Product Associations (PMPAs) coordinate cosmetic sampling on an annual basis. This is a significant part of post-market surveillance. Manufacturers must take measures to pull or correct failing products. Sampling is organized at the national level and focuses on various products that could be higher risk due to the nature of the product, its prior record, or the breadth of its circulation. Sampling is done by different levels of local governments (province, municipality) and includes both onsite sampling in stores and sampling of online products.[60] The latter have been increasing cause for concern as several adverse

[58] National People's Congress Standing Committee (2018).

[59] State Council of the People's Republic of China (2020), Article 55.

[60] National Medical Products Administration (2022a) 2021 National Annual Cosmetic Sampling Report.

reactions have been linked to products purchased online. NMPA recently special rules to surveil cosmetics sold through domestic e-commerce (see above for cross-border e-commerce, which is handled separately).[61] Despite this concern, China reported a 98% pass rate for sampling in 2021.[62]

Sampling occurs in accordance with a sampling standard issued in 2017. The sampling standard covers both testing of product and reviewing labels and credentials to ensure, for example, that all products are appropriately approved via registration and filing procedures and that the label is consistent with that approval. It can be for targeted reasons, such as when there is a suspicion of a quality or safety issue, including illegal ingredients or contamination.[63] Under CSAR, NMPA began to reshape and further establish the sampling program through new rules. In 2022, NMPA issued a procedural rule to replace the 2017 standard and govern the sampling process, such as retesting and subsequent measures to remediate product issues.[64] The new rules provide that sampling should focus on (1) cosmetics formulated for children (which are subject to special rules in China), (2) those with new ingredients, (3) those that have a record of safety and quality problems, (4) those with higher rates of fails in prior sampling, and (5) those with a wider circulation. The extent of prior issues and circulation that would qualify a product for this are not specified.

NMPA also issued separate procedures to create supplemental testing measures when these were insufficient under national standards to meet the demands of post-market testing.[65] And, in 2022, it established new testing methods for the following eight components: neoconazole, benzidazole, clometronidazole, ornidazole, tinidazole, ipronidazole, ronidazole, and hydroxymetronidazol.[66]

CSAR also calls for post-market inspections of manufacturing and distribution facilities. Local medical-products authorities have the right to enter those facilities as part of their supervision work. This area is still developing and, at the time of this writing, comprehensive national inspection rules are still being made and considered, including those for overseas cosmetic facilities inspections. Even in the absence of rules, these types of inspections continue to happen at the local level of government. For example, Shanghai has released a rule on the quality and safety grading of cosmetics-manufacturing enterprises for purposes of determining the

[61] National Medical Products Administration (2023b) Cosmetics Internet Distribution Supervision Measures.

[62] National Medical Products Administration (2022a) 2021 National Annual Cosmetic Sampling Report.

[63] China Food and Drug Administration (2017a) Cosmetics Supervision Sampling Work Standard.

[64] National Medical Products Administration (2023a) Administrative Measures on Cosmetics Sampling and Testing.

[65] National Medical Products Administration (2021b) Announcement on Issuing the Regulations on the Management of Supplementary Testing Methods for Cosmetics and the Technical Guidelines for Research and Drafting of Supplementary Testing Methods for Cosmetics.

[66] National Medical Products Administration (2022e) Notice Establishing Testing for 8 Components of Neoconazole in Cosmetics.

level and frequency of different types of government supervision, including inspections.[67]

9.15.5 Enforcement

Like other areas, cosmetics enforcement is a mixture of administrative enforcement through agencies for what are mainly strict liability prohibited acts, as they investigate complaints and safety signals and test products and facilities, and civil and criminal enforcement. CSAR provides for a standard range of penalties, including monetary fines, warnings, interviews (to educate the business as to compliance), and product seizures and, in more serious cases, cancellation of product and facility licenses, debarment of entities and fines and debarment of individuals responsible for violations (see Chap. 3). As is the trend with drug and medical device regulation, the penalties in CSAR are substantially more severe than in its predecessor.

CSAR provides that whoever causes damage to human health or injury in violation of the CSAR shall be liable for compensation.[68] However, unlike the Food Safety Law, it does not provide for nominal damages in the event of a violation of any relevant standard, and it does not provide for punitive damages. The CSAR essentially echoes compensation rights that are also available under the Civil Code when products cause damage. Consumer complaints against various aspects of cosmetics through litigation or reports to administrative agencies are not uncommon.

References

China Food and Drug Administration (2014) Administrative Measures on Cosmetic Labeling (Draft for Comments). https://www.ccfdie.org/zryyxxw/hzp/jgdt/webinfo/2014/11/1485613785277072.htm

China Food and Drug Administration (2015a) Announcement on Matters Concerning Cosmetics Production License. https://www.nmpa.gov.cn/hzhp/hzhpggtg/hzhpqtgg/20151215120001817.html?type=pc&m=

China Food and Drug Administration (2015b) Cosmetics Manufacturing Licensing Work Guideline (Trial Implementation). https://www.nmpa.gov.cn/directory/web/nmpa/xxgk/zhqyj/zhqyjhzhp/20150317145901794.html

China Food and Drug Administration (2015c) Cosmetics Safety and Technical Standard. https://www.nmpa.gov.cn/hzhp/hzhpfgwj/hzhpgzwj/20151223120001986.html

China Food and Drug Administration (2017a) Cosmetics Supervision Sampling Work Standard. https://www.nmpa.gov.cn/directory/web/nmpa/images/u6Xsca3vOC2vbPpvOy5pNf3uea3ti5kb2N4.docx

China Food and Drug Administration (2017b) Shanghai Pudong New Area Imported Non-Special Use Cosmetics Filing Management. https://www.nmpa.gov.cn/hzhp/hzhpfgwj/hzhpgzwj/20170118174101168.html?type=pc&m=

[67] Shanghai Medical Products Administration (2022).

[68] State Council of the People's Republic of China (2020), Article 76.

Ernst and Young (2021) How Should Cross-Border E-Commerce on the Racing Track of Investment Hotspots Get Prepared for the Capital Markets. https://assets.ey.com/content/dam/ey-sites/ey-com/en_cn/topics/consumer-products/ey-how-cross-border-e-commerce-in-china-prepare-move-to-capital-markets.pdf?download

General Office of the National Health Commission (2018) Disinfectant Measures. http://www.nhc.gov.cn/zhjcj/gongwen1/201806/047c54980196495ab95856cc4839f3cc.shtml

Ministry of Commerce (2018) Circular of the Ministry of Commerce, National Development and Reform Commission, Ministry of Finance, General Administration of Customs, State Administration of Taxation and State Administration for Market Regulation on Improving the Supervision of Retail Imports in Cross-border Electronic Commerce. http://english.mofcom.gov.cn/article/policyrelease/announcement/201812/20181202813328.shtml

National Health and Family Planning Commission (2009) Catalogue of Disinfectants. http://dghb.dg.gov.cn/attachment/cmsfile/dgwsjd/qtxz/201606/daofile/13512398391270b62e043bf091d20f85f68f56042_492.doc

National Medical Products Administration (2018) Implementation of Record-filing Management of Imported Non-Special-Purpose Cosmetics Nationwide. https://www.nmpa.gov.cn/hzhp/hzhpfgwj/hzhpgzwj/20181109112801602.html?type=pc&m=

National Medical Products Administration (2019a) Cosmetics Registration and Record-Filing Testing Work Standard. https://www.nmpa.gov.cn/directory/web/nmpa/xxgk/ggtg/qtggtg/20190910153001302.html

National Medical Products Administration (2019b) Rules on Cosmetics Overseas Inspections (Draft for Comment). https://www.nmpa.gov.cn/directory/web/nmpa/xxgk/zhqyj/zhqyjhzhp/20191122150401393.html

National Medical Products Administration (2021a) Announcement on Further Clarifying the Transitional Management of Original Special Purpose Cosmetics and Other Related Matters. https://www.nmpa.gov.cn/directory/web/nmpa/images/localgov_h4/gov_1640282661028.pdf

National Medical Products Administration (2021b) Announcement on Issuing the Regulations on the Management of Supplementary Testing Methods for Cosmetics and the Technical Guidelines for Research and Drafting of Supplementary Testing Methods for Cosmetics. https://www.nmpa.gov.cn/xxgk/ggtg/qtggtg/20210428140400103.html

National Medical Products Administration (2021c) Cosmetics Categorization Principles and Categorization Catalogue. https://www.nmpa.gov.cn/xxgk/ggtg/qtggtg/20210409160151122.html?type=pc&m=

National Medical Products Administration (2021d) Cosmetics Claims Evaluation Standard. https://www.nmpa.gov.cn/xxgk/fgwj/xzhgfxwj/20210409160321110.html

National Medical Products Administration (2021e) Cosmetics Ingredient Safety Information Registration Platform Online. https://www.nmpa.gov.cn/directory/web/nmpa/hzhp/hzhpjgdt/20211230165501124.html

National Medical Products Administration (2021f) Inventory of Existing Cosmetics Ingredients. https://www.nmpa.gov.cn/xxgk/ggtg/qtggtg/20210430162707173.html?type=pc&m=

National Medical Products Administration (2021g) Measures on Cosmetics Labeling. https://www.nmpa.gov.cn/xxgk/fgwj/xzhgfxwj/20210603171933181.html

National Medical Products Administration (2021h) Provisions on Cosmetics Registration and Record-Filing Materials. https://www.nmpa.gov.cn/xxgk/fgwj/xzhgfxwj/20210304140747119.html

National Medical Products Administration (2021i) Provisions on the Registration and Filing Materials for New Cosmetic Ingredients. https://www.nmpa.gov.cn/xxgk/fgwj/xzhgfxwj/20210304140454159.html

National Medical Products Administration (2021j) Q&A on the Registration and Filing Policy for New Cosmetic Ingredients. https://www.nmpa.gov.cn/directory/web/nmpa/xxgk/zhcjd/zhcjdhzhp/20211111150042125.html

National Medical Products Administration (2021k) Administrative Provisions for the Supervision of Children's Cosmetics. https://www.nmpa.gov.cn/directory/web/nmpa/hzhp/hzhpjgdt/20211008171226187.html

National Medical Products Administration (2021l) Cosmetics Registration and Filing Dossier Provisions, Appendix 10. https://www.nmpa.gov.cn/xxgk/fgwj/xzhgfxwj/20210304140747119.html

National Medical Products Administration (2022a) 2021 National Annual Cosmetic Sampling Report. https://www.nmpa.gov.cn/directory/web/nmpa/hzhp/hzhpjgdt/20220321165109138.html

National Medical Products Administration (2022b) Announcement of the National Cosmetics Reaction Monitoring System. https://www.nmpa.gov.cn/xxgk/ggtg/qtggtg/20220920170353136.html

National Medical Products Administration (2022c) Cosmetics Good Manufacturing Practice. https://www.nmpa.gov.cn/hzhp/hzhpfgwj/hzhpgzwj/20220107101645162.html?type=pc&m=

National Medical Products Administration (2022d) Cosmetics GMP Inspection Main Points and Determination Principles. https://www.nmpa.gov.cn/xxgk/ggtg/qtggtg/jmhzhptg/20221025104946190.html

National Medical Products Administration (2022e) Notice Establishing Testing for 8 Components of Neoconazole in Cosmetics. https://www.nmpa.gov.cn/xxgk/ggtg/qtggtg/jmhzhptg/20220913105611132.html

National Medical Products Administration (2022f) R/F Cosmetic Adverse Reaction Collection and Reporting Guiding Principles (Trial Implementation) (Draft for Comment). https://www.ccfdie.org/cn/yjxx/hzp/webinfo/2022/12/1668876200094474.htm

National Medical Products Administration (2022g) Policy Interpretation—Cosmetics GMP Freshly Released—Main Responsibilities of Entrusting Parties (Cosmetics Registrants and Filers). https://www.nmpa.gov.cn/xxgk/zhcjd/tjzhc/tjzhchzhp/20220107183349115.htm

National Medical Products Administration (2023a) Administrative Measures on Cosmetics Sampling and Testing https://www.nmpa.gov.cn/xxgk/ggtg/qtggtg/jmhzhptg/20230112162311197.html

National Medical Products Administration (2023b) Cosmetics Internet Distribution Supervision Measures. http://www.gov.cn/zhengce/zhengceku/2023-04/04/content_5750049.htm

National Medical Products Administration (2023c) Optimizing the Management Measures for Filing and Inspection of Ordinary Cosmetics. https://www.nmpa.gov.cn/xxgk/ggtg/qtggtg/jmhzhptg/20230118093634177.html

National People's Congress Standing Committee (2018) Product Quality Law

Shanghai Drug Administration (2022) Explanation of the Cosmetics Manufacturing Enterprise Quality and Safety Risk Grading Management Measures. http://www.zhuceabc.com/2022/08/17/%E3%80%8A%E4%B8%8A%E6%B5%B7%E5%B8%82%E5%8C%96%E5%A6%86%E5%93%81%E7%94%9F%E4%BA%A7%E4%BC%81%E4%B8%9A%E8%B4%A8%E9%87%8F%E5%AE%89%E5%85%A8%E9%A3%8E%E9%99%A9%E5%88%86%E7%BA%A7%E7%AE%A1%E7%90%86%E5%8A%9E/

Sina (May 25, 2021) Over 3000 Cosmetics ARs in Shanghai and 62% From Online Shopping. https://finance.sina.cn/2022-05-25/detail-imizmscu3364753.d.html?oid=4062650155088394&vt=4&cid=76675&node_id=76675

State Administration for Industry and Commerce (1993) Administrative Measures on Cosmetics Advertisements. http://tradeinservices.mofcom.gov.cn/article/zhengce/flfg/201710/1232.html

State Administration for Market Regulation (2021a) Cosmetics Manufacturing and Distribution Supervision Administrative Measures. https://gkml.samr.gov.cn/nsjg/fgs/202108/t20210806_333443.html

State Administration for Market Regulation (2021b) Cosmetics Registration and Record-Filing Measures. https://gkml.samr.gov.cn/nsjg/fgs/202101/t20210112_325127.html

State Council (1989) Cosmetics Hygiene Management Regulation

State Council of the People's Republic of China (2020) Cosmetics Supervision and Administration Regulation (CSAR)
State Food and Drug Administration (2010) Cosmetics Naming Provisions and Naming Guidelines. http://www.gov.cn/gzdt/2010-02/09/content_1531311.htm
State Food and Drug Administration (2013a) Circular on Adjusting Cosmetics Registration and Filing Management. https://www.nmpa.gov.cn/hzhp/hzhpfgwj/hzhpgzwj/20131216084301437.html
State Food and Drug Administration (2013b) Letter on Strictly Circumscribing Documentation of Proof of Approval and Sale for Imported Cosmetics. https://www.nmpa.gov.cn/xxgk/fgwj/gzwj/gzwjhzhp/20131216120001584.html
State Food and Drug Administration (2013c) Notice on Further Improving Work Related to Cosmetics Production Licenses. https://www.nmpa.gov.cn/directory/web/nmpa/xxgk/fgwj/gzwj/gzwjhzhp/20131011120001445.html
State Food and Drug Administration (2022) Cosmetic Adverse Reaction Administrative Measures. https://www.nmpa.gov.cn/xxgk/fgwj/xzhgfxwj/20220221165805149.html
U.S. Department of Commerce, International Trade Administration (2020) China – Country Commercial Guide, China Cosmetics and Toiletries Industry. https://www.trade.gov/country-commercial-guides/china-cosmetics-and-toiletries-industry
Xinhua News Agency (April 26, 2022) China's Cross-Border E-commerce Gains Momentum, Bolsters Consumption. http://english.www.gov.cn/news/topnews/202204/26/content_WS62675588c6d02e5335329edd.html

Advertising and Promotion for Regulated Products 10

10.1 Introduction

Now that we have examined the components of the legal system relevant to FDD regulation and each of the regulated spaces—drugs, medical devices, cosmetics, and food—we know that there are various ways in which the government regulates the claims and statements that can be made about products. Those ways include:

- Product-specific statutes to control what information appears on labels (including the label on the box, the package insert, and, in some cases, electronic content, e.g., via QR code)
- The Advertisement Law, which governs both what is on the label of FDD products and everything that is said in advertisements outside of the label
- The Law Against Unfair Competition and other consumer protection laws that control the fairness and non-misleading nature of product publicity generally

While the National Medical Products Administration (NMPA) and local Medical Product Associations (MPAs) control the review of labels and package inserts during applicable pre-market approval processes and the approval of advertisements (if required), the State Administration for Market Regulation (SAMR) and its local administrations for market regulation (AMRs) control enforcement of the broader laws on advertising, consumer protection, and fair competition. As this chapter will discuss, there is a fair amount of discretion in the assessment of content and enforcement and agencies do not always align with each other or with the courts.

These laws and their implementing rules share various general principles and content restrictions in common. This area of regulation controls not only the technical aspects of what can be said about these products, but it also provides a tool for local regulators to reflect elements of the political and cultural context in China in regulated product discussions. The Chinese government has a number of other ways to control what companies or individuals can say about these areas. This may include general policy, but other methods are non-legal and even invisible—such as

J. C. Balzano, *China Food and Drug Law*, Law for Professionals,
https://doi.org/10.1007/978-3-031-61901-4_10

discussions and influence applied outside of the public eye. For example, the government can use broadly written laws against making trouble to punish content. Critics of the handling of the response to SARS and COVID have been told certain speech was unacceptable, forced to issue apology letters—or detained.[1]

10.2 General Principles of Product Promotion

Chinese law and policy do not draw practical distinctions between objectively presented information for scientific exchange (e.g., about the specifics of product development), objective drug/device and healthcare information, including that on labels and product detailing materials (i.e., a type of informational material about drugs), and advertisements, whether through traditional publication or other forms of publicity and promotion (e.g., verbal conversations or presentations).

Advertisement and promotion law does not have a safe harbor for puffery, meaning even flowery language can be taken literally. If the advertisement associates a product with luck or good fortune or a healthy life, a consumer could bring a claim that they did not receive these benefits after use of the product. No provision of the Advertisement Law defines what a reasonable consumer would understand not to be literally true. And, while some courts or agencies have applied reasonableness standards (i.e., the consumer would reasonably understand that certain words are just for advertising purposes), in these cases this determination is at the relevant court's or authority's discretion.

Advertisement law also prohibits content that undermines good social customs or morals, but does not define those morals, leaving decisions as to what might be against the spirit of social and political policy at a given time to discretion of local officials based on the policies of time. For example, superstitious statements about spirits and myths can be prohibited, and yet various products with dragon or other traditional imagery can be found on the market.

China further prohibits cultural or political appropriation for commercial purposes: It is not permitted to use the names or likenesses of leaders or of the government to promote products, although this does not stop many companies from labeling products as approved by government agencies, including foreign government agencies. One company was fined for claiming that Chinese leaders drank its coffee and for using their images.[2] During the pandemic, distributors of various personal protective equipment products claimed to have U.S. government or FDA approval—even offering printed certificates with purported U.S. government seals—which was certainly not accurate in many cases.

Statements that contradict the government's policies or goals could also be the subject of enforcement. For example, one popular medical website posted criticisms of Chinese medicine as a treatment for COVID and hypothesized about the origins of

[1] Buckley (February 6, 2020); Weinland (February 3, 2020).

[2] Sina Finance (May 23, 2022).

the pandemic in China. Both areas were considered unacceptable from a policy perspective, and the site was temporarily shut down.[3]

10.3 Entities Subject to this Law

The Advertisement Law creates different obligations for advertisers, advertising agents, advertising publishers, and endorsers. The drug, device, and health food license holders and their registration agents may be "advertisers," which is defined as an entity or person that designs, produces, and publishes advertisements or authorizes others to do so on his or her behalf. For overseas companies that hold product permissions, their regulatory agents or distributors, which in some cases must submit advertisement approvals, can be the object of enforcement.

Agents for the purposes of the Advertisement Law are a bit different: third-party entities that provide design and production services. Publishers are a separate class of entity that publishes the advertisement in the press or online. Publishers routinely have their own set of internal standards and supervisory personnel to help to ensure compliance with the law, given that they may also be held liable for certain actions. If the advertisement relates to a product that affects the health or life of the consumer, the publisher can be held jointly liable with the advertiser itself.[4]

Endorsers are another regulated entity under the Advertisement Law. They are either natural persons, entities, or organizations that use their own name or image to issue a testimonial to a product or service as part of an advertisement.[5] They are also required to observe content standards and must have actually used the product that they are endorsing. In the cases of drugs and devices, endorsers are forbidden.

The Anti-Unfair Competition Law and its ban on misleading publicity applies to "operators," which includes "natural persons, legal persons, and unincorporated organizations engaged in the production and distribution of commodities or the provision of services. . ."[6]

10.4 What Constitutes an Advertisement

What content makes something an advertisement? This question is important because for drugs, medical devices, certain special foods, and medical services, the distinction between advertisement and non-advertisement means the difference between cumbersome pre-release approval processes and strict content requirements and simply releasing informational product-related content into the media, subject to

[3] Zhou (August 18, 2022).

[4] National People's Congress Standing Committee (2018) Advertisement Law. Article 56.

[5] Ibid., Article 2.

[6] National People's Congress Standing Committee (2019a) Anti-Unfair Competition Law. Article 2.

any requirements or content restrictions that are the policies of the website or television channel.

Perhaps one reason that the Advertisement Law is not always strictly enforced is that it sweeps too broadly and could potentially cover all media or even private content. Advertisement activities are defined as those activities of sellers or service providers to promote their goods and services to consumers directly or indirectly through a defined medium and format.

So, what does this mean? In terms of the *form or medium,* practice has taught that it covers television (there are some special rules for TV commercials), radio spots, and print news that aim to promote goods or services for sale.[7] It covers outdoor advertisements, such as billboards and signs, which are also subject to subsets of local advertisement rules.

A 2004 implementing rule of the Advertisement Law (now defunct) offered the following helpful list:

> Publishing advertisements in newspapers, periodicals, books, directories, [and other such media].
>
> Using radio, television, film, video, slideshow, [and other such methods] to broadcast advertisements.
>
> Using buildings or spaces such as streets, squares, airports, stations, docks, [and other such methods] to set up signs, neon lights, electronic display boards, windows, light boxes, walls and other [such] advertisements.
>
> Using theaters, stadiums, cultural centers, exhibition halls, hotels, restaurants, playgrounds, shopping malls and other [such] places to set up and post advertisements inside and outside.
>
> Releasing, drawing, and posting advertisements by means of vehicles, boats, and airplanes.
>
> Mailing of different types of advertising materials through the post.
>
> Giving in-kind items as a form of advertising.
>
> Using other media and forms to publish, set up, and post advertisements.[8]

There was some debate about what web-based or email content was covered, but in 2016 the State Administration for Industry and Commerce (now SAMR) released Tentative Measures on Internet Advertising that clearly stated that words and images online could constitute advertisements and specially regulated their content and posting.[9] These measures followed on the heels of a scandal, in which a 21-year-old alleged that he had located what turned out to be questionable immunotherapy treatments to treat his cancer (soft tissue carcinoma) online, via a hospital's

---

[7] Ministry of Justice (2011).

[8] State Administration for Industry and Commerce (2004).

[9] State Administration for Industry and Commerce (2016).

advertisement. The patient ultimately died. The posts related to this incident sparked outrage that the main search engine had allowed such questionable medical advertising to remain unchecked, giving license for doctors and the hospital to promote this medical treatment.[10]

The internet measures apply not only to standard webpages, but also pop-ups and emails used to promote items. The internet rules required that online promotional content be clearly marked as "advertisement," in Chinese characters.

Less guidance exists regarding the meaning of directly or indirectly *promoting* a product or service. This distinction relates to the line between information and scientific exchange and promotional activity. For example, does merely posting a label constitute advertising? The answer should be at least "not necessarily," given that NMPA itself posts drug labels on its website, and most would agree it is not advertising those products. But some regulators take the position that a manufacturer of a drug or device can only advertise by using language that it takes from the label itself and places into an advertisement format.

There is also potential overlap between regulatory categories. For example, as discussed in Chap. 5 on Drug Regulation, websites that purvey information about drugs (or medical devices) receive record-filing numbers. Until recently, the National Health Commission (NHC) also supervised the implementation of rules regulating the establishment and content of health information websites that provide preventive and health care knowledge for free or a fee, provided that the company or institution releasing the website had required personnel with expertise to support the type of information it seeks to provide.[11] It may appear that approval of such information would mean that the website content is non-promotional, but that is not the case. Regulators can take the position that an informational website can also contain advertisements. And the record-filing process has not been designed to weed out promotional content.

For drugs, devices, and health foods, for which advertising approvals are required, an approval can help extinguish any doubt about the category of content. However, those approvals are not easy to obtain, as officials may strictly limit the content (to what is included on the label) and, at least for drugs, the placement of that advertisement. For example, prescription-drug content may only be advertised to healthcare professionals in journals that are accredited by the NMPA and NHC. Furthermore, MPA authorities must approve these advertisements; if they do not deem it to be promotional in nature, they will not examine the submission, leaving the applicant to post the content at risk. The problem is that the local counterparts of SAMR may disagree with the conclusion that the content is non-promotional and decide to enforce against it as an unapproved advertisement. Some flexibility was recently introduced by the establishment of NMPA's Medical Representative system for drug companies (see Chap. 5), allowing these representatives to be authorized by the drug-marketing-authorization holder to engage in an informational dialogue

[10] Phillips (May 3, 2016).

[11] Ministry of Health (2009).

about the product and its features with healthcare professionals; this form of dialogue is called "academic promotion."[12]

The Drug Administration Law contains a provision that states that "non drug advertisements may not contain drug promotion."[13] Although this is not often cited, it may reasonably be read to mean companies may not use another type of advertisement or information content to indirectly promote a drug.

10.5 Intersection with Labeling

Product-label content is governed by separate rules and standards in each of the food, drug, device, and cosmetics spaces. But, in some cases, the labeling rules, standards, or product-naming guidelines administered by NMPA or SAMR, such as those for food and cosmetics, contain the same or similar prohibitions as those in the Advertisement Law (e.g., bans on superstitious content, superlatives, or hyperbole). As discussed, prototypes of the labels and package inserts are submitted and reviewed as part of pre-market registration or record-filing processes (except for unapproved products, such as general foods), and must also contain a set of mandatory items, including ingredients, instructions, warning, safety information, and, in some cases, data.

This approval process can also provide valuable input into the permissibility of certain claims on the label if the regulator proposes revisions or rejects claims on the label, as may occur in some cases. Adhering closely to the approved content can be a way to reduce risk post-market. However, because enforcement in the market against content that potentially violates the Advertisement or Anti-Unfair Competition Laws comes from SAMR, authorities can disagree and reliance on NMPA- or PMPA-approved content is not a guarantee.

10.6 General and Specific Content Requirements

The Advertisement Law articulates several content requirements for all product and service advertisements and rules against false or misleading content. The more specific rules on content revolve around puffery, government endorsements, superstitious, sexual or violent content, and political issues, such as advertisements that harm the state, national minorities or the environment.

Article Nine of the Advertisement Law contains the following list of general prohibitions that are repeated throughout other advertising and labeling regulations:

> [Express] or disguised use of the national flag, national anthem, national emblem[s], military flag, military song, and military emblem of the People's Republic of China;

[12] National Medical Products Administration (2020).

[13] National People's Congress Standing Committee (2019b) Drug Administration Law. Article 90.

[Express] or disguised use of the name or image of state organs or staff of state organs;

Use terms such as "national level," "superlative," "best";

[Content that] harms the dignity or interests of the State, or content that divulges state secrets;

[Content that] obstructs social stability and damages public interests;

[Content that] endangers personal and property safety, or releases personal privacy [information];

[Content that] obstructs social public order or violates good social customs;

[Content that] is obscene, pornographic, related to gambling, superstitious, related to terror, and[/or] violent;

[Content that] discriminates against ethnicity, race, religion, or sex;

[Content that] obstructs the protection of the environment, natural resources, or cultural heritage;

Some enforcement cases revolve around the restriction on puffery. As noted above, the Advertisement Law and related regulations do not have safe harbors for puffery, although some court cases have recognized that consumers should understand that certain statements are not intended to be taken literally. Specifically, courts have analyzed the "context" to determine whether the content at issue "was enough to induce a rational consumer to make a wrongful decision to purchase the products."[14] Still, enforcement against puffery is sufficiently prevalent that one can often find advertisements that contain disclaimers that glossy pictures on the packaging are "for reference only," meaning they are not intended to necessarily represent what is in the package.

A common basis for enforcement is the ban on superlatives. These statements are prohibited under Article 9 of the Advertisement Law, but certain commentary, including that endorsed by local officials, indicates that some "absolute" terms can be used if they are supported by facts.[15] Certain provinces may also have rules that assist with understanding the contours of enforcement related to these restrictions. For example, in Jiangsu Province the following types of language do not fall within the ban on superlatives:

- Terms expressing the order of time and space;
- Awards and titles judged in accordance with laws and regulations;

[14]Taizhou City Intermediate Court (2017) Consumer Rights Trial Cases.

[15]Ningyang Market Supervision Bureau (2017) Policy Interpretation: Extreme Language is not Absolutely Prohibited Under the New Advertisement Law.

- Categories identified in accordance with national standards for specific industries and fields;
- Grading terms expressing an advertiser's [internal] self-comparison;
- Terms used to express the pursuit of the advertiser's goals;
- Factual information, such as sales volume, sales revenue, and market share that are objectively stated and verifiable.[16]

In 2023, SAMR issued guidelines on absolute terms that exempted terms or content from these Article 9 restrictions, for different reasons, such as because the content is not an advertisement of a product or because the term does refer to a product but is not objectively misleading (e.g., comparing products of the same company or the product has received an aware in accordance with state regulation. The Guidelines also direct authorities to assess penalties in light of the harm to consumers and if there is no or slight harm no penalty may be imposed. However, absolute terms used in relation to drugs, devices, or health or medical foods and their safety and effectiveness will not be considered minor violations.[17]

Other provincial rules, documents, and commentary are also important in interpretations of the Advertisement Law because provinces may regulate more specific areas. For example, some provinces have issued rules governing the placement and establishment of outdoor advertisements.[18] In some cases, these outdoor advertisement rules have regulated content, such as prohibiting advertisements with sexual content.[19]

Again, local officials have a significant amount of discretion to determine how the broad restrictions in Article 9 apply. In a recent case, a local government determined that an advertisement for "secret testicle medicine" was illegal—not because of the potentially misleading and confusing language describing it as a "secret" medicine, but because it was considered obscene and pornographic by local officials.[20]

There are also sweeping prohibitions on false and misleading advertisements. These come in various shapes and forms. In the Advertisement Law, there is a general provision prohibiting "false or misleading content" or content with the effect of "defrauding or misleading consumers."[21] This occurs if the advertisement is (1) for goods or services that do not exist, (2) if the advertisement presents content that is different from the actual situation (e.g., with respect to quality, price, or function) and that misrepresentation has an impact on consumer behavior or (3) the advertisement simply mispresents data or statistics underlying the assertions or the

[16] Jiangsu Provincial People's Congress Standing Committee (2010), Article 11.

[17] State Administration for Market Regulation (2023).

[18] Shanghai Government (2017).

[19] China Youth Daily (October 9, 2014).

[20] FJ Daily (December 16, 2020).

[21] National People's Congress Standing Committee (2018) Advertisement Law. Advertisement Law, Article 28.

product's effectiveness.[22] The broad prohibition on misleading advertisements goes hand-in-hand with other articles in the Law. For example, there is another article of the Advertisement Law that requires that all content in an advertisement concerning the "performance, function, place of origin, use, quality, composition, price, producer, expiration date, promise, etc. of a commodity or about the content, provider, form, quality, price, promise..." be accurate *and* understandable.[23] Another provision requires that any statistics or data presented be supported with citations.[24] A recent advertisement that stated that a product was 99.9% effective at sterilizing and "effectively killed coronavirus" was found to violate this provision for failing to indicate the source of these statements.[25]

A similarly broad provision exists under the Anti-Unfair Competition Law, which prohibits "false or misleading commercial publicity on the performance, function, quality, sales status, user evaluation, honors received, [among other aspects] of their products to deceive or mislead consumers."[26] This provision can be used to capture a number of violations, including when off-label claims go beyond the labeling.

The Supreme People's Court defines misleading publicity under the Anti-Unfair Competition Law as (1) biased comparisons to other goods, (2) assertions of facts that are not scientifically supported, (3) use of ambiguous language in commercial advertising, and (4) any other act of commercial publicity that is misleading. To determine whether the advertising is misleading the Supreme People's Courts to examine daily experience of consumers, public attention to the content, and the differences between actual product advertised and any misunderstanding of consumers.[27] Individuals can seek compensation under this provision and bear the burden of showing they have suffered harm as a result of the misleading advertisement.

In addition, there is a provision under the Law for the Protection of the Rights and Interests of Consumers that prohibits inducement of purchase decisions by using fraud. It requires that manufacturers and distributors "shall provide consumers with information on the quality, performance, use, and expiration date of relevant goods or services, which shall be true and comprehensive, and shall not be false or misleading publicity."[28]

---

[22] National People's Congress Standing Committee (2018) Advertisement Law. Advertisement Law, Article 28.

[23] National People's Congress Standing Committee (2018) Advertisement Law. Advertisement Law, Article 8.

[24] National People's Congress Standing Committee (2018) Advertisement Law. Advertisement Law, Article 11.

[25] FJ Daily (December 16, 2020).

[26] National People's Congress Standing Committee (2019a) Anti-Unfair Competition Law. Article 8.

[27] Supreme People's Court (2022), Articles 17–18.

[28] State Administration for Market Regulation (2019b) Law for the Protection of the Rights and Interests of Consumers. Article 20.

A detailed analysis of the application of these provisions is beyond the scope of this discussion, except to note that there they give both officials and private consumers the ability to question product claims and seek enforcement and/or compensation. There is not clear guidance on the difference between the Anti-Unfair Competition Law's prohibition on misleading content and that in the Advertisement Law and the Consumer Protection Law, but courts may distinguish them based on somewhat different purposes: The former is regulating the adverse effect of this content on the market, whereas the latter two laws are more consumer protection focused.

What law is chosen is often based on the amount of penalty or compensation available. For example, the Law for the Protection of the Rights and Interests of Consumers may be chosen because punitive damages, depending on the circumstance, might equal up to three times the cost of the good or service or twice the loss.[29] As discussed below, the Anti-Unfair Competition Law makes higher penalties available from officials from the outset.

10.7 Overlapping Claims and Categories

Unlike in other jurisdictions, China does not generally recognize overlap between therapeutic categories and consumer products. In other words, a product is not usually permitted to be both a cosmetic and a drug. As discussed above, there are some ingredients that are permitted in both foods and drugs (e.g., traditional medicine ingredients and ingredients such as caffeine), and there are foods formulated for medical uses, but there are not products that are regulated as both a food and a drug. If a product is considered a drug, it requires pre-market approval and all advertisements require approval. In the case of a drug-device combination product, the approval would be based on the license issued. In the case of advertising a companion diagnostic (See Chap. 6) with the drug with which it is used approvals for both the drug and device components may be required.

Various laws and regulations indicate that only drug, medical device, and medical service advertisements by hospitals may make therapeutic claims, subject to pre-approval. But there is not a clear definition as to what constitutes a therapeutic claim and the laws have taken different positions on how to identify offending language. Some guidance is very specific: words like "doctor" or "clinical," and symbols associated with hospitals, such as a red cross, can be viewed as medical terms, regardless of the context. That is why in the various spaces there have often been lists of prohibited words that are considered associated with therapeutic products. For example, the naming guidance for health foods (as discussed above, products regulated as a food) may not use words like "organs, tissue, or cells" or

[29] National People's Congress Standing Committee (2019b) Law for the Protection of the Rights and Interests of Consumers. Articles 55. Standing Committee of the National People's Congress (2019).

"health function or words related to the description of the health function of the product (approximate, homonymous, implied, similar)" in trademarks. In a proposed rule in 2014, NMPA proposed an extensive list of prohibited medical terms:

(1) Terms related to indications, including but not limited to: stretchmarks, pregnancy spots, chloasma, pathological alopecia, alopecia areata, alopecia totalis, alopecia universalis, scars, scars, rashes, sores, boils, scabies, furuncles, carbuncles, ringworm, inflammation, injury, pain, swelling, pus, blisters, spasm, convulsions, rosacea, beriberi, infection, anti-hair loss, hair growth, spot removal.
(2) Terms related to medical professions, including but not limited to: medical*, special titles for medical personnel (such as doctor, doctor, doctor, Chinese medicine doctor, military doctor, pharmacist, doctor, nurse, etc.), specialist, patient, targeted, medical treatment, treatment, treatment, cure, healing, endocrine, virus, bacteria (pathogenic bacteria), fungi (pathogenic bacteria), immunity, detoxification, desensitization, anti-allergic, anti-allergic, sterilization, sterilization, anti-bacterial, antibacterial, antibacterial, sterilization, disinfection.
(3) Terms related to pharmacy, including but not limited to: medicine*, prescription-related titles (such as medicine prescription, Chinese prescription, Korean prescription, Tibetan prescription, Miao prescription, Mongolian prescription, Uyghur prescription, ancient prescription, secret prescription, proven prescription, ancestral prescription, Single prescription, compound prescription, prescription, prescription, etc.), medicinal, drug, traditional Chinese medicine, Chinese herbal medicine, growth factor, hormone, hormone, antibiotic.
(4) Other terms: gene, factor, stem cell, interferon, capillary, lymph, central nervous system, cell repair, red blood streak, dark circle, cosmeceutical.[30]

Companies can maintain lists of prohibited words that indicate health functions and other prohibited content (e.g., superlatives) to help avoid consumer complaints and enforcement inquiries. Some government documents provide more general guidance on the breadth of enforcement. For example, the following statement is from the Shanghai Administration for Industry and Commerce (now the Shanghai AMR) in response to a question about when certain language in a cosmetics advertisement could confuse consumers into thinking that the product is a drug:

> Whether advertising terms used to describe food, alcohol, or cosmetics are medical terms or terms that are easily confused with drugs shall be determined by whether they may produce the effect of making the advertising audience believe that certain products have the effect of treating a certain type of disease. [In this case,] the phrase "comprehensively adjust the endocrine balance of the human body..." may make consumers believe that this [] pure pollen [product] has a therapeutic effect on endocrine imbalances in the body.[31]

As indicated above, in practice, AMRs may employ a common-sense or subjective analysis when reviewing advertisement copy, meaning if there is any basis for a consumer to misunderstand (and the consumer did in fact misunderstand) the advertisement could be problematic, even if the context is sufficiently clear to most reasonable consumers. In other words, even if someone could understand

[30] Zhang (2014).

[31] State Administration for Industry and Commerce (1996).

that the word "doctor" in a food or cosmetics advertisement refers to a scientific or medical professional, who is evaluating the nutrition or cosmetic efficacy of the product, the fact that "doctor" may be more associated with the prescription of drugs could be enough to conclude that the content is misleading as to the true product category.

10.8 Other Product Restrictions

In addition to the general prohibitions on content (e.g., no superlatives) and misleading language, both the Advertisement Law and its implementation rules include more specific safeguards to prevent misleading advertising language on regulated products.

This includes prohibitions on endorsements by hospitals and doctors and assertions of the product's safety or efficacy, including comparisons to the efficacy of other drugs. Advertisements also may not assert that products are essential or represent "cutting edge" science or developments. Importantly, regulated products like drugs, medical devices, and health foods may not have content in their advertisements that exceeds the scope of their approvals, i.e., drug indications, device scope of use, and registered health food claims.[32] The Advertisement Law and related regulations also mandate instructions to consumers for different types of advertisements. For example, drug and device advertisements must remind consumers to follow the instructions on the label and health food advertisements must remind consumers that the products are not substitutes for drugs.

10.9 Advertising and Promotion Enforcement

China's advertising regime is strict regarding content, but the focus and intensity of enforcement can vary depending on a range of factors, including general governance policies at a given them. Enforcement under the Advertisement Law can have penalties ranging from tens to hundreds of thousands of dollars, but it does not often venture into millions. The Advertisement Law underwent a major revision in 2015, which substantially increased penalties. The fines can be substantial for smaller companies in China, such as $30,000 for use of words that are superlative in nature. In some cases, courts have reduced penalties when they determined that the fine was disproportionate with the harm caused by the offending advertisement.[33] Penalties under the Anti-Unfair Competition Law start higher, and that law has also often been used to penalize advertising and promotion violations.

[32] State Administration for Market Regulation (2019a) Interim Measures for Advertisement Review and Administration of Drugs, Medical Devices, Health Foods, and Formula Foods for Special Medical Purposes. Articles 5–7.

[33] China Supervision News (2020).

Specifically, under the Advertising Law the penalties can be three to five times the advertising cost; if that is difficult to calculate, the penalties can start at $30,000 and rise into the hundreds of thousands (or millions of RMB). Under the Anti-Unfair Competition Law, the penalties can start at the $30,000 mark.[34]

Because most public content that is arguably promotional will trigger some Advertisement Law violation, SAMR has emphasized that officials need to exercise reasonable discretion in investigating and punishing these violations. For example, not every use of a superlative will violate Article 9 of the Advertisement Law. Some provinces have established rules that expressly exempt certain terms.[35] Commentary has indicated that the harm of the violation must be considered when imposing penalties, e.g., how broadly the offending content was disseminated, for how long, and by what company with what kind of reputation.

Given the uncertainty in this area, companies must consider these penalties when determining their risk tolerance and internal policies. In airports or in malls in China, one can find many advertisements that have questionable compliance with some of the broad and vague principles in Article 9 of the Advertisement Law and with that Law's other content restrictions and the implementing regulations for various FDD spaces. Enforcement does occur (there are many examples online), but it sometimes appears selective and inconsistent, perhaps because SAMR and NMPA have not established the resources to police all content the way that China polices the internet and media for other politically or socially sensitive content. For example, the Drug Administration Law only allows advertisements for prescription drugs in pre-approved journals and, even then, only after the specific content has been approved.

But the internet in China contains information in the form of websites and press releases that are publicly available and speak to product approvals and the results of clinical trials, touting new forms of treatment or healthy foods. In the end, the decision to publish any content entails some risk and requires a fact-specific analysis of the content, the policies of the time, and the trends in enforcement, including local attitudes.

References

Buckley C (February 6, 2020) Chinese Doctor, Silenced After Warning of Outbreak, Dies from Coronavirus. The New York Times. https://www.nytimes.com/2020/02/06/world/asia/chinese-doctor-Li-Wenliang-coronavirus.html?login=email&auth=login-email

China Youth Daily (October 9, 2014) Remediation of Bad Advertisements Requires National Standards. People.com.cn. http://opinion.people.com.cn/n/2014/1009/c159301-25792546.html

FJ Daily (December 16, 2020) Fujian Provincial Market Supervision Bureau announced 17 typical cases of false and illegal advertisements. https://www.fjdaily.com/app/content/2020-12/16/content_892162.html

[34] The Paper (2020).

[35] State Administration for Market Regulation (2020).

He M, China Supervision News (2020) Correctly Exercising Administrative Punishment Discretion for Advertisement Violations No. 1. Republished on SAMR Website. https://www.samr.gov.cn/ggjgs/sjdt/gzdt/202008/t20200814_320760.html
Jiangsu Provincial People's Congress Standing Committee (2010) Jiangsu Advertisement Provisions. http://scjgj.jiangsu.gov.cn/art/2019/4/1/art_78977_9243623.html
Ministry of Health (2009) Internet Medical and Health Care Information Service Measures. http://www.gov.cn/flfg/2009-06/23/content_1347818.htm
Ministry of Justice (2011) Administrative Measures on the Broadcasting of Radio and Television Advertisements. http://www.gd.gov.cn/zwgk/wjk/zcfgk/content/post_2532399.html
National Medical Products Administration (2020) Administrative Measures on Record-Filing of Medical Representatives (Trial Implementation). https://www.nmpa.gov.cn/xxgk/ggtg/qtggtg/20200930163955170.html
National People's Congress Standing Committee (2018) Advertisement Law. http://www.gov.cn/guoqing/2020-12/24/content_5572939.htm
National People's Congress Standing Committee (2019a) Anti-unfair Competition Law. https://gkml.samr.gov.cn/nsjg/fgs/201906/t20190625_302771.html
National People's Congress Standing Committee (2019b) Drug Administration Law
Ningyang Market Supervision Bureau (2017) Policy Interpretation: Extreme Language is not Absolutely Prohibited Under the New Advertisement Law. http://www.ny.gov.cn/art/2017/5/18/art_171925_9939923.html
Phillips T (May 3, 2016) China investigates Baidu after death of student who sought cancer cure on internet. The Guardian. https://www.theguardian.com/world/2016/may/03/baidu-investigated-in-china-after-death-of-student-who-sought-cancer-cure-on-internet
Shanghai Government (2017) Shanghai Administrative Measures on Outdoor Facility Advertisements. https://www.shqp.gov.cn/shqp/shqp/upload/202109/0903_145209_383.pdf
Sina Finance (May 23, 2022) Do not use words at will in communication, beware of being punished for breaking the law, the new "Advertising Law" complete list of prohibited words and cases. Sina.com.. https://finance.sina.com.cn/jjxw/2022-05-23/doc-imizmscu2851374.shtml
Standing Committee of the National People's Congress (2019) Law for the Protection of the Rights and Interests of Consumers
State Administration for Industry and Commerce (1996) Answers of the State Administration for Industry and Commerce on the Identification of Medical Terms or Terms that are Easily Confused with Drugs. http://scjgj.jiangsu.gov.cn/art/2008/11/17/art_78976_9315674.html
State Administration for Industry and Commerce (2004) Advertisement Law Implementing Measures. https://baike.baidu.com/item/%E5%B9%BF%E5%91%8A%E7%AE%A1%E7%90%86%E6%9D%A1%E4%BE%8B%E6%96%BD%E8%A1%8C%E7%BB%86%E5%88%99/7995877
State Administration for Industry and Commerce (2016) Tentative Measures on Internet Advertising. http://www.gov.cn/gongbao/content/2016/content_5120707.htm
State Administration for Market Regulation (2019a) Interim Measures for Advertisement Review and Administration of Drugs, Medical Devices, Health Foods, and Formula Foods for Special Medical Purposes. https://gkml.samr.gov.cn/nsjg/fgs/201912/t20191227_309564.html
State Administration for Market Regulation (2019b) Law for the Protection of the Rights and Interests of Consumers. https://gkml.samr.gov.cn/nsjg/fgs/201906/t20190625_302783.html#:~:text=%E4%B8%AD%E5%8D%8E%E4%BA%BA%E6%B0%91%E5%85%B1%E5%92%8C%E5%9B%BD%E6%B6%88%E8%B4%B9%E8%80%85%E6%9D%83%E7%9B%8A%E4%BF%9D%E6%8A%A4%E6%B3%95&text=%E7%AC%AC%E4%B8%80%E6%9D%A1%E4%B8%BA%E4%BF%9D%E6%8A%A4,%E5%85%B6%E4%BB%96%E6%9C%89%E5%85%B3%E6%B3%95%E5%BE%8B%E3%80%81%E6%B3%95%E8%A7%84%E4%BF%9D%E6%8A%A4%E3%80%82
State Administration for Market Regulation (2020) Correctly Exercising the Discretionary Power of Advertising Administrative Punishment No. 2. https://www.samr.gov.cn/ggjgs/sjdt/gzdt/202008/t20200814_320761.html

State Administration for Market Regulation (2023) Guidelines for the Enforcement Against Absolute Terms in Advertising. https://gkml.samr.gov.cn/nsjg/ggjgs/202303/t20230320_353975.html
Supreme People's Court (2022) Interpretation of the Supreme People's Court on Several Issues Concerning the Anti-Unfair Competition Law. https://www.court.gov.cn/fabu-xiangqing-351291.html
Taizhou City Intermediate Court (2017) Consumer Rights Trial Cases. https://tzcourt.taizhou.gov.cn/art/2017/10/14/art_56702_2234350.html
The Paper (2020) How to Clarify the Boundary Between the Advertising Law and the Anti-Unfair Competition Law? https://www.thepaper.cn/newsDetail_forward_8657689
Weinland D (February 3, 2020) Tale of two doctors reveals how China controls the narrative. The Financial Times. https://www.ft.com/content/cf59b132-43d7-11ea-a43a-c4b328d9061c
Zhang Y (2014) List of prohibited words, cosmetics labeling regulation. Zhihu.com. https://www.zhihu.com/column/p/20686705
Zhou V (August 18, 2022) china censored a top health site after it challenged traditional medicine. Rest of World. https://restofworld.org/2022/china-health-platform-censored

MIX
Papier aus verantwortungsvollen Quellen
Paper from responsible sources
FSC® C105338

If you have any concerns about our products,
you can contact us on
ProductSafety@springernature.com

In case Publisher is established outside the EU,
the EU authorized representative is:
Springer Nature Customer Service Center GmbH
Europaplatz 3, 69115 Heidelberg, Germany

Printed by Libri Plureos GmbH
in Hamburg, Germany